AF843357

Modelling the Evolution of Natural Fracture Networks

Michael John Welch · Mikael Lüthje ·
Simon John Oldfield

Modelling the Evolution of Natural Fracture Networks

Methods for Simulating the Nucleation, Propagation and Interaction of Layer-Bound Fractures

Springer

Michael John Welch
Danish Hydrocarbon Research and Technology Centre
Technical University of Denmark
Kongens Lyngby, Denmark

Mikael Lüthje
Danish Hydrocarbon Research and Technology Centre
Technical University of Denmark
Kongens Lyngby, Denmark

Simon John Oldfield
Danish Hydrocarbon Research and Technology Centre
Technical University of Denmark
Kongens Lyngby, Denmark

ISBN 978-3-030-52413-5   ISBN 978-3-030-52414-2 (eBook)
https://doi.org/10.1007/978-3-030-52414-2

© The Editor(s) (if applicable) and The Author(s), under exclusive license to Springer Nature Switzerland AG 2020
This work is subject to copyright. All rights are solely and exclusively licensed by the Publisher, whether the whole or part of the material is concerned, specifically the rights of translation, reprinting, reuse of illustrations, recitation, broadcasting, reproduction on microfilms or in any other physical way, and transmission or information storage and retrieval, electronic adaptation, computer software, or by similar or dissimilar methodology now known or hereafter developed.
The use of general descriptive names, registered names, trademarks, service marks, etc. in this publication does not imply, even in the absence of a specific statement, that such names are exempt from the relevant protective laws and regulations and therefore free for general use.
The publisher, the authors and the editors are safe to assume that the advice and information in this book are believed to be true and accurate at the date of publication. Neither the publisher nor the authors or the editors give a warranty, expressed or implied, with respect to the material contained herein or for any errors or omissions that may have been made. The publisher remains neutral with regard to jurisdictional claims in published maps and institutional affiliations.

This Springer imprint is published by the registered company Springer Nature Switzerland AG
The registered company address is: Gewerbestrasse 11, 6330 Cham, Switzerland

Preface

Natural fracture networks are important in controlling the mechanical behaviour and flow of fluids through geological formations. It is, therefore, essential to include fracture networks in the static geomodels that are used to model such behaviour, in applications ranging from tunnel excavation and mining to groundwater management, hydrocarbon exploration and production, geothermal energy extraction and CO_2 sequestration. However, fractures cannot normally be mapped directly in the subsurface, as they are below the resolution of geophysical data, and boreholes provide only very limited data coverage. Traditional solutions to this problem include building stochastic fracture models, in which fractures of arbitrary size are placed at random locations in an attempt to match the fracture densities measured in the boreholes, or using numerical methods such as the finite element method to simulate the nucleation and growth of the fractures. However, the former method produces inaccurate and geologically unrealistic fracture models, while the latter is too computationally expensive to be practical for any but the simplest fracture systems.

In this book, we present a new method of simulating the growth of layer-bound fracture networks, which is based on fundamental geomechanical principles, but is simple enough to model large networks containing hundreds of thousands of fractures across major geological structures such as anticlines and diapirs. This is achieved by combining the established theories of subcritical fracture propagation, linear elastic fracture mechanics and fracture distribution to derive quantitative expressions describing the evolution of the fracture network, based on the mechanical properties of the host rock and the deformation history.

We start by modelling the growth of small circular fractures within a homogeneous layer. We then focus on layer-bound fractures that are confined within brittle competent layers (which often act as aquifers or reservoirs) sandwiched between more ductile layers (which often act as seals). By applying the expressions for fracture growth to cumulative distribution functions describing the fracture populations as a whole, we can model the evolution of these populations through time without needing to model the growth of each fracture individually. To do this accurately, we must also model the effects of fracture interaction; this includes both intersection with perpendicular or oblique fractures and stress shadow interaction with parallel fractures. The resulting expressions can predict the density of fractures of different sizes, and in different orientations, at different locations across the geological structure we are modelling, as well as predicting important properties of the fracture network such as fracture porosity and connectivity. Finally, we can use these results to generate a geomechanically consistent explicit Discrete Fracture Network (DFN) model.

This method can be used to explore the controls on different aspects of the fracture networks that develop under different conditions. We show that the mean linear density (P_{32}) of layer-bound fractures is mainly controlled by the presence of stress shadows and the layer thickness, while the volumetric density (P_{30}) and mean length of layer-bound fractures are controlled by many factors, including the duration of deformation, the subcritical fracture propagation index (which controls the fracture propagation rate), the initial microfracture population and the interactions between the fractures (which are in turn dependent on anisotropy in the applied strain). We show that fracture networks may not start to develop immediately when horizontal strain is applied, but that once they do start to develop they grow very rapidly, often reaching "saturation" (when no further fractures can nucleate or propagate) within tens of thousands of years or less. On a large scale, the fracture network often appears first at the location of greatest strain on the large-scale geological structure (e.g. on the crest of an anticline or diapir), and then sweeps outwards across the structure as a sharp "deformation front". If we look at the large-scale structure at any specific point in time, therefore, we can divide it into unfractured and fully fractured (saturated) zones separated by a clearly defined boundary, rather than seeing a gradual variation in fracture density across the structure. We also show that mode of the fractures (i.e. Mode 1 dilatant fractures verse Mode 2 shear fractures) is dependent not only on the mechanical properties and fluid overpressure (which we might expect), but also the propagation rate (Mode 1 fractures are more likely to form if the subcritical fracture propagation index is high). Finally, we investigate ways of characterising the anisotropy and connectivity of the fracture network and show how these properties evolve as the fracture network develops.

A method of this complexity requires calibration against real fracture networks before we can apply it with confidence to the subsurface. We have selected three fractured outcrops from the UK for this purpose. The Nash Point outcrop in south Wales exposes a uniaxial fracture network (i.e. comprising a single set of parallel fractures) in thin brittle limestone beds sandwiched between ductile shales (Maerten et al. 2016). We show that the thickness of the limestone beds controls the observed fracture spacing, and we show how to replicate the observed fracture lengths by adjusting the subcritical fracture propagation index and the initial microfracture density. The Robin Hood's Bay outcrop in northeast England exposes an orthogonal fracture network, comprising two perpendicular sets of layer-bound fractures, again confined within thin brittle limestone beds sandwiched between ductile shales (Rawnsley et al. 1993). The anisotropy of this network varies laterally across the outcrop; we show that this variation in fracture anisotropy is the result of lateral variation in the local strain around the km-scale Peak Fault, which lies to the east of the outcrop. Finally, the Pegwell Bay outcrop in southeast England exposes a series of fracture corridors propagating outwards from strike-slip faults in chalk (Souque et al. 2019). We show that it is possible to reproduce the growth of these fracture corridors, but only in conditions of high fluid overpressure, high subcritical fracture propagation index and without the development of stress shadows around the fractures.

Finally we use this new method to predict fracture networks in two subsurface examples. The Kraka field is a hydrocarbon field offshore Denmark, producing from a fractured chalk reservoir overlying a salt pillow. We show that the fracture network generated by a combination of growth of the salt pillow and local strain around a set of seismically-mapped faults gives a good match for the fractures observed on borehole images and in core (described by Aabø et al. 2019). The Anloo salt diapir near Drenthe in the Netherlands is considered a potential prospect for geothermal energy extraction, utilising several fractured reservoir layers overlying the diapir. Although the large-scale structure has been mapped out from seismic data, there is very little well data available with which to constrain the fracture population. We show how the new method can be used to identify the areas most likely to be fractured and predict the most likely fracture geometries at different points around the structure. We then show how to quickly build multiple, geologically realistic fracture models from this limited dataset for use in uncertainty and risk analysis.

Kongens Lyngby, Denmark
July 2020

Michael John Welch
Mikael Lüthje
Simon John Oldfield

Acknowledgements

The authors kindly acknowledge the Danish Underground Consortium (Total E&P Denmark, Noreco & Nordsøfonden) for providing data for the Kraka field and granting the permission to publish this work. This research has received funding from the Danish Hydrocarbon Research and Technology Centre (DHRTC) under the Advanced Water Flooding programme.

We would also like to thank Total E&P Denmark, and especially Amit Singh, Alain Lejay, Maurits de Heer, Klaas Kostro and Alan Cunningham for their help in testing the model output in their dynamic simulations, providing feedback and good discussions.

We would like to thank everyone at DHRTC and partner institutions who have contributed to this research project. In particular we would like to thank Florian Smit providing us with input and help building the Drenthe model. Thanks also goes to Tala Aabø, Jesper Dramsch and Solomon Seymun for providing us with results from their analysis of borehole images, seismic data and core for the Kraka field. We would like to acknowledge Frederic Amour letting us use results from his extensive work characterising the elastic moduli of the chalk in Kraka. We would like to give a special thanks to Aslaug Clemmensen Glad for her work on the sensitivity analyses.

We would like to thank Ole Rønø Clausen and Kenni Petersen at Aarhus University for providing us with strain data for the Kraka field, and for allowing us to use the strain history modelling software that they developed to generate strain data for Drenthe.

We would like to thank Bertrand Gauthier and his colleagues in the Naturally Fractured Reservoir Team at Total for helpful feedback and discussion.

We would like to acknowledge Ordnance Survey data and style sheets shared under the Open Government Licence.

We would like to acknowledge data shared by the Geologische Dienst Nederland, part of TNO, through dinoloket.nl.

Finally, we would also like to acknowledge our colleagues, students, families and friends who have often contributed to our work through discussion and questioning of the processes and concepts behind this book.

Contents

Constants and Symbols

The following fundamental geometric and material properties are user-specified input parameters in the calculations. Where appropriate, typical values are given for consolidated sedimentary rocks at c.2km burial depth.

Table 1 List of fundamental geometric and material properties used in these calculations

Symbol	Property	Typical value (SI Unit)
A	Rupture velocity in the rock matrix; similar to the sonic velocity	2000 m/s
b	Subcritical fracture propagation index	10–50
G_c	Crack surface energy of the rockmass	1000 J/m^2
E_r	Young's Modulus of the intact rock	10^{10} Pa (=10 GPa)
ν_r	Poisson's ratio of the intact rock	0.25
υ	Initial compaction factor	0–1
μ_{fr}	Friction coefficient of fracture	0.5
B	Initial microfracture density coefficient	0.001 fractures/m$^{3\text{-}c}$
c	Initial microfracture distribution coefficient	2
h	Layer thickness	0.1–100 m
z	True Vertical Depth (TVD)	0–5000 m
g	Gravitational constant	9.81 m/s^2
ρ_b	Bulk rock density	2200–2500 kg/m^3
ρ_f	Pore fluid density	1000 kg/m^3
t_r	Strain relaxation time constant	0.1 ma $\approx 3.15 \times 10^{11}$ s

The following symbols represent derived variables that are parameters of or output from the calculations described in this document. In some cases, modifiers (subscripts or superscripts) are applied to constrain the scope of the variables (for example, so they apply only to a specific fracture type or set, or a specific timestep). Where these modifiers are absent, the variables can be assumed to apply generally or in situations in which the scope is implied or not defined; e.g. if we are considering only one fracture type or set, if the timestep is undefined.

Table 2 List of variables used in these calculations

Symbol	Property	SI Unit
N	Current timestep	none
M	Timestep in which specified macrofracture nucleated	none
J,K	Counters used for iteration through timesteps	none
I,J	Represent two perpendicular-striking fracture sets	none
t_0	Initial time	s
t_n	A time during timestep **N**	s
t_N	Time at the end of timestep **N**	s
Δt	Time increment (e.g. a timestep duration)	s
ω	Fracture dip	rad
r	Microfracture radius	m
L	Macrofracture length (along strike)	m
ℓ	Half-macrofracture length (along strike)	m
K_c	Critical stress intensity (also known as fracture toughness)	$Nm^{3/2}$
K_f	Stress intensity at the tip of a fracture	$Nm^{3/2}$
$\sigma_v^{'}$	Vertical effective stress (+ve for compressive stress)	Pa
$\sigma_n^{'}$	Effective normal stress acting on a fracture	Pa
τ	Shear stress acting on a fracture	Pa
σ_{dN}	Driving stress acting on a fracture during timestep **N**	Pa
ε_{hmax}	Horizontal strain in the maximum (most compressional) strain direction (+ve for compressional strain)	ratio
ε_{hmin}	Horizontal strain in the minimum (most extensional) strain direction	ratio
$_{MF}\varepsilon_h$	Horizontal strain accommodated by elastic displacement on a set of macrofractures	ratio
α_{MF}	Macrofracture propagation rate factor	variable
β	Factor related to subcritical fracture propagation index	none
γ	Instantaneous microfracture growth factor	variable
Γ_N	Cumulative microfracture growth factor at the end of timestep **N**	variable
$\Lambda_{N,M}$	Cumulative macrofracture growth factor between the end of timestep **M** and the end of timestep **N**	variable

(continued)

Table 2 (continued)

Symbol	Property	SI Unit
$_{\mu F}P_{30}$	Volumetric microfracture density (number of microfractures per unit volume)	fracs/m^3
$a_{\mu F}P_{30}$	Volumetric density of active microfractures	fracs/m^3
$s_{\mu F}P_{30}$	Volumetric density of static microfractures	fracs/m^3
$_{\mu F}P_{30}(r,t)$	Volumetric density of microfractures with radius $\geq$ r at time t	fracs/m^3
$_{\mu F}P_{32}$	Mean linear microfracture density (microfracture area per unit volume)	m^2/m^3 = fracs/m
$_{MF}P_{30}$	Volumetric macrofracture density	fracs/m^3
$_{MF}P_{30}(L, t)$	Volumetric density of macrofractures with length $\geq$ L at time t	fracs/m^3
$_{MF}P_{32}$	Mean linear macrofracture density	fracs/m
$_{\underline{MF}}P_{30}$	Volumetric half-macrofracture density	fracs/m^3
$_{\underline{MF}}P_{30}(\ell, t)$	Volumetric density of half-macrofractures with half-length $\geq \ell$ at time t	fracs/m^3
$_{\underline{MF}}P_{32}$	Mean linear macrofracture density	fracs/m
$\mathbb{I}_{30}$	Anisotropy of the fracture network, calculated from volumetric macrofracture density	ratio
$\mathbb{I}_{32}$	Anisotropy of the fracture network, calculated from mean linear macrofracture density	ratio
$\mathbb{V}_{MF}(\mathbf{L})$	Internal volume of a Mode 1 dilatant macrofracture with length L	m^3
ϕ_{MF}	Total macrofracture porosity (for Mode 1 macrofractures)	ratio
η_{MF}	Total macrofracture volumetric heave (for Mode 2 macrofractures)	ratio
W	Macrofracture stress shadow width	m
ψ_{MF}	Total macrofracture stress shadow volume	ratio
χ_{MF}	Total macrofracture exclusion zone volume	ratio
q	Instantaneous probability of microfracture deactivation	s^{-1}
$\bar{q}_{\mathbf{N}}$	Mean probability of microfracture deactivation during timestep **N**	s^{-1}
$\theta_{\mathbf{N}}$	Cumulative microfracture activation probability to the end of timestep **N**	ratio
$\theta_{\mathbf{N}}'$	Effective cumulative microfracture activation probability to the end of timestep **N**	ratio
$F_{I,I,\mathbf{N}}$	Instantaneous probability of deactivation for a macrofracture of set I due to stress shadow interaction during timestep **N**	s^{-1}
$\Phi_{I,I,\mathbf{N}}$	Cumulative macrofracture activation probability for a macrofracture of set I due to stress shadow interaction during timestep **N**	ratio
$\bar{F}_{I,J,\mathbf{N}}$	Mean probability of deactivation for a macrofracture of set I due to intersection with a macrofracture of set J during timestep **N**	s^{-1}
$\Phi_{I,J,\mathbf{N}}$	Cumulative macrofracture activation probability for a macrofracture of set I due to intersection with a macrofracture of set J during timestep **N**	ratio

(continued)

Table 2 (continued)

Symbol	Property	SI Unit
$\bar{F_N}$	Mean probability of macrofracture deactivation for any reason during timestep **N**	s^{-1}
$\Phi_{\mathbf{N,M}}$	Cumulative macrofracture activation probability from the end of timestep **M** to the end of timestep **N**	ratio
$\xi(t)$	Stress shadow interaction box growth rate at time t	$m^{-2}s^{-1}$
${}^{I,I}\varsigma$	Stress shadow obstruction modifier for static half-macrofracture tip density	ratio
${}^{I,J}\varsigma$	Intersection obstruction modifier for static half-macrofracture tip density	ratio
${}_{IB}P_{32}(t + dt)$	Total macrofracture area per unit volume at time t plus the cross-sectional area of the stress shadow interaction boxes for the interval dt	ratio
$\mathcal{X}_{IB}(t + dt)$	Total macrofracture exclusion zone volume at time t plus the volume of the interaction boxes for the interval dt	ratio
S(x)	Cumulative macrofracture spacing distribution function, for macrofracture spacing x	m^{-1}
$\mathbb{A}$	Coefficient for cumulative macrofracture spacing distribution function	m^{-1}
$\mathbb{B}$	Exponent for cumulative macrofracture spacing distribution function	m^{-1}
M_{hh}	Fracture mode factor relating horizontal stress to horizontal strain	none
M_{vh}	Fracture mode factor relating vertical stress to horizontal strain	none
M_{hv}	Fracture mode factor relating horizontal stress to vertical strain	none
M_{vv}	Fracture mode factor relating vertical stress to vertical strain	none

Chapter 1
Introduction

Abstract In this chapter, we review previous approaches to modelling natural fracture networks in the subsurface and explain the drawbacks of each. The most common method is implicit stochastic modelling, where the fracture population is represented statistically by one or more spatially variable properties. The main drawback is that there is generally very limited data available to populate such models, so they are usually incomplete and unreliable. Another approach is to build an explicit Discrete Fracture Network (DFN) model in which fractures are represented as geometric objects; however, because fractures cannot be imaged directly in the subsurface, these models also suffer from the drawback of uncertainty due to limited data. A solution to this problem is to simulate the growth of the fracture network dynamically. This has the advantage that it will generate geologically realistic models based only on the geometry of the larger geological structure and the mechanical properties, which are generally known. However, conventional numerical methods such as the finite element method are far too computationally expensive to simulate a network of hundreds of thousands of fractures across a large geological structure. We therefore lay out the requirement and specifications for a new geomechanically-based method of simulating the growth of large fracture networks. We explain that this could be achieved by combining the theories of subcritical fracture propagation, linear elastic fracture mechanics and fracture size distributions. Such a method would be able to generate both implicit fracture models and explicit DFNs.

Keywords Fracture modelling · Discrete fracture networks · Naturally fractured reservoir · Geomechanics · Reservoir modelling

The purpose of this book is to present a new method for simulating the growth of a network of layer-bound fractures during an episode of geological deformation. Our method provides detailed statistical information on the development of the fracture network as a whole, as well as a means for visualising this development in the form of an explicit geometric model of the fractures. It is hoped that this method will help in understanding the controls on the development and resulting geometry of natural fracture networks in real geological formations, as well as provide improved

© The Editor(s) (if applicable) and The Author(s), under exclusive license to Springer Nature Switzerland AG 2020

M. J. Welch et al., *Modelling the Evolution of Natural Fracture Networks*,
https://doi.org/10.1007/978-3-030-52414-2_1

tools for modelling geomechanical behaviour and fluid flow in fractured layers in the subsurface.

1.1 Background and Previous Work

Many geological formations contain extensive networks of natural fractures that developed in response to an externally-applied horizontal strain. This strain can be generated by various mechanisms: regional tectonic deformation, flexure over large folds or diapirs, local deformation around larger faults, or high pressure fluids injected into the formation. The fractures that develop in response to the strain may be vertical Mode 1 dilatant fractures or inclined Mode 2 shear fractures, depending on the in situ stress state and fluid pressure at the time of deformation (Anderson 1951). If the formation comprises interbedded layers with contrasting mechanical properties, the fractures may often be confined within brittle layers sandwiched between more ductile layers, forming "layer-bound fractures" (Gross and Engelder 1995). In such cases, the spacing of the layer-bound fractures may be controlled by the "stress shadows" that develop around them. These stress shadows are zones, surrounding each fracture, in which the applied horizontal strain is accommodated by displacement on the fracture rather than by elastic strain in the host rock. The in situ stress that drives fracture propagation is much reduced within these stress shadows, and therefore other fractures cannot propagate into them (Pollard and Segall 1987; Bai and Pollard 2000; Bai et al. 2000a, b).

The overall fracture population within each brittle layer can typically be subdivided into sets of approximately parallel-striking fractures. If the applied horizontal strain was biaxial, we may see two orthogonal fracture sets, one striking perpendicular to the maximum horizontal strain and the other striking perpendicular to the minimum horizontal strain (Pollard and Aydin 1988; Lorenz et al. 1991); outcrop examples include those described by Rawnsley et al. (1993), Engelder and Peacock (2000), and Gillespie et al. (2001).

Such fracture networks will have a major impact on fluid flow through a fractured layer, especially when the host rock has low permeability, as they will provide important long-distance fluid conduits. They will also have a significant impact on the strength and elastic properties of the fractured layer as a whole (the "bulk rock" properties), potentially introducing planes of weakness and elastic anisotropy to the layer. It is, therefore, important to incorporate the fracture network into any 3D models of the subsurface geology ("static geomodels") that may be used for fluid flow modelling, geomechanical modelling or assessing rock stability.

Direct evidence of fractures in the subsurface is, however, often elusive. Their impact on seismic anisotropy is known, but may be hard to differentiate from stress effects, and they are typically both too small and lack sufficient contrast in elastic properties to be resolved or detected by seismic imaging. There may be some limited data from boreholes: fractures can be identified and logged in core, where this is available, and those fractures which have acoustic or electrical properties with sufficiently contrast to the host rock may be visible on borehole image logs. It may also be possible to identify fractures indirectly through their impact on bulk rock properties (e.g. on bulk density or sonic velocity logs) or on fluid flow (e.g. from production logging tools or temperature logs).

In outcrop, fractures are usually visible, but may be subject to preferential erosion. Outcrops often give us a good understanding of the overall geometry of a fracture network, although this is normally limited to two-dimensions. At a smaller scale, some laboratory analyses (e.g. CT scanning) provide three-dimensional analysis of microfracture networks. In both cases, the fracture geometries and properties may differ from those expected at depth since any outcrop will, by definition, have experienced a different geological history to a layer in the subsurface.

1.1.1 Impact of Fracture Networks on Bulk Rock Properties: The "TM" Approach

The conventional solution to the problem of modelling subsurface fracture networks, described comprehensively in Nelson (2001) and Mäkel (2007), is to try to characterise the subsurface fracture network as best we can from available data, and then extrapolate this to build a stochastic fracture model. In the simplest stochastic models, the fractures are not represented individually, but rather the fracture network is described statistically by one or more spatially variable properties, such as fracture density, fracture orientation or fracture connectivity (which describes proportion of fractures that are connected to other fractures in the network, versus the proportion of isolated fractures). These properties are then used to directly modify key bulk rock properties, such as porosity and permeability (for flow modelling) or failure strength and elastic moduli (for geomechanical modelling). In the case of permeability, we can define a multiplier for the transmissibility between each gridblock in the static geomodel based on the estimated fracture density in that gridblock, known as a Transmissibility Multiplier or TM. Models of this type are often called "implicit" fracture models; examples are given in Conway et al. (1996) and Gauthier et al. (2000).

The properties that can be measured and extrapolated in these models are usually very limited. From borehole data, we can only measure the density of fractures along a one-dimensional transect through the fractured layer, although we can correct this for borehole orientation (as described in Terzaghi 1965) to obtain the density of fractures along a linear transect perpendicular to the fracture planes (the "mean linear fracture density" or P_{32}). This is equivalent to the total fracture area per unit volume and can thus be used as a proxy for fracture porosity ϕ, if the fracture aperture is known or can be estimated. The boreholes also give us data on fracture orientation. The mean linear fracture density and orientation of each fracture set can then be interpolated between boreholes, with guidance from surface attributes or seismic anisotropy data if available (see e.g. Hart 2006; Singh et al. 2008; Chopra et al.

2009; Donati et al. 2016; Williams et al. 2017). However, the boreholes provide no data on the fracture sizes or connectivity, so we cannot determine the fracture size distribution, or determine how many of the fractures are connected to other fractures and how many are isolated. Nor can we measure the total number of fractures per unit volume (the "volumetric fracture density" or P_{30}). This lack of data makes it difficult to calculate the fracture permeability.

1.1.2 Modelling Fractures Explicitly: The "DFN" Approach

A more sophisticated approach is, therefore, to build an "explicit" stochastic model, comprising a series of geometric objects representing individual fractures, also known as a Discrete Fracture Network model or DFN (Dershowitz et al. 2000; also described by Bowen et al. 2013). Examples of this approach are given in Fonta et al. (2005), Barr et al. (2007), Casabianca et al. (2007), Rawnsley et al. (2007), Rogers et al. (2007), and Souche et al. (2012).

The fractures in the DFN do not correspond directly with actual fractures in the subsurface; instead, we aim to generate a DFN which has the same overall statistical and topological properties as the real fracture network. Key properties that we would like to replicate are the fracture density (both volumetric fracture density and mean linear fracture density), the fracture size distribution and the fracture connectivity.

The fracture density and size distribution can both be characterised using the cumulative fracture density distribution functions $P_{30}(S)$ or $P_{32}(S)$, which give the volumetric or mean linear density of fractures of size S or greater, as a continuous function of S. The fracture size S can be represented by a variety of parameters, but typically we use the fracture radius for circular fractures and the fracture length for layer-bound fractures. Many studies have suggested that natural fracture populations often follow fractal (power-law) cumulative fracture density distributions (e.g. Marrett and Allmendinger 1991, 1992; Westaway 1994; Ouillon et al. 1996; Marrett 1996), although this is by no means universal and other types of cumulative density distributions have also been recorded (e.g. Nicol et al. 1996; Bonnet et al. 2001).

The fracture connectivity is partly dependent on the fracture density and size distribution (Odling 1997; Manzocchi 2002), but also reflects other geometric attributes of the fracture network such as the fracture spacing distribution and fracture anisotropy. Fracture connectivity has a significant (and non-linear) impact on the bulk rock permeability, since fluid can flow much more easily through a network of connected fractures than it can through isolated fractures (Balberg et al. 1991; Berkowitz 1995; Odling et al. 1999).

Unfortunately neither the fracture size distribution nor the fracture connectivity can be measured directly in the subsurface, even from seismic anisotropy data. We must, therefore, rely on measurements from outcrop analogues to condition our stochastic DFN (e.g. De Kejzer et al. 2007; Wennberg et al. 2007; Bisdom et al. 2014). However, both are emergent properties that derive from the growth of the fracture network. An alternative approach is, therefore, to generate the DFN using a simple kinematic growth model. In this approach, fractures nucleate and grow at arbitrarily specified rates. New fractures are placed at random locations, with orientations corresponding to the orientations of observed fractures, and they interact following simple rules. A DFN generated in this way should exhibit a more realistic fracture size distribution and connectivity than a DFN generated by simply placing fractures of arbitrary size in arbitrary locations. This technique is explored in detail by Bonneau et al. (2013). However, we should bear in mind that these are not dynamic models, i.e. they do not take into account the in situ stress, applied strain or mechanical properties of the fractured layer. Furthermore they do not give direct expressions for the fracture size distribution or connectivity; rather these properties must be calculated empirically from the explicit DFN.

1.1.3 Simulating Fracture Growth: The Mechanical Approach

A third approach is to generate a fracture model by simulating the growth of the fracture network dynamically, based on fundamental geomechanical principles. The main advantage of this approach is that we do not require a priori knowledge of the fracture density, size distribution, connectivity or any of the other statistical parameters that are so difficult to measure in the subsurface, and nor do we need to specify arbitrary fracture nucleation and propagation rates; all we need is a knowledge of the geomechanical properties and deformation history of the fractured layer. This information is usually readily available: the geomechanical properties can be taken from mechanical tests on core samples, wireline log geophysical data or published data for similar lithologies, while the deformation history is often described in studies of the regional geology, or can be inferred from the geometry of larger-scale geological structures (e.g. major faults, folds or diapirs). DFNs generated in this way are more realistic than stochastic DFNs as they are by definition consistent with the geomechanical processes of fracture development, and hence honour the geology of the fractured layer and the larger-scale geological structure. The statistical properties of the fracture network emerge from the simulation, unlike the stochastic modelling approaches where they are specified in advance. These simulations, therefore, generate new information, unlike the stochastic DFNs which simply reflect the (often unreliable) data used to create them.

In most cases, this approach involves using some numerical method to simulate the propagation and interaction of individual fractures dynamically, and thereby generate an explicit DFN. The finite element method is most commonly used, but other methods such as the boundary element method or the discrete element method have also been used for this purpose. This approach has been used to model both natural and artificially induced fracture networks; examples include Advani et al. (1987), Sousa et al. (1993), Huang and Desroches (2004), Bunger et al. (2007),

Adachi et al. (2007), Welch et al. (2015), Guo et al. (2017), Huang et al. (2017), Paluszny et al. (2018), Salimzadeh et al. (2018), Vik et al. (2018).

Dynamic numerical methods allow detailed modelling of local stress anomalies that build up around fractures and faults, as well as local mechanical heterogeneities (e.g. cemented nodules, stiffer or more ductile layers or previous fractures). They can also incorporate thermal or chemical controls on fracturing. They can thus predict the effect that such heterogeneities will have on the ultimate geometry of the fracture network. However, numerical methods are computationally expensive, as the models must have sufficient resolution to capture the local stress anomalies around all of the fractures. It is, therefore, rarely practical to use them to model more than a few fractures in a restricted area. Certainly it is not possible to generate DFNs across large geological structures, which may contain hundreds of thousands of individual fractures, using dynamic numerical methods.

Some studies have, therefore, tried to use simplified (but still geomechanically consistent) rules governing fracture nucleation, propagation and interaction to generate larger DFNs. The best example of this is a series of papers by Olson and co-workers (Olson 1993, 2004; Olson et al. 2001, 2007, 2009), in which they generate explicit DFNs comprising a single set of layer-bound fractures by applying a horizontal strain to a layer containing an initial population of "seed" microfractures, in order to demonstrate the impact of fracture propagation rate on fracture geometry. Their method is based on subcritical fracture propagation theory, which broadly speaking distinguishes slow "subcritical" fracture propagation at low stress, and fast "critical" fracture propagation at high stress (see Sect. 1.2 and Sect. 3.1 for a more detailed review of subcritical fracture propagation theory).

The Olson studies show that, for a single set of parallel fractures, slow subcritical fracture propagation results in a random fracture length and spacing distribution, while intermediate rates of fracture propagation result in a series of regularly-spaced fractures of similar lengths, and fast critical fracture propagation leads to a fractal population of fractures following a power-law length and spacing distribution. It should be noted, however, that their method does not give the cumulative fracture density distribution functions directly; instead, these must be calculated empirically from the explicit DFNs. Therefore, in order to build an implicit fracture model using the Olson method (or other similar methods), it is necessary to first build an explicit DFN which includes all fractures within the size range we are interested in.

If we want to avoid this step and generate the implicit fracture model directly, we must apply the mechanical laws governing fracture propagation to functions describing the fracture population as a whole (i.e. the cumulative fracture density distribution functions), rather than to individual fractures. This is much more complicated as it requires combining statistical and mechanical laws, but it would allow large fracture populations to be modelled very quickly, since the computation time for such calculations will not be proportional to the number of fractures. Such a method could be applied to problems that are very difficult to solve by conventional numerical methods, for example, predicting the variation in size distribution and connectivity of small, centimetre-scale fractures across a large geological structure such as a thrust sheet or a salt dome.

However, the only significant developments in this technique to date are those of Davy and co-workers (Davy et al. 2010, 2013; Maillot et al. 2014), who use it to model the evolution and size distribution of a population of randomly oriented circular fractures in the subsurface. No previous studies have applied this technique to interacting sets of layer-bound fractures, or to examine the effect of fracture stress shadows, and none have derived expressions for the cumulative fracture density distribution functions that take into account fracture deactivation. These steps are necessary if we want to generate realistic implicit fracture models of subsurface layers acting as reservoirs.

1.1.4 Implementation and Practical Application of Fracture Models

As we have previously noted, fracture models (implicit or explicit) are mostly used to enable more accurate simulation of subsurface fluid flow, geomechanical response or both (for example, to predict subsidence or fracture reactivation in response to changes in subsurface fluid pressure). The approach taken will depend on the input data available, the time and resources available for modelling and the accuracy required.

Often, when an approximate solution is all that is required or when time and input data are limited, an implicit fracture model may be sufficient. The simplest implicit models may comprise only one parameter, for example, total mean linear density or total porosity of all fractures, extrapolated laterally across the fractured layer. This property is then scaled to generate a transmissibility multiplier for each gridblock in the static geomodel. Where greater accuracy is required, dual permeability or dual porosity models may be used, in which the flow through the fractures and/or storage of fluid in the fractures is calculated separately from the flow of fluid through the pores of the host rock. Such models require an estimate of the rate of fluid transfer between the pores and the fractures, which is dependent on the fracture spacing and geometry, as well as an estimate of the fracture orientation and connectivity (Cottereau et al. 2010). We must, therefore, build a more sophisticated implicit fracture model that includes these parameters.

To obtain a full statistical description of the fracture population in a specific layer, we must subdivide all the fractures in the layer into sets of approximately parallel fractures. Each set is then described statistically by two cumulative density distribution functions:

- the volumetric fracture density distribution function, $P_{30}(S)$, is the number of fractures of any specified size S or larger, per unit volume of rock and
- the mean linear fracture density distribution function, $P_{32}(S)$, is the total area of all fractures of any specified size or larger within a unit volume of rock,

where the fracture size S is expressed as fracture radius for circular fractures and length for layer-bound fractures. These functions can be defined as continuous analytical functions or as piecewise or discretised functions. The total volumetric fracture density and total mean linear fracture density, P_{30} and P_{32} respectively, are scalar parameters that represent the total number and total area of all fractures within a unit volume of rock, regardless of fracture size. Other statistical measurements, such as the fracture spacing distribution, the total fracture porosity ϕ (i.e. the combined volume of all fractures per unit volume of rock) or the fracture connectivity (i.e. the proportion of fractures that are connected to other fractures) can also be used to enhance the statistical description of the fracture population.

However it is very difficult to obtain all this data for a subsurface fractured layer. The required data can be obtained more easily from an explicit DFN model, so a standard approach is to build a DFN based on the limited statistical data available (e.g. total mean linear fracture density from a wellbore, plus an approximate breakdown into fracture sets), and then extract the more detailed statistical data from the DFN. This is in fact the most common use for explicit DFN models. The drawback with this approach is that a stochastic DFN, generated with no geomechanical constraint, cannot really give out any more information than was used to generate it; any other statistical measurements derived from it must be regarded as extremely unreliable.

More sophisticated flow models are able to simulate fluid flow through individual fractures, as well as fluid exchange between the fractures and the host rock pores (e.g. Sarda et al. 2002; Matthäi et al. 2007; Paluszny et al. 2007). This approach may be particularly useful when fluid flow is concentrated in a few large fractures, which provide pressure communication over long distances. Obviously, an explicit DFN is required to do this.

A major issue when building these more detailed fracture models is the enormous complexity of natural fracture networks in the subsurface. Such networks may contain hundreds of thousands of fractures at a whole range of scales, from mm- or cm-sized microfractures, up to fractures or small faults that are hundreds of metres long and are on the limits of seismic resolution. We must be aware, when comparing (for example) fractures observed in core or borehole image data with fractures mapped in outcrop or inferred from seismic anisotropy analysis, that we may be comparing fractures of very different sizes. Fractures on all these scales may have a significant impact on fluid flow or geomechanical properties in the subsurface: the smaller microfractures may provide additional porosity, or significant local permeability enhancement (or permeability reduction, if they are cemented or filled), or they may reduce the bulk rock strength and elastic moduli; while the largest fractures may provide long-distance fluid flow pathways, or fluid baffles (if they are cemented or filled), and may be reactivated in response to in situ stress or fluid pressure changes. Ideally, therefore, a comprehensive fracture model should include fractures on all scales.

It is impractical, however, to build a DFN across an entire geological structure that includes fractures on all these scales, as it will contain too many fractures to calculate within a reasonable timeframe. One solution is, therefore, to generate a combined implicit and explicit fracture model, in which only the larger fractures that provide long-distance flow pathways are represented in the explicit DFN, while the smaller fractures that provide local permeability enhancement are represented implicitly by one or more parameters in the static geomodel gridblocks. Such an approach is adopted by Fonta et al. (2005) and Souche et al. (2012) to model giant fractured carbonate fields.

It is possible to include a greater range of fracture scales in a DFN that covers a smaller area. For example, DFNs are often used in sector or near wellbore models; an added advantage here is that well test data can be used to calibrate the DFN (Casabianca et al. 2007; Rogers et al. 2007). Alternatively, we can build DFNs for each gridblock in the static geomodel and use them in sensitivity analyses to determine the overall permeability tensor of the gridblock. This data can then be used to populate the flow model (e.g. Jonoud et al. 2011).

Numerical geomechanical methods are even more impractical than stochastic methods for building large DFNs, as considerable runtime is required to simulate the propagation of even a single fracture, and each fracture in the network must be simulated to build the full DFN. For this reason, they are mostly used for simulating the propagation of hydraulically induced fractures, which can be modelled singly or as very small networks (Sousa et al. 1993; Huang and Desroches 2004; Bunger et al. 2007; Adachi et al. 2007; Salimzadeh et al. 2018; Vik et al. 2018). When they are applied to real natural fracture networks, only small patches of the network are fully simulated; results from these patches are then extrapolated across the fractured layer (Huang et al. 2017).

1.2 Objectives of This Study

The objective of this study is to develop a geomechanically-based method that is capable of generating a combined implicit and explicit model comprising both circular and layer-bound fractures, through a full range of scales, across large geological structures, within a practical timeframe (ideally minutes to hours of computational time) and using commonly available subsurface data. This method should be capable of modelling both Mode 1 dilatant fractures and Mode 2 shear fractures and should be capable of realistically recreating the geometry of the fracture network developed by the interactions between two sets of fractures, striking perpendicular to the minimum and maximum horizontal strain orientations. It should also be capable of generating a purely implicit fracture model in which the fracture sets are described by cumulative fracture density distribution functions, to allow rapid generation of preliminary models. Finally it should be capable of generating multiple, geologically-consistent fracture models for use in uncertainty risk analysis.

We will do this by combining three established theories:

- **Subcritical fracture propagation theory** describes the rate of fracture propagation. It is based on rock mechanics experiments carried out by Atkinson (1984)

and Swanson (1984), who show that the rate of fracture propagation is proportional to the stress intensity at the fracture tip raised to a power b, known as the subcritical fracture propagation index. The subcritical fracture propagation index is a mechanical property of the rock which can be measured in the laboratory and which varies between different rock types. Broadly speaking there are two types of fracture propagation: subcritical propagation, which occurs in rocks with low subcritical fracture propagation index (c.3−10) and critical fracture propagation, which occurs in rocks with high subcritical fracture propagation index ($\geq$10). Subcritical fracture propagation is characterised by low propagation rates but can occur at low stress; it is analogous to fatigue processes in engineering materials. Critical fracture propagation is characterised by very high propagation rates (close to the sonic velocity of the rock) but can only occur when the stress concentration at the fracture tip reaches a critical value; it is analogous to brittle shattering.
- **Linear elastic fracture mechanics theory** describes the stress intensity at the fracture tip, as a function of the fracture size and geometry, the far-field stress (i.e. the in situ stress in a rock layer) and the mechanical properties of the material containing the fracture. It was originally developed by Griffith (1921), who derived an analytical expression for the stress intensity at the tip of a laterally extensive fracture (i.e. a fracture with much greater length than height) in a homogeneous elastic material. The theory was further developed by Sneddon (1946) and Sack (1946) to give the stress intensity at the tip of a circular fracture in a homogeneous elastic material.
- **The fractal theory of fracture size distribution** states that genetically related fracture populations follow a power-law size distribution (and also a power-law displacement distribution). This theory is supported by numerous field studies of fracture sizes, e.g. Marrett and Allmendinger (1991, 1992), Westaway (1994), Watterson et al. (1996).

Combining these three theories allows us to apply the mechanical laws governing fracture propagation to the functions describing fracture populations as a whole and to generate expressions for the evolution of those fracture populations in response to the in situ stress (which can be calculated as a function of the depth of burial, the fluid pressure, the applied horizontal strain and the mechanical properties of the fractured layer). Unlike the models of Davy and co-workers (Davy et al. 2010, 2013; Maillot et al. 2014), our method applies to both circular and layer-bound fractures, it includes the effect of fracture interactions on the evolving fracture populations in a fully deterministic manner (both intersection and stress shadow interaction), and it can calculate both the mean linear fracture density and the volumetric fracture density.

This latter feature is important, as it allows us to determine the impact of the growing fracture network on the bulk rock elastic properties, which will in turn affect the rate of fracture nucleation and propagation as the fracture network grows. We can then use this dynamic data from the implicit model, combined with the rules governing fracture interaction, to generate a geomechanically-based explicit DFN without having to simulate the propagation of each fracture numerically. We can even generate implicit fracture models independently for every gridblock in a static geomodel, using mechanical property and horizontal strain data specific to that gridblock, and then combine these results to generate a single, integrated explicit DFN covering the entire geomodel. This is a key development as it allows us to generate large, geomechanically consistent DFNs that reflect the variability in geometry, mechanical properties and in situ stress across a large geological structure, for example, a DFN containing all the layer-bound fractures across an entire salt diapir or a near wellbore DFN containing the larger microfractures that develop in the damage zone around a nearby fault. This cannot be done using previous techniques.

Other key advantages of this method over previous methods are as follows:

- The combined implicit and explicit geomechanically-based approach allows us to include fractures through a full range of scales, from mm- or cm-size microfractures to small faults with lengths of hundreds of metres, in a single, geomechanically consistent model.
- Only this combined implicit and explicit geomechanically-based approach can generate geomechanically-based DFNs that cover large geological structures such as salt domes or large anticlines; dynamic numerical methods (even those using simplified mechanical rules) cannot run fast enough to generate such models within a reasonable time.
- The implicit component of this method can calculate cumulative density distribution functions for the volumetric fracture density $P_{30}(S)$, the mean linear fracture density $P_{32}(S)$, the fracture porosity $\phi(S)$ and the total stress shadow volume $\psi(S)$, for each fracture set, over a full range of scales from mm-scale microfractures to large layer-bound fractures and even small faults. Note that, even if the initial microfracture population has a power-law size distribution, the final cumulative density distribution functions will not be simple power-law functions, due to the effects of variable fracture propagation rates and fracture interaction.
- Since fracture interactions are included in the implicit component of the method, we can calculate the fracture connectivity for the implicit fracture populations directly, without needing to generate an explicit DFN.
- By generating implicit fracture models independently for every gridblock in the static geomodel, based on mechanical property and horizontal strain data specific to that gridblock, and combining the results, we can generate a single, integrated explicit DFN where fracture density and orientation varies laterally, reflecting local variations in the lithology or applied strain. Thus it is possible to generate a DFN in which the longest layer-bound fractures follow curves in the situ stress orientation around larger faults or other geological structures.
- We can also use this method to study the development of the fracture networks through geological time. We can follow the growth of individual fractures, from small circular microfractures to large layer-bound fractures and small faults, and we can also follow the evolution of the cumulative fracture density distribution functions and fracture connectivity values through time. We can, therefore,

quickly generate multiple, geologically-consistent fracture models, each representing a different stage in the growth of the fracture network, which can be used for uncertainty and risk analysis.
- Finally, we can calibrate the method by comparing the implicit cumulative fracture density distribution functions against the cumulative fracture density distribution functions calculated for the explicit DFN. A close match will demonstrate that the implicit component of the method is correctly predicting the fracture populations, based on the premises concerning fracture nucleation and interaction.

1.3 Structure of This Book

Due to the complexity of the method, we describe its derivation in stages:

- In Chap. 2, we outline the conceptual model of fracture growth on which the method is based, and detail the simplifications and assumptions it includes.
- In Chap. 3, we consider the growth of circular microfractures, and derive simple expressions for the evolution of the cumulative density distribution functions describing the microfracture population, ignoring fracture interaction.
- In Chap. 4, we examine how circular microfractures can develop into layer-bound macrofractures, and derive expressions for the rate of macrofracture nucleation and growth. We use these to calculate the cumulative density distribution functions describing the layer-bound macrofracture population, again ignoring fracture interaction.
- In Chap. 5, we study the interaction between fractures. We start by considering the different ways that fractures may interact, and derive expressions for the probability that a propagating fracture will interact in each of these ways. We then modify the expressions for the evolution of the cumulative microfracture and macrofracture density distribution functions, derived in Chaps. 3 and 4 respectively, to include the effects of fracture interaction.
- In Chap. 6, we examine the impact of the growing fracture network on the bulk rock elastic properties and show how to calculate the in situ stress at any time based on depth of burial, fluid pressure and applied horizontal strain rate.
- In Chap. 7, we combine all these components to create a workflow that can be used to generate combined implicit and explicit fracture models, as described in Sect. 1.1.2. We also run some simple test models to show that the implicit cumulative fracture density distribution functions generated by this workflow are consistent with the explicit DFNs.
- In Chap. 8, we use the method to examine some of the controls on the evolution and geometry of layer-bound fracture networks.
- In Chap. 9, we apply the method to fractured outcrops and show that it is able to replicate the observed fracture patterns.
- In Chap. 10, we investigate how the method can be applied in the subsurface.
- Finally in Chap. 11, we will discuss possible further developments of the technique.

Note that a further study of controls on fracture propagation rate and geometry using this method is presented in Welch et al. (2019).

References

Adachi, J., E. Siebrits, A. Peirce, and J. Desroches. 2007. Computer simulation of hydraulic fractures. *International Journal of Rock Mechanics and Mining Sciences* 44: 739–757.

Advani, S.H., J.S. Torok, J.K. Lee, and S. Choudhry. 1987. Explicit time-dependent solutions and numerical evaluations for penny-shaped hydraulic fracture models. *Journal of Geophysical Research* B92: 8049–8055.

Anderson, E.M. 1951. *The dynamics of faulting and Dyke formation with applications to Britain*, 2nd ed. Edinburgh: Oliver & Boyd.

Atkinson, B.K. 1984. Subcritical crack growth in geological materials. *Journal of Geophysical Research* 89 (B6): 4077–4114.

Bai, T., and D.D. Pollard. 2000. Fracture spacing in layered rocks: a new explanation based on the stress transition. *Journal of Structural Geology* 22: 43–57.

Bai, T., D.D. Pollard, and H. Gao. 2000a. Explanation for fracture spacing in layered materials. *Nature* 403: 753–756.

Bai, T., D.D. Pollard, and M.R. Gross. 2000b. Mechanical prediction of fracture aperture in layered rocks. *Journal of Geophysical Research* 105 (B1): 707–721.

Balberg, I., B. Berkowitz, and G.E. Draschler. 1991. Application of a percolation model to flow in fractured hard rocks. *Journal of Geophysical Research* 96B: 10015–10021.

Barr, D., K.E. Savory, S.R. Fowler, K. Arman, and J.P. McGarrity. 2007. Pre-development fracture modelling in the Clair field, west of Shetland. In *Fractured reservoirs*, vol. 270, ed. L. Lonergan, R.J.H. Jolly, K. Rawnsley, and D.J. Sanderson, 205–225. London: Geological Society, Special Publications.

Berkowitz, B. 1995. Analysis of fracture network connectivity using percolation theory. *Mathematical Geology* 27: 467–483.

Bisdom, K., B.D.M. Gauthier, G. Bertotti, and N.J. Hardebol. 2014. Calibrating discrete fracture-network models with a carbonate three-dimensional outcrop fracture network: Implications for naturally fractured reservoir modelling. *American Association of Petroleum Geologists Bulletin* 98: 1351–1376.

Bonneau, F., V. Henrion, G. Caumon, P. Renard, and J. Sausse. 2013. A methodology for pseudo-genetic stochastic modelling of discrete fracture networks. *Computers & Geosciences* 56: 12–22.

Bonnet, E., O. Bour, N.E. Odling, P. Davy, I. Main, P. Cowie, and B. Berkowitz. 2001. Scaling of fracture systems in geological media. *Reviews of Geophysics* 39: 347–383.

Bowen, G., T.W. Stone, D.C. Bradley, and N. Morozov. 2013. Multisegment fractures. United States Patent Application, US2013/0124178 A1.

Bunger, A.P., E. Detournay, D.I. Garagash, U. Clarkson, and A.P. Pierce. 2007. Numerical simulation of hydraulic fracturing in the viscosity-dominated regime. SPE106115, presented at SPE Hydraulic Fracturing Technology Conference, College Station, Texas.

Casabianca, D., R.J.H. Jolly, and R. Pollard. 2007. The machar oil field: Waterflooding a fractured chalk reservoir. In *Fractured reservoirs*, vol. 270, ed. L. Lonergan, R.J.H. Jolly, K. Rawnsley, and D.J. Sanderson, 171–191. London: Geological Society, Special Publications.

Chopra, S., K.J. Marfurt, and H.T. Mai. 2009. Using automatically generated 3D rose diagrams for correlation of seismic fracture lineaments with similar lineaments from attributes and well log data. *First Break* 27: 37–42.

Conway, P.I., M. Damm, and P.M. Andersen. 1996. Full field dual porosity modelling to both gas and water injection. SPE36931, presented at SPE European Petroleum Conference, Milan, Italy.

Cottereau, N., M.H. Garcia, O.R. Gosselin, and L. Vigier. 2010. Effective fracture network permeability: comparative study of calculation methods. SPE131126, presented at SPE Europec/EAGE Annual Conference and Exhibition, Barcelona, Spain.

Davy, P., R. Le Goc, C. Darcel, O. Bour, D.R. de Dreuzy, and R. Munier. 2010. A likely universal model of fracture scaling and its consequence for crustal hydromechanics. *Journal of Geophysical Research Solid Earth* 115: 10411–10424.

Davy, P., R. Le Goc, and C. Darcel. 2013. A model of fracture nucleation, growth and arrest, and consequences for fracture density and scaling. *Journal of Geophysical Research Solid Earth* 118: 1393–1407.

De Kejzer, M., H. Hillgartner, S. Al Dhahab, and K. Rawnsley. 2007. A surface-subsurface study of reservoir-scale fracture heterogeneities in Cretaceous carbonates, North Oman. In *Fractured reservoirs*, vol. 270, ed. L. Lonergan, R.J.H. Jolly, K. Rawnsley and D.J. Sanderson, 227–244. London: Geological Society, Special Publications.

Dershowitz, B., P. LaPointe, T. Eiben, and L. Wei. 2000. Integration of discrete feature network methods with conventional simulator approaches. SPE49069, presented at SPE Annual Technical Conference and Exhibition, New Orleans, Louisiana.

Donati, M., J.L. Piazza, A. Rollet, S. Baillon, D. Marin, V. Belz, H. Toubiana, J. Castro, A. Bouheouira, T. Belhouchet, and M. Raha. 2016. 3D-3C multicomponent seismic–A successful fracture characterization case study in Algeria. *First Break* 34: 35–47.

Engelder, T., and D.C.P. Peacock. 2000. Joint development normal to regional compression during flexural-flow folding: The lilstock buttress anticline, Somerset, England. *Journal of Structural Geology* 23: 259–277.

Fonta, O., H. Al-Ajmi, N.K. Verma, and S. Matar. 2005. The fracture characterization and fracture modelling of a tight carbonate reservoir: The Najmah Sargelu of West Kuwait. SPE93557, presented at 14th SPE Middle East Oil and Gas Show and Conference, Bahrain.

Gauthier, B.D.M., A.M. Zellou, A. Toublanc, M. Garcia, and J.-M. Daniel. 2000. Integrated fractured reservoir characterisation: A case study in a North Africa field. SPE 65118, presented at SPE European Petroleum Conference, Paris, France.

Gillespie, P.A., J.J. Walsh, J. Watterson, C.G. Bonson, and T. Manzocchi. 2001. Scaling relationships of joint and vein arrays from The Burren, Co. Clare, Ireland. *Journal of Structural Geology* 23: 183–201.

Griffith, A.A. 1921. The phenomena of rupture and flow in solids. *Philosophical Transactions of the Royal Society* A221: 163–197.

Gross, M.R., and T. Engelder. 1995. Strain accommodated by brittle failure in adjacent units of the Monterey Formation, U.S.A.: Scale effects and evidence for uniform displacement boundary conditions. *Journal of Structural Geology* 17: 1303–1318.

Guo, L., J.-P. Latham, and J. Xiang. 2017. A numerical study of fracture spacing and through-going fracture formation in layered rocks. *International Journal of Solids and Structures* 110–111: 44–57.

Hart, B.S. 2006. Seismic expression of fracture-swarm sweet spots, upper cretaceous tight-gas reservoirs, San Juan Basin. *American Association of Petroleum Geologists Bulletin* 90: 1519–1534.

Huang, H., J. Desroches. 2004. A PKN hydraulic fracturing model with piecewise fluid rheology. ARMA/NARMS04–560, presented at Gulf Rocks 2004, the 6th North America Rock Mechanics Symposium, Houston Texas.

Huang, H., G.D. Becker, and R. Myers. 2017. Method and system for characterizing fractures in a subsurface region. United States Patent Application, US2017/0316128 A1.

Jonoud, S., O.P. Wennberg, A. Larsen, and G. Casini. 2011. Capturing the effect of fracture heterogeneity on multiphase flow. SPE147834, presented at SPE Reservoir Characterisation and Simulation Conference and Exhibition, Abu Dhabi, UAE.

Lorenz, J.C., L.W. Teufel, and N.R. Warpinski. 1991. Regional fractures I: A mechanism for the formation of regional fractures at depth in flat-lying reservoirs. *American Association of Petroleum Geologists Bulletin* 75: 1714–1737.

Maillot, J., P. Davy, J.R. de Dreuzy, R. Le Goc, C. Darcel and M. Stiggson. 2014. Comparison between "Poissonian" and "mechanically-oriented" DFN models for the prediction of permeability and flow channeling. In International Conference on Discrete Fracture Network Engineering, Vancouver, Canada, 114.

Mäkel, G.H. 2007. The modelling of fractured reservoirs: Constraints and potential for fracture network geometry and hydraulics analysis. In *Structurally complex reservoirs*, vol. 292, ed.

S.J. Jolley, D. Barr, J.J. Walsh, and R.J. Knipe, 375–403. London: Geological Society, Special Publications.

Manzocchi, T. 2002. The connectivity of two-dimensional networks of spatially correlated fractures. *Water Resources Research* 38: 1162–1181.

Marrett, R. 1996. Aggregate properties of fracture populations. *Journal of Structural Geology* 18: 169–178.

Marrett, R., and R.W. Allmendinger. 1991. Estimates of strain due to brittle faulting: sampling of fault populations. *Journal of Structural Geology* 13: 735–738.

Marrett, R., and R.W. Allmendinger. 1992. Amount of extension on "small" faults: An example from the Viking graben. *Geology* 20: 47–50.

Matthäi, S.K., S. Geiger, S.G. Roberts, A. Paluszny, M. Belayneh, A. Burri, A. Mezentsev, H. Lu, D. Coumou, T. Dreisner, and C.A. Heinrich. 2007. Numerical simulation of multi-phase flow in structurally complex reservoirs. In *Structurally complex reservoirs*, vol. 292, ed. S.J. Jolley, D. Barr, J.J. Walsh, and R.J. Knipe, 405–429. Geological Society, London, Special Publications.

Nelson, R.A. 2001. *Geological analysis of naturally fractured reservoirs*, 2nd ed. Boston: Butterworth-Heinemann.

Nicol, A., J.J. Walsh, J. Watterson, and P.A. Gillespie. 1996. Fault size distributions—are they really power law? *Journal of Structural Geology* 18: 191–197.

Odling, N.E. 1997. Scaling and connectivity of joint systems in sandstones from western Norway. *Journal of Structural Geology* 19: 1257–1271.

Odling, N.E., P. Gillespie, B. Bourgine, C. Castaing, J.-P. Chiles, N.P. Christensen, E. Fillion, A. Genter, C. Olsen, L. Thrane, R. Trice, E. Aarseth, J.J. Walsh, and J. Watterson. 1999. Variations in fracture system geometry and their implications for fluid flow in fractured hydrocarbon reservoirs. *Petroleum Geology* 5: 373–384.

Olson, J.E. 1993. Joint pattern development: Effects of subcritical crack growth and mechanical crack interaction. *Journal of Geophysical Research* B98: 12251–12265.

Olson, J.E. 2004. Predicting fracture swarms—The influence of subcritical crack growth and the crack-tip process zone on joint spacing in rock. In *The initiation, propagation and arrest of joints and other fractures*, vol. 231, ed. J.W. Cosgrove, and T. Engelder, 73–88. London: Geological Society, Special Publications.

Olson, J.E., Y. Qiu, J. Holder, and P. Rijken. 2001. Constraining the spatial distribution of fracture networks in naturally fractured reservoirs using fracture mechanics and core measurements. SPE71342, presented at SPE Annual Technical Conference and Exhibition, New Orleans, Louisiana.

Olson, J.E., S.E. Laubach, and R.H. Lander. 2007. Combining diagenesis and mechanics to quantify fracture aperture distributions and fracture pattern permeability. In *Fractured reservoirs*, vol. 270, ed. L. Lonergan, R.J.H. Jolly, K. Rawnsley, and D.J. Sanderson, 101–116. London: Geological Society, Special Publications.

Olson, J.E., S.E. Laubach, and R.H. Lander. 2009. Natural fracture characterization in tight gas sandstones: Integrating mechanics and diagenesis. *American Association of Petroleum Geologists Bulletin* 93: 1535–1549.

Ouillon, G., C. Castaing, and D. Sornette. 1996. Hierarchical geometry of faulting. *Journal of Geophysical Research* 101 (B3): 5477–5487.

Paluszny, A., S.K. Matthäi, and M. Hohmeyer. 2007. Hybrid finite element-finite volume discretization of complex geologic structures and a new simulation volume workflow demonstrated on fractured rocks. *Geofluids* 7: 186–208.

Paluszny, A., S. Salimzadeh, and R.W. Zimmerman. 2018. Finite-element modeling of the growth and interaction of hydraulic fractures in poroelastic rock formations. In *Hydraulic fracture modelling*, ed. Y.-S. Wu, 1–19. Colorado: Colorado School of Mines, Golden.

Pollard, D.D., and A. Aydin. 1988. Progress in understanding jointing over the past century. *Geological Society of America Bulletin* 100: 1181–1204.

Pollard, D.D., and P. Segall. 1987. Theoretical displacements and stresses near fractures in rocks: with applications to faults, joints, veins, dikes and solution surfaces. In *Fracture mechanics of rock*, ed. B.K. Atkinson, 277–349. London: Academic Press.

Rawnsley, K., M. De Kejzer, L. Wei, S. Bettembourg, W. Asyee, J.-L. Massaferro, P. Swaby, D. Drysdale, and D. Boettcher. 2007. Characterizing fracture and matrix heterogeneities in folded Devonian carbonate thrust sheets, Waterton tight gas fields, Western Canada. In *Fractured reservoirs*, vol. 270, ed. L. Lonergan, R.J.H. Jolly, K. Rawnsley, and D.J. Sanderson, 265–279. London: Geological Society, Special Publications.

Rogers, S., C. Enachescu, R. Trice, and K. Buer. 2007. Integrating discrete fracture network models and pressure transient data for testing conceptual fracture models of the Valhall chalk reservoir, Norwegian North Sea. In *Fractured reservoirs*, vol. 270, ed. L. Lonergan, R.J.H. Jolly, K. Rawnsley and D.J. Sanderson, 193–204. London: Geological Society, Special Publications.

Sack, R.A. 1946. Extension of Griffith's theory to three dimensions. *Proceedings of the Physical Society, London* 58: 729–736.

Salimzadeh, S., A. Palszny, H.M. Nick, and R.W. Zimmerman. 2018. A three-dimensional coupled thermo-hydro-mechanical model for deformable fractured geothermal systems. *Geothermics* 71: 212–224.

Sarda, S., L. Jeannin, R. Basquet, B. Bourbiaux. 2002. Hydraulic characterisation of fractured reservoirs: Simulation on discrete fracture models. SPE77300, presented at SPE Reservoir Simulation Symposium, Houston, Texas.

Singh, S.K., H. Abu-Habbiel, and B. Khan. 2008. Mapping fracture corridors in naturally fractured reservoirs: an example from Middle East carbonates. *First Break* 26: 109–113.

Sneddon, I.N. 1946. The distribution of stress in the neighbourhood of a crack in an elastic solid Proceedings of the Royal Society A187, 229–260.

Souche, L., D. Astratti, V. Aarre, N. Clerc, A. Clark, T.N.A. Al Dayyni, and S.L. Mahmoud. 2012. A dual representation of multiscale fracture network modelling: Application to a giant UAE carbonate field. *First Break* 30: 43–52.

Sousa, J.L., B.J. Carter, and A.R. Ingraffea. 1993. Numerical simulation of 3D hydraulic fracture using Newtonian and power-law fluids. *International Journal of Rock Mechanics and Mining Sciences and Geomechanics Abstracts* 30: 1265–1271.

Swanson, P.L. 1984. Subcritical crack growth and other time- and environment-dependent behaviour in crustal rocks. *Journal of Geophysical Research* 89: 4137–4152.

Terzaghi, R.D. 1965. Sources of errors in joint surveys. *Geotechnique* 15: 287–304.

Vik, H.S., S. Salimzadeh, and H.M. Nick. 2018. Heat recovery from multiple-fracture enhanced geothermal systems: The effect of thermoelastic fracture interactions. *Renewable Energy* 121: 606–622.

Watterson, J., Walsh, J.J., Gillespie, P.A. and S. Easton. 1996. Scaling systematics of fault sizes on a large-scale range fault map. *Journal of Structural Geology*, 18: 199–214. https://doi.org/10.1016/S0191-8141(96)80045-9.

Welch, M.J., C. Souque, R.K. Davies, and R.J. Knipe. 2015. Using mechanical models to investigate the controls on fracture geometry and distribution in chalk. In *Fundamental controls on fluid flow in carbonates*, vol. 406, ed. S.M. Agar, and S. Geiger, 281–309. London: Geological Society, Special Publications.

Welch, M.J., M. Lüthje, and A.C. Glad. 2019. Influence of fracture nucleation and propagation rates on fracture geometry: Insights from geomechanical modelling. *Petroleum Geoscience* 25: 470–489.

Wennberg, O.P., M. Azizzadeh, A.A.M. Aqrawi, E. Blanc, P. Brockbank, K.B. Lyslo, N. Pickard, L.D. Salem, T. Svånå. 2007. The Khaviz anticline: An outcrop analogue to giant fractured Asmari Formation reservoirs in SW Iran. In *Fractured reservoirs*, vol. 270, ed. L. Lonergan, R.J.H. Jolly, K. Rawnsley, and D.J. Sanderson, 23–42. London: Geological Society, London, Special Publications.

Westaway, R. 1994. Quanititative analysis of population of small faults. *Journal of Structural Geology* 16: 1259–1273.

Williams, R.M., E. Pascual_Cebrian, J.C. Gutmanis, and G.S. Paton. 2017. Delineating fractures in the sub-seismic domain through seismic and image log analysis: A North Sea case study. *First Break* 35: 49–57.

Chapter 2
Conceptual Model for Fracture Network Growth

Abstract In this chapter, we describe a conceptual model for the growth of a network of layer-bound fractures in a brittle layer sandwiched between more ductile layers. We assume that the layer contains an initial population of small, circular microfractures. A horizontal extensional strain is applied to the layer, which causes the microfractures to grow. Initially the microfractures grow at an equal rate in all directions, maintaining a circular profile. Eventually, however, they reach the upper and lower boundaries of the layer, which they cannot cross. Once the microfractures span the layer, they can only propagate laterally (i.e. parallel to the layer), becoming layer-bound macrofractures. As these propagate, they start to interact: they may intersect a perpendicular macrofracture, or they may propagate into the stress shadow of another parallel macrofracture; in both cases they cease propagating. The fracture network will continue to grow until all macrofractures have ceased propagating. The density, size distribution, connectivity and anisotropy of the resulting fracture network will depend on the rates of fracture growth and interaction. Therefore, if we can describe these processes quantitatively, we should be able to predict the eventual fracture patterns. This conceptual model provides a framework to develop algorithms that can do this.

Keywords Fracture modelling · Discrete fracture networks · Naturally fractured reservoir · Subcritical crack propagation · Fracture anisotropy

In this chapter we describe a conceptual model for the evolution of layer-bound fracture networks that will be used as the basis for the quantitative models derived in subsequent chapters.

In the simplest version of the conceptual model, we consider a flat, homogeneous and isotropic brittle poroelastic layer of uniform thickness h, sandwiched between two ductile layers, as illustrated in Fig. 2.1. Fractures are confined within the brittle layer and cannot propagate into the surrounding ductile layers. The brittle layer is subjected to a constant and uniform vertical effective stress σ_v', and to extensional horizontal principal strains $\varepsilon_{h\min}$ and $\varepsilon_{h\max}$. By convention, compressional stresses and strains are positive so the extensional horizontal strains will be negative; $\varepsilon_{h\min}$

© The Editor(s) (if applicable) and The Author(s), under exclusive license to Springer Nature Switzerland AG 2020

M. J. Welch et al., *Modelling the Evolution of Natural Fracture Networks*,
https://doi.org/10.1007/978-3-030-52414-2_2

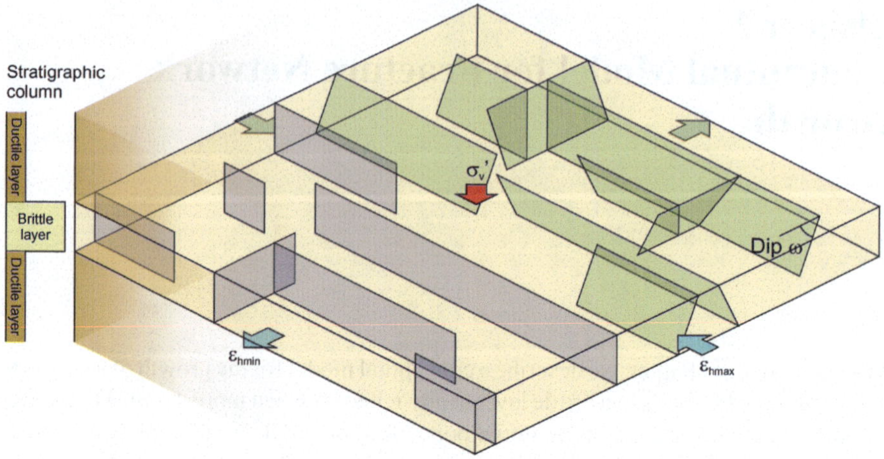

Fig. 2.1 Schematic illustration of the layer-bound fracture model showing Mode 1 (blue) and Mode 2 (green) fractures confined within a brittle layer (in yellow), sandwiched between two ductile layers (in brown). In this example, the minimum effective horizontal strain $\varepsilon_{h\min}$ is extensional while the maximum effective horizontal strain $\varepsilon_{h\max}$ is compressive

represents the most extensional, and hence most negative horizontal strain, while $\varepsilon_{h\max}$ represents the least extensional or most compressional strain. Two sets of fractures develop in response to the applied horizontal strains, striking perpendicular to $\varepsilon_{h\min}$ and $\varepsilon_{h\max}$, respectively (see Fig. 2.1). These can be either Mode 1 vertical dilatant fractures or Mode 2 inclined shear fractures, depending on the in situ stress state. Note that the Mode 2 fractures are assumed to dip in either direction with equal probability, so the net strain on the whole fracture set is horizontal pure shear with no component of simple shear. Note also that when we refer to the height and area of Mode 2 fractures, we will take this to mean the height and area projected onto a vertical plane parallel to fracture strike.

The host rock is assumed to be a perfectly elastic material, so the applied horizontal strain can be accommodated either by elastic strain in the host rock or by elastic displacement on the fractures. The stress and strain tensors are related by the three-dimensional form of Hooke's Law. However, while the elastic properties of the host rock are assumed to be isotropic, the fracture network will generally not be, although since the two fracture sets strike perpendicular to the principal horizontal strains, the principal stresses and strains will always be coaxial. We can, therefore, use the orthotropic version of Hooke's Law to relate the stress and strain tensors, when the fractures have a significant impact on the bulk rock elastic properties of the layer.

We assume that the layer contains an initial "seed" population of small circular microfractures, following a power-law cumulative density distribution function, as shown by the red circles in Fig. 2.2. These microfractures will grow in response to the in situ stress in the layer, induced by the applied strain. Initially they are contained entirely within the layer, and they propagate at an equal rate in all directions,

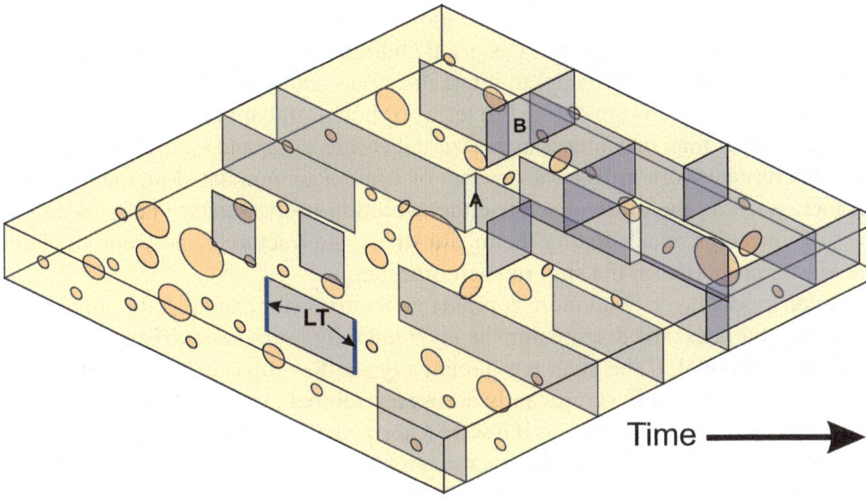

Fig. 2.2 Schematic illustration showing the growth of a fracture network through time. Microfractures are shown as red circles and layer-bound macrofractures are shown as blue rectangles. The labels LT indicate the two lateral, propagating tips of one of the layer-bound macrofractures. A and B indicate instances of fracture termination due to stress shadow interaction and due to intersection with an orthogonal fracture, respectively

maintaining a circular shape, until they span the entire layer. The propagation rates are given by subcritical fracture propagation theory and are thus dependent on the subcritical fracture propagation index of the host rock, as well as the in situ stress. For simplicity we assume that the microfractures will span the layer when their radius reaches half the layer thickness $h/2$, although in practice they may be centred anywhere in the layer, so will often not intersect the top and bottom of the layer simultaneously.

At this point they become layer-bound macrofractures. Layer-bound macrofractures are assumed to be rectangular, with a uniform and constant height equal to the layer thickness h, as shown by the blue rectangles in Fig. 2.2. Although they cannot propagate vertically, they can propagate laterally (parallel to the layer) in response to the in situ stress. The macrofractures will, therefore, have different lengths, depending on when they nucleated. Each macrofracture has two lateral tips (labelled LT in Fig. 2.2), representing the two opposite ends of the fracture, propagating in opposite directions parallel to the layer. The two tips of each macrofracture propagate independently of each other, and each will cease propagating if they fall into the stress shadow of another parallel fracture (shown by A in Fig. 2.2), or if they intersect with an orthogonal fracture (shown by B in Fig. 2.2). Eventually, all the fractures in the layer will cease propagating; at this point the fracture network will be "fully saturated" and will not develop further, although the fractures themselves can continue to accommodate strain by accumulating displacement. When this occurs we can end the simulation.

In more sophisticated versions of the model, the layer can also deform by time-dependent inelastic processes such as grain sliding, creep and pressure solution, so we must use a viscoelastic model to relate the stress and strain tensors. In this case, the boundary conditions are defined in terms of horizontal applied strain rates $\dot{\varepsilon}_{h\min}$ and $\dot{\varepsilon}_{h\max}$, rather than instantaneous horizontal strains $\varepsilon_{h\min}$ and $\varepsilon_{h\max}$. Furthermore, as the fracture network grows, the amount of strain accommodated on the fractures will increase. We, therefore, discretise the evolution of the fracture network into a series of timesteps, recalculating the in situ stress, the fracture density and the bulk rock elastic properties at the start of each timestep.

We can model layers with more complex geometries (for example, dipping layers of variable thickness) by discretising the layer laterally into a series of gridblocks, as in a static geomodel. If the layer has a relatively shallow dip (<c.20°), and its depth and thickness do not vary significantly across a gridblock, we can obtain reasonable results by treating the gridblock as if it were a horizontal and planar, using the average depth and thickness to calculate the fracture growth, and then projecting the resulting fracture network onto the true top and bottom surfaces of the gridblock. However, we should ensure that the length and width of the gridblocks are significantly greater than their height and (ideally) greater than the typical length of the layer-bound macrofractures (see Sect. 7.1).

More steeply-dipping layers can also be modelled, as long as we assume that the fracture network developed when they were relatively flat-lying and they were only subsequently tilted. This is often likely to be a reasonable assumption, since fracture networks usually develop early in the deformation episode, at relatively low strains (see Sect. 8.1.4 and Welch et al. 2019). In this case, we must first reorient the gridblock back to horizontal and calculate the fracture growth in a horizontal layer, before rotating the resulting fracture network back to its current orientation.

Reference

Welch, M.J., M. Lüthje, and A.C. Glad. 2019. Influence of fracture nucleation and propagation rates on fracture geometry: insights from geomechanical modelling. *Petroleum Geoscience* 25: 470–489.

Chapter 3
Modelling Microfractures

Abstract In this chapter, we derive equations to describe the growth of microfractures in response to an applied horizontal strain. We show that by combining subcritical fracture propagation theory and linear elastic fracture mechanics, we can obtain a differential equation describing microfracture growth. We then combine the solution to this equation with an equation describing the initial, fractal microfracture population to derive equations for the volumetric microfracture density (${}_{\mu F}P_{30}$) and the mean linear microfracture density (${}_{\mu F}P_{32}$). These can be expressed in the form of cumulative density distribution functions that describe the size distribution of the microfracture populations through time.

Keywords Fracture modelling · Discrete fracture networks · Naturally fractured reservoir · Subcritical crack propagation · Microfractures

Microfractures are defined as small circular fractures that are entirely contained within the fractured layer. When a horizontal strain is applied to the layer, they will propagate at an equal rate in all directions. They therefore remain circular until they reach the top and the bottom of the layer. We assume that this happens when the fracture radius r reaches half the layer thickness $h/2$, although in practice the microfractures may be centred anywhere in the layer, so will often not intersect the top and bottom of the layer simultaneously.

We use the label μF to denote microfractures in all subsequent equations.

3.1 Microfracture Propagation Rate

The microfracture propagation rate is calculated by combining subcritical fracture propagation theory and linear elastic fracture mechanics (see Sect. 1.2). Subcritical fracture propagation theory is based on the results of various experimental studies,

© The Editor(s) (if applicable) and The Author(s), under exclusive license to Springer Nature Switzerland AG 2020
M. J. Welch et al., *Modelling the Evolution of Natural Fracture Networks*,
https://doi.org/10.1007/978-3-030-52414-2_3

e.g. Atkinson (1984) and Swanson (1984), which showed that the fracture propagation rate *dr/dt* follows a power-law relationship with the stress acting on the fracture, given by

$$\frac{dr}{dt} = A\left(\frac{K_f}{K_c}\right)^b \tag{3.1}$$

A and K_c are mechanical properties of the host rock in the fractured layer: A is the rupture velocity (typically similar to the sonic velocity, c.2000 m/s for consolidated sedimentary rock), and K_c is the critical stress intensity (also known as the fracture toughness). For a circular fracture, this is given by

$$K_c = \sqrt{\frac{G_c E}{1 - \nu^2}} \tag{3.2}$$

where G_c is the crack surface energy of the material (typically c.1000 J/m^2), E is Young's Modulus (c.10 GPa) and ν Poisson's ratio (c.0.25).

The exponent b in (3.1) represents the subcritical fracture propagation index, which controls the rate of fracture propagation. For low values of b (<c.10), slow subcritical fracture propagation will predominate, whereas at high values of b ($\geq$10), fast critical fracture propagation will predominate.

K_f represents the stress intensity at the fracture tip, which can be calculated using linear elastic fracture mechanics theory. The stress intensity factor at the tip of a circular Mode 1 dilatant fracture, K_I, is derived by Sneddon (1946), Sneddon and Elliott (1946) and Sack (1946) as

$$K_1 = -2\sigma'_n\sqrt{\frac{r}{\pi}} \tag{3.3}$$

when the effective normal stress acting on the fracture, $\sigma_n{}'$, is tensile (i.e. negative). The stress intensity factor for a circular Mode 2 shear fracture is more complicated, as it is dependent on the magnitude of the shear stress acting on the fracture, τ, and the frictional traction on the fracture surface. However, Griffith (1924), Brace (1960) and McClintock and Walsh (1962) showed that the analytical solutions derived for stress and strain around Mode 1 fractures are also valid for Mode 2 fractures, if we substitute the shear stress minus the frictional traction for the tensile normal stress in (3.3). The frictional traction is given by the effective normal stress on the fracture multiplied by the fracture friction coefficient, μ_{fr}, which is a function of the surface roughness. We can, therefore, express the stress intensity factor for a circular Mode 2 shear fracture, K_2, as

$$K_2 = 2\left(|\tau| - \mu_{fr}\sigma'_n\right)\sqrt{\frac{r}{\pi}} \tag{3.4}$$

Following Pollard and Segall (1987), we can simplify subsequent calculations by defining a fracture driving stress for each facture mode, which we use to write a generalised expression for the stress intensity factor for any circular microfracture K_f:

$$K_f = 2\sigma_d \sqrt{\frac{r}{\pi}} \tag{3.5}$$

where σ_d is the fracture driving stress, given by

$$\sigma_d = \begin{cases} \max\begin{Bmatrix} -\sigma_n' \\ 0 \end{Bmatrix} & \text{for Mode 1 fractures} \\ |\tau| - \max\begin{Bmatrix} \mu_{fr}\sigma_n' \\ 0 \end{Bmatrix} & \text{for Mode 2 fractures} \end{cases} \tag{3.6}$$

The magnitude of the effective normal stress and shear stress acting on a specific fracture are of course dependent on the fracture orientation, and the orientation and magnitude of the in situ stress field.

3.2 Microfracture Growth

Since K_r is a function of the fracture radius r, combining (3.1) and (3.5) gives a differential equation

$$\frac{dr}{dt} = A\left(\frac{2}{\sqrt{\pi}K_c}\right)^b \sigma_d^b r^{\frac{b}{2}} \tag{3.7}$$

A solution to (3.7) (when $b > 2$) is given by

$$r(t) = \gamma(|t_i - t|)^\beta \tag{3.8}$$

where

$$\beta = \frac{2}{2-b}$$

$$\gamma = \left(\frac{A}{|\beta|}\right)^\beta \left(\frac{2\sigma_d}{\sqrt{\pi}K_c}\right)^{\beta b}$$

$$t_i = \left(\frac{r_0}{\gamma}\right)^{\frac{1}{\beta}} \tag{3.9}$$

γ is hereafter referred to as the microfracture growth factor, and controls the microfracture growth rate. β is a factor derived from the subcritical propagation index; note that β is negative when $b > 2$. The form of the solution to (3.7) is slightly different when $b = 2$ or $b < 2$ (see Appendix A).

We can rearrange (3.8) to calculate the microfracture radius r_0 at time $t = 0$ if we know the radius r_1 at time $t = t_1$:

$$r_0 = \left(r_1^{\frac{1}{\beta}} + \gamma^{\frac{1}{\beta}} t_1\right)^{\beta} \tag{3.10}$$

Equation (3.10) assumes that the driving stress, and hence γ, is constant. In general, however, the driving stress will vary with time. If we break the time down into a series of discrete timesteps and assume the driving stress is constant within each timestep but can vary between timesteps, we can derive a form of (3.10) valid for variable driving stress:

$$r_0 = \left(r_N^{\frac{1}{\beta}} + \Gamma_N\right)^{\beta} \tag{3.11}$$

Γ_N is hereafter referred to as the cumulative microfracture growth factor. This will later be used to calculate macrofracture nucleation rate and density. It is defined as

$$\Gamma_N = \sum\nolimits_{J=1}^{N} \left[\gamma_J^{\frac{1}{\beta}} (t_J - t_{J-1})\right] \tag{3.12}$$

for a series of timesteps up to the present timestep $\boldsymbol{N}$, where γ_J and σ_{dJ} represent the values of γ and σ_d during each timestep $\boldsymbol{J}$.

3.3 Volumetric Microfracture Density

We can now use the microfracture growth Eq. (3.11) to derive expressions for the cumulative microfracture density distribution functions. As discussed in Sect. 1.1.2, these are defined in terms of the microfracture radius; however, since we are modelling the evolution of the microfracture populations, we also need to include a parameter for the time t. The cumulative volumetric microfracture density and cumulative mean linear microfracture density distribution functions will, therefore, be denoted by ${}_{\mu F}P_{30}(r,t)$ and ${}_{\mu F}P_{32}(r,t)$, respectively.

If we assume that the initial microfracture population has a power-law distribution, as suggested by various studies (e.g. Marrett and Allmendinger 1991, 1992; Westaway 1994; Ouillon et al. 1996; Marrett 1996; Bonnet et al. 2001), then the initial cumulative microfracture density distribution function will be

$${}_{\mu F}P_{30}(r, 0) = Br^{-c} \tag{3.13}$$

where B is the initial microfracture density coefficient and c is the microfracture size distribution exponent. These studies suggest that typically $c \approx 2$, although it may have values up to 4 (see Sect. 8.3.2 for a more detailed discussion of initial microfracture populations). We can combine (3.13) with (3.11) to derive an expression for the evolution of the volumetric microfracture density distribution function over time:

$$_{\mu F}\mathrm{P}_{30}(r_N, t_N) = Br_0^{-c} = B\left(r_N^{\frac{1}{\beta}} + \Gamma_N\right)^{-\beta c} \tag{3.14}$$

where r_N represents the radius of a microfracture at time t_N, and t_N represents the end of timestep $\boldsymbol{N}$. A similar expression is derived by Davy et al. (2013).

Note that (3.14) will give an infinite result if we set $\mathrm{r_N}$ to zero. We must, therefore, define a minimum cut-off radius in order to calculate the total volumetric density of all microfractures in the layer. We will discuss the minimum microfracture radius further in Sect. 8.3.2. Note also that the volumetric microfracture density given by (3.14) includes those microfractures that have already reached the layer boundaries and become layer-bound macrofractures; to calculate the population of microfractures with radius less than $h/2$ we must subtract the layer-bound macrofracture population from (3.14). An expression for the volumetric density of layer-bound macrofractures will be derived in Sect. 4.2.

3.4 Mean Linear Microfracture Density

The total area of all microfractures in a unit volume can be calculated by integrating the number of microfractures of a given radius r multiplied by the area of a microfracture with radius r, for all radii between 0 and $h/2$ (the maximum radius of a microfracture in the layer). As we noted previously, the total area of microfractures per unit volume is equivalent to the number of microfractures that will be encountered along a linear transect perpendicular to the fractures, otherwise known as the mean linear microfracture density. Hereafter we will use the term total mean linear microfracture density to refer to the mean linear density of all microfractures, irrespective of size, i.e. $_{\mu F}\mathrm{P}_{32}(0,t)$.

The number of microfractures with radius between r and $r + dr$ can be obtained by differentiating the cumulative volumetric microfracture density distribution function $_{\mu F}\mathrm{P}_{30}(r,t)$ with respect to r. Thus the total mean linear microfracture density is given by

$$_{\mu F}P_{32}(0, t) = \int_0^{\frac{h}{2}} -\frac{d_{\mu F}P_{30}(r, t)}{dr}\pi r^2 dr \tag{3.15}$$

Assuming $b > 2$, $d_{\mu F}\mathrm{P}_{30}/\mathrm{dr}$ at time t_N can be obtained by differentiating (3.14) with respect to r:

$$\frac{d_{\mu F}P_{30}(r, t_N)}{dr} = \frac{d}{dr}\left[B\left(r^{\frac{1}{\beta}} + \Gamma_N\right)^{-\beta c}\right] = -cBr^{\frac{1}{\beta}-1}\left(r^{\frac{1}{\beta}} + \Gamma_N\right)^{-(\beta c+1)} \tag{3.16}$$

So (3.15) becomes

$$_{\mu F}P_{32}(0, t_N) = \pi cB \int_0^{\frac{h}{2}} r^{\frac{1}{\beta}+1}\left(r^{\frac{1}{\beta}} + \Gamma_N\right)^{-(\beta c+1)} dr \tag{3.17}$$

Unfortunately there is no analytical solution to (3.17). However, the limits of the integral (0 and *h*/2) are well defined and constant, so it is possible to obtain an approximate solution for (3.17) numerically, e.g.:

$$_{\mu F}P_{32}(0, t_N) = \pi B \int_0^{\frac{h}{2}} cr^{\frac{1}{\beta}+1}\left(r^{\frac{1}{\beta}} + \Gamma_N\right)^{-(\beta c+1)} dr = \pi B \int_0^{\frac{h}{2}} r^2 \frac{d}{dr}\left[\left(r^{\frac{1}{\beta}} + \Gamma_N\right)^{-\beta c}\right] dr$$

$$\approx \pi \left(\frac{h}{2}\right)^2 B \sum_{i=1}^{I-1}\left[\left(\frac{i}{I}\right)^2\left(\left(\left(\frac{ih}{2I}\right)^{\frac{1}{\beta}} + \Gamma_N\right)^{-\beta c} - \left(\left(\frac{(i+1)h}{2I}\right)^{\frac{1}{\beta}} + \Gamma_N\right)^{-\beta c}\right)\right] \tag{3.18}$$

Sensitivity studies showed that in most cases a reasonably accurate result can be obtained by setting the number of integration points $I = 10$, although this can be increased if a greater accuracy or resolution is required. The cumulative mean linear microfracture density for any microfracture radius r_N, $_{\mu F}P_{32}(r_N, t_N)$ can be calculated by adjusting the lower limit of the numerical integration in (3.18).

References

Atkinson, B.K. 1984. Subcritical crack growth in geological materials. *Journal of Geophysical Research* 89 (B6): 4077–4114.

Bonnet, E., O. Bour, N.E. Odling, P. Davy, I. Main, P. Cowie, and B. Berkowitz. 2001. Scaling of fracture systems in geological media. *Reviews of Geophysics* 39: 347–383.

Brace, W.F. 1960. An extension of the Griffith theory of fracture to rocks. *Journal of Geophysical Research* 65: 3477–3480.

Davy, P., R. Le Goc, and C. Darcel. 2013. A model of fracture nucleation, growth and arrest, and consequences for fracture density and scaling. *Journal of Geophysical Research Solid Earth* 118: 1393–1407.

Griffith, A.A. 1924. Theory of rupture. In *Proceedings of the first international congress on applied mechanics*, eds. Biezeno, Cornelis Benjamin, Johannes M. Burgers, 55–63, Delft.

Marrett, R., and R.W. Allmendinger. 1991. Estimates of strain due to brittle faulting: sampling of fault populations. *Journal of Structural Geology* 13: 735–738.

Marrett, R., and R.W. Allmendinger. 1992. Amount of extension on "small" faults: an example from the Viking graben. *Geology* 20: 47–50.

Marrett, R. 1996. Aggregate properties of fracture populations. *Journal of Structural Geology* 18: 169–178.

McClintock, F.A., J.B. Walsh. 1962. Frictional of Griffith cracks under pressure. In *Proceedings of the fourth U.S. national congress of applied mechanics*, 1015–1021.

Ouillon, G., C. Castaing, and D. Sornette. 1996. Hierarchical geometry of faulting. *Journal of Geophysical Research* 101 (B3): 5477–5487.

Pollard, D.D., and P. Segall. 1987. Theoretical displacements and stresses near fractures in rocks: with applications to faults, joints, veins, dikes and solution surfaces. In *Fracture mechanics of rock*, ed. B.K. Atkinson, 277–349. London: Academic Press.

Sack, R.A. 1946. Extension of Griffith's theory to three dimensions. *Proceedings of the Physical Society, London* 58: 729–736.

Sneddon, I.N. 1946. The distribution of stress in the neighbourhood of a crack in an elastic solid. *Proceedings of the Royal Society* A187: 229–260.

Sneddon, I.N., and H.A. Elliot. 1946. The opening of a Griffith crack under internal pressure. *Quarterly of Applied Mathematics* 4(3): 262–267

Swanson, P.L. 1984. Subcritical crack growth and other time- and environment-dependent behaviour in crustal rocks. *Journal of Geophysical Research* 89: 4137–4152.

Westaway, R. 1994. Quanititative analysis of population of small faults. *Journal of Structural Geology* 16: 1259–1273.

Chapter 4
Modelling Layer-Bound Macrofractures

Abstract In this chapter, we derive equations to describe the growth of layer-bound macrofractures in response to an applied horizontal strain. These form when microfractures grow large enough to span the entire layer, so we can calculate the rate of macrofracture nucleation using the equations for microfracture growth. The rate of macrofracture propagation is controlled by the layer thickness. We can, therefore, derive expressions for the total macrofracture area at any time based on the macrofracture nucleation and propagation rates. Using this information we can derive cumulative density distribution functions that describe the volumetric macrofracture density ($_{MF}P_{30}$) and mean linear macrofracture density ($_{MF}P_{32}$) through time. Finally we show how to use these results to calculate important parameters such as the fracture porosity, the total strain accommodated by fracture displacement and the total volume of the stress shadows surrounding the fractures.

Keywords Fracture modelling · Discrete fracture networks · Naturally fractured reservoir · Subcritical crack propagation · Layer-bound fractures

Once the radius of a propagating microfracture reaches half the layer thickness, the fracture will span the entire fractured layer and can no longer propagate vertically. It can, however, continue to propagate laterally, parallel to the layer. It can thus be represented as a rectangular layer-bound fracture. We will use the term macrofracture to describe such fractures, and we will denote them with the label MF in subsequent equations. Both tips of a macrofracture will propagate independently until they cease propagating for some reason, for example, they intersect an orthogonal fracture or they enter the stress shadow of another parallel fracture. The key parameter characterising macrofracture size is thus the macrofracture length, L, and the cumulative volumetric macrofracture density and cumulative mean linear macrofracture density functions will be denoted by $_{MF}P_{30}(L,t)$ and $_{MF}P_{32}(L,t)$, respectively.

© The Editor(s) (if applicable) and The Author(s), under exclusive license to Springer Nature Switzerland AG 2020

M. J. Welch et al., *Modelling the Evolution of Natural Fracture Networks*,
https://doi.org/10.1007/978-3-030-52414-2_4

4.1 Macrofracture Propagation Rate

We assume that the macrofracture length is significantly greater than the layer thickness, h, which is also the macrofracture height. The stress intensity at the lateral tips of the macrofracture, K_f, is thus dependent only on h and not on L:

$$K_f = 2\sigma_d \sqrt{\frac{h}{2\pi}} \tag{4.1}$$

where the driving stress, σ_d, is as defined in (3.6). We can use (4.1) to derive an equation for the lateral macrofracture propagation rate, dL/dt, that is similar to (3.7):

$$\frac{dL}{dt} = 2A\left(\frac{\sqrt{2h}}{\sqrt{\pi}K_C}\right)^b \sigma_d^b = 2\alpha_{\mathrm{MF}}\sigma_d^b \tag{4.2}$$

where

$$\alpha_{\mathrm{MF}} = A\left(\frac{\sqrt{2h}}{\sqrt{\pi}K_C}\right)^b \tag{4.3}$$

The factor of 2 is included in (4.2) because the macrofracture propagates at both tips.

The key point to note about (4.2) is that the macrofracture propagation rate is independent of the macrofracture length. Therefore, all macrofractures will propagate at the same rate, regardless of their size, although this rate will vary with time as the driving stress changes. We do not, therefore, need to solve a differential equation to calculate macrofracture length, which makes the calculation of the cumulative macrofracture density distribution functions much easier. We do, however, need to bear in mind that macrofractures nucleate continuously through time, as microfractures grow to the maximum radius $h/2$.

In this chapter, we will show how to calculate the cumulative macrofracture density distribution functions when all macrofractures remain active and do not cease propagating for any reason. In Chap. 5, we will extend the method to calculate the density distribution functions for macrofractures that can cease propagating (for example, if they intersect with an orthogonal fracture or they enter the stress shadow of another parallel fracture).

4.2 Volumetric Macrofracture Density

The volumetric macrofracture density distribution function, ${}_{\mathrm{MF}}\mathrm{P}_{30}(L_N, t_N)$, gives the number of macrofractures of length L_N or greater per m^3 volume at the current time t_N. We can calculate ${}_{\mathrm{MF}}\mathrm{P}_{30}(L_N, t_N)$ by extrapolating back to the time t_m at which a macrofracture with current length L_N would have nucleated. Because all macrofractures propagate at the same rate, the number of macrofractures of length L_N or greater at time t_N is equal to the total number of macrofractures at time t_m. This in turn is equal to the total number of microfractures with radius $h/2$ or greater at time t_m, ${}_{\mu F}P_{30}(h/2, t_m)$, as given by (3.14):

$$ {}_{MF}P_{30}(L_N, t_N) = {}_{MF}P_{30}(0, t_m) = {}_{\mu F}P_{30}\left(\frac{h}{2}, t_m\right) \tag{4.4} $$

If the driving stress varies through time then so will the macrofracture propagation rate, and we must take this into account when determining t_m. If we split time into a series of timesteps $\boldsymbol{J}$, each of duration $t_J - t_{J-1}$ and with driving stress σ_{dJ}, then the maximum macrofracture propagation distance $\Delta L_{N,M}$ between the end of timestep $\boldsymbol{M}$ and the end of timestep $\boldsymbol{N}$ will be given by

$$ \Delta L_{N,M} = \sum_{J=M+1}^{N} \left[2\alpha_{MF} \sigma_{dJ}^{b} (t_J - t_{J-1}) \right] \tag{4.5} $$

If t_N represents at the end of timestep $\boldsymbol{N}$ and t_m lies within timestep $\boldsymbol{M}$ ($t_{M-1} \leq t_m < t_M$) then

$$ L_N = 2\alpha_{MF} \sigma_{dM}^{b} (t_M - t_m) + \sum_{J=M+1}^{N} \left[2\alpha_{MF} \sigma_{dJ}^{b} (t_J - t_{J-1}) \right] \tag{4.6} $$

$$ \Rightarrow t_m = t_M - \frac{L_N - \sum_{J=M+1}^{N} \left[2\alpha_{MF} \sigma_{dJ}^{b} (t_J - t_{J-1}) \right]}{2\alpha_{MF} \sigma_{dM}^{b}} \tag{4.7} $$

It is helpful to define a function $\Lambda_{N,M}$ such that

$$ \Lambda_{N,M} = \sum_{J=M+1}^{N} \left[\sigma_{dJ}^{b} (t_J - t_{J-1}) \right] \tag{4.8} $$

So that (4.7) becomes

$$ t_m = t_M - \frac{L_N - 2\alpha_{MF} \Lambda_{N,M}}{2\alpha_{MF} \sigma_{dM}^{b}} \tag{4.9} $$

Substituting (4.9) into (3.14) gives:

$$
\begin{aligned}
{}_{MF}P_{30}(L_N, t_N) &= B\left(\left(\frac{h}{2}\right)^{\frac{1}{\beta}} + \Gamma_M + \frac{1}{2\beta}\left(\frac{2}{h}\right)^{\frac{b}{2}}\left(L_N - 2\alpha_{MF}\Lambda_{N,M}\right)\right)^{-\beta c} \\
&= B\left(\left(\frac{h}{2}\right)^{\frac{1}{\beta}} + \Gamma_N + \left(\frac{2}{h}\right)^{\frac{b}{2}}\frac{L_N}{2\beta}\right)^{-\beta c}
\end{aligned} \tag{4.10}
$$

where timestep $\boldsymbol{M}$ is defined such that

$$
2\alpha_{MF}\Lambda_{N,M} \leq L_N < 2\alpha_{MF}\Lambda_{N,M-1} \tag{4.11}
$$

Note that the total volumetric density of all macrofractures at time t_N, ${}_{MF}P_{30}(0,t_N)$, is simply given by

$$
{}_{MF}P_{30}(0, t_N) = B\left(\left(\frac{h}{2}\right)^{\frac{1}{\beta}} + \Gamma_N\right)^{-\beta c} \tag{4.12}
$$

4.3 Mean Linear Macrofracture Density

We can calculate the mean linear macrofracture density distribution function, ${}_{MF}P_{32}(L_N,t_N)$, in the same way that we did for microfractures, i.e. by differentiating the volumetric macrofracture density distribution function ${}_{MF}P_{30}(L,t_N)$ with respect to L to obtain the number of macrofractures with length between L and $L + dL$, multiplying the result by fracture area (given by hL), and integrating across the full range of possible macrofracture lengths. However for macrofractures, the fracture height is constant, so the macrofracture area is linearly proportional to the macrofracture length. This simplifies the calculation of ${}_{MF}P_{32}(L_N,t_N)$, as there is an analytical solution to the resulting differential equation:

$$
\begin{aligned}
{}_{MF}P_{32}(L_N, t_N) &= \int_{L_N}^{L_{max}} -\frac{d\,{}_{MF}P_{30}(L, t_N)}{dL} hLdL \\
&= \sum_{M=0}^{N}\left[\int_{\max\left\{\begin{matrix}2\alpha_{MF}\Lambda_{N,M} \\ L_N\end{matrix}\right.}^{\max\left\{\begin{matrix}2\alpha_{MF}\Lambda_{N,M-1} \\ L_N\end{matrix}\right.} -\frac{d}{dL}\left[B\left(\left(\frac{h}{2}\right)^{\frac{1}{\beta}} + \Gamma_N + \left(\frac{2}{h}\right)^{\frac{b}{2}}\frac{L}{2\beta}\right)^{-\beta c}\right] hLdL\right]
\end{aligned} \tag{4.13}
$$

Equation (4.13) can be solved using the technique of integration by parts, to give

$$
\begin{aligned}
{}_{\mathrm{MF}}\mathrm{P}_{32}(\mathrm{L}_{N},\mathrm{t}_{N}) = 2\mathrm{Bh}\sum_{M=1}^{K-1}&\left[\left(\begin{array}{l}\frac{-\beta}{(1-\beta c)}\left(\frac{h}{2}\right)^{\frac{b}{2}}\left(\left(\left(\frac{h}{2}\right)^{\frac{1}{\beta}}+\Gamma_{M}\right)^{1-\beta c}-\left(\left(\frac{h}{2}\right)^{\frac{1}{\beta}}+\Gamma_{M-1}\right)^{1-\beta c}\right)\\ +\left(\left(\left(\frac{h}{2}\right)^{\frac{1}{\beta}}+\Gamma_{M}\right)^{-\beta c}\alpha_{\mathrm{MF}}\Lambda_{N,M}-\left(\left(\frac{h}{2}\right)^{\frac{1}{\beta}}+\Gamma_{M-1}\right)^{-\beta c}\alpha_{\mathrm{MF}}\Lambda_{N,M-1}\right)\end{array}\right)\right]\\
+2\mathrm{Bh}&\left(\begin{array}{l}\frac{-\beta}{(1-\beta c)}\left(\frac{h}{2}\right)^{\frac{b}{2}}\left(\left(\left(\frac{h}{2}\right)^{\frac{1}{\beta}}+\Gamma_{N}+\left(\frac{2}{h}\right)^{\frac{b}{2}}\frac{L_N}{2\beta}\right)^{1-\beta c}-\left(\left(\frac{h}{2}\right)^{\frac{1}{\beta}}+\Gamma_{K-1}\right)^{1-\beta c}\right)\\ +\left(\left(\left(\frac{h}{2}\right)^{\frac{1}{\beta}}+\Gamma_{N}+\left(\frac{2}{h}\right)^{\frac{b}{2}}\frac{L_N}{2\beta}\right)^{-\beta c}\frac{L_N}{2}-\left(\left(\frac{h}{2}\right)^{\frac{1}{\beta}}+\Gamma_{K-1}\right)^{-\beta c}\alpha_{MF}\Lambda_{N,K-1}\right)\end{array}\right)
\end{aligned}
\tag{4.14}
$$

where timestep $\boldsymbol{K}$ represents the timestep of nucleation of a macrofracture with current length L_N. $\boldsymbol{K}$ is defined such that

$$2\alpha_{\mathrm{MF}}\Lambda_{N,K} \le L_N < 2\alpha_{\mathrm{MF}}\Lambda_{N,K-1} \tag{4.15}$$

Equation (4.14) can be simplified by cancelling terms across the timesteps, to obtain

$$
{}_{\mathrm{MF}}\mathrm{P}_{32}(\mathrm{L}_{N},\mathrm{t}_{N}) = 2\mathrm{Bh}\left(\begin{array}{c}\left(\left(\frac{h}{2}\right)^{\frac{1}{\beta}}+\Gamma_{N}+\left(\frac{2}{h}\right)^{\frac{b}{2}}\frac{L_N}{2\beta}\right)^{-\beta c}\frac{L_N}{2}\\ -\frac{\beta}{(1-\beta c)}\left(\frac{h}{2}\right)^{\frac{b}{2}}\left(\left(\left(\frac{h}{2}\right)^{\frac{1}{\beta}}+\Gamma_{N}+\left(\frac{2}{h}\right)^{\frac{b}{2}}\frac{L_N}{2\beta}\right)^{1-\beta c}-\left(\frac{h}{2}\right)^{\frac{1-\beta c}{\beta}}\right)\end{array}\right)
\tag{4.16}
$$

By setting $L_N = 0$, we can also derive an expression for the total mean linear density of all macrofractures at time $\mathrm{t}_{\mathbf{N}}$:

$$
{}_{\mathrm{MF}}\mathrm{P}_{32}(0,\mathrm{t}_{N}) = 2\mathrm{Bh}\frac{-\beta}{(1-\beta c)}\left(\frac{h}{2}\right)^{\frac{b}{2}}\left(\left(\left(\frac{h}{2}\right)^{\frac{1}{\beta}}+\Gamma_{N}\right)^{1-\beta c}-\left(\frac{h}{2}\right)^{\frac{1-\beta c}{\beta}}\right)
\tag{4.17}
$$

As we noted in Chap. 2, strictly speaking the mean linear macrofracture density ${}_{\mathrm{MF}}\mathrm{P}_{32}$ as defined above represents the total fracture area projected onto a vertical plane parallel to fracture strike, or alternatively the mean density along a horizontal transect perpendicular to the fracture strike; to obtain the true fracture area, we must divide ${}_{\mathrm{MF}}\mathrm{P}_{32}$ by the sine of the fracture dip.

4.4 Calculating Macrofracture Porosity and Stress Shadow Volume

We can convert (4.16) into an expression for the total porosity of a set of Mode 1 dilatant macrofractures, ϕ_{MF}, if we know the mean fracture aperture. Griffith (1921) and Sneddon and Elliott (1946) showed that if we take a cross section of a laterally extensive elastic Mode 1 fracture of height h (i.e. a 2D plane strain approximation), it will dilate with an elliptical profile. The maximum aperture at the centre of the fracture, $a_{\max}$, will be given by

$$a_{\max} = \frac{2\sigma_{dN}\left(1 - \nu^2\right)}{E}h \tag{4.18}$$

Therefore, the total volume of a macrofracture of length L, $\mathbb{V}_{\mathrm{MF}}(L)$, is given by

$$\mathbb{V}_{\mathrm{MF}}(L) = \frac{\pi\sigma_{dN}\left(1 - \nu^2\right)}{2E}h^2 L \tag{4.19}$$

and the total macrofracture porosity at time t_N, $\phi_{\mathrm{MF}}(t_N)$, is given by

$$\phi_{MF}(t_N) = {}_{\mathrm{MF}}\mathrm{P}_{32}(0, t_N)\frac{\pi\sigma_{dN}\left(1 - \nu^2\right)}{2E}h \tag{4.20}$$

The total fracture porosity for a set of vertical Mode 1 dilatant macrofractures also represents the total horizontal extensional strain accommodated by displacement on the fracture set, ${}_{\mathrm{MF}}\varepsilon_{\mathrm{h}}(t)$.

For a set of Mode 2 shear fractures, the total horizontal extensional strain accommodated by displacement on the fractures in a unit volume is represented by the total volumetric heave, $\eta_{\mathrm{MF}}(t_N)$. Griffith (1924), Brace (1960) and McClintock and Walsh (1962) showed that the displacement of a Mode 2 fracture follows an identical profile to that of a Mode 1 fracture, and hence we can calculate the total volumetric heave by simply adding an additional term to the formula for total macrofracture porosity given by (4.20), to account for the fracture dip, ω:

$$\eta_{\mathrm{MF}}(t_N) = {}_{\mathrm{MF}}\mathrm{P}_{32}(0, t_N)\frac{\pi\sigma_{dN}\left(1 - \nu^2\right)}{2E}h\cos\omega \tag{4.21}$$

We can, therefore, derive an expression for the total horizontal strain accommodated by a set of macrofractures in terms of the mean linear macrofracture density ${}_{\mathrm{MF}}\mathrm{P}_{32}$:

$${}_{\mathrm{MF}}\varepsilon_{\mathrm{h}}(t_N) = \begin{cases} -\phi_{MF}(t_N) = -\frac{\pi}{2}\frac{(1-\nu^2)}{E}h\sigma_{dN}\,{}_{\mathrm{MF}}\mathrm{P}_{32}(0, t_N) & \text{for Mode 1 fractures} \\ -\eta_{MF}(t_N) = -\frac{\pi}{2}\frac{(1-\nu^2)}{E}h\sigma_{dN}\,{}_{\mathrm{MF}}\mathrm{P}_{32}(0, t_N)\cos\omega & \text{for Mode 2 fractures} \end{cases} \tag{4.22}$$

where the minus signs indicate that the strain is extensional.

Fractures are often surrounded by a stress shadow, which comprises the zone in which an increment of applied horizontal strain is accommodated by an elastic displacement on the fracture, rather than by an increase in the elastic strain (and hence stress) in the host rock (Pollard and Segall 1987; Bai and Pollard 2000; Bai et al. 2000a, b). Since the elastic strain within the stress shadow does not change in response to an applied external strain, neither will the in situ stress or the fracture driving stress (hence the name).

To calculate the width of the stress shadow around a fracture, we must, therefore, divide the change in the fracture heave (the horizontal displacement on the fracture) by the increment in applied horizontal strain. We can, therefore, combine (4.22) with an expression for the fracture driving stress as a function of applied horizontal strain, to obtain an expression for the stress shadow width around a macrofracture. The relationship between driving stress and applied horizontal strain in an elastic rockmass can be obtained by combining (3.6) with the three-dimensional form of Hooke's Law, giving

$$\sigma_d = \max\left\{ \begin{array}{c} -\frac{E}{1-\nu^2}(\varepsilon_{h\min} + \nu\varepsilon_{h\max}) - \frac{\nu}{1-\nu}\sigma'_v \\ 0 \end{array} \right\} \tag{4.23}$$

for a Mode 1 macrofracture perpendicular to the minimum horizontal applied strain $\varepsilon_{h\min}$ and

$$\begin{aligned} \sigma_d = &\left(\sin\omega\cos\omega - \mu_{fr}\cos^2\omega\right)\sigma'_v \\ &- \left(\sin\omega\cos\omega + \mu_{fr}\sin^2\omega\right)\left(\frac{E}{1-\nu^2}(\varepsilon_{h\min} + \nu\varepsilon_{h\max}) + \frac{\nu}{1-\nu}\sigma'_v\right) \end{aligned} \tag{4.24}$$

for a Mode 2 macrofracture striking perpendicular to $\varepsilon_{h\min}$, with $\varepsilon_{\mathrm{hmax}}$ representing the maximum horizontal strain, σ'_v the effective vertical stress and E and ν Young's Modulus and Poisson's ratio of the bulk rock.

The total volume of the stress shadows around all macrofractures, ψ_{MF}, is thus given by

$$\psi_{\mathrm{MF}}(\mathrm{t}_N) = -\frac{d\phi_{\mathrm{MF}}}{d\varepsilon_{h\min}} = -\frac{d\phi_{\mathrm{MF}}}{d\sigma_d}\frac{d\sigma_d}{d\varepsilon_{h\min}} = \frac{\pi}{2}h_{\mathrm{MF}}\mathrm{P}_{32}(0, \mathrm{t}_N) \tag{4.25}$$

for Mode 1 dilatant fractures and

$$\begin{aligned} \psi_{\mathrm{MF}}(\mathrm{t}_N) &= -\frac{d\eta_{MF}}{d\varepsilon_{hmin}} = -\frac{d\eta_{MF}}{d\sigma_d}\frac{d\sigma_d}{d\varepsilon_{hmin}} \\ &= \frac{\pi}{2}h\cos\omega\left(\sin\omega\cos\omega + \mu_{fr}\sin^2\omega\right)_{\mathrm{MF}}\mathrm{P}_{32}(0, \mathrm{t}_N) \end{aligned} \tag{4.26}$$

for Mode 2 shear fractures. To calculate the width of the stress shadow around individual macrofractures, W, we simply divide (4.25) and (4.26) by the mean linear macrofracture density, to obtain

$$W = \frac{\pi}{2}h \tag{4.27}$$

for Mode 1 macrofractures and

$$W = \frac{\pi}{2} h \cos\omega \left(\sin\omega\cos\omega + \mu_{fr}\sin^2\omega\right) \tag{4.28}$$

for Mode 2 shear fractures.

Note that the width of the macrofracture stress shadows is independent of the macrofracture length, the driving stress and the elastic moduli of the host rock. In the case of Mode 1 fractures, it is dependent only on the layer thickness, although for Mode 2 fractures it will also depend on the fracture dip and the friction coefficient.

References

Bai, T., and D.D. Pollard. 2000. Fracture spacing in layered rocks: a new explanation based on the stress transition. *Journal of Structural Geology* 22: 43–57.

Bai, T., D.D. Pollard, and H. Gao. 2000a. Explanation for fracture spacing in layered materials. *Nature* 403: 753–756.

Bai, T., D.D. Pollard, and Michael R. Gross. 2000b. Mechanical prediction of fracture aperture in layered rocks. *Journal of Geophysical Research* 105 (B1): 707–721.

Brace, W.F. 1960. An extension of the Griffith theory of fracture to rocks. *Journal of Geophysical Research* 65: 3477–3480.

Griffith, A.A. 1921. The phenomena of rupture and flow in solids. *Philosophical Transactions of the Royal Society* A221: 163–197.

Griffith, A.A. 1924. Theory of rupture. In *Proceedings of the first international congress on applied mechanics*, eds. Biezeno, Cornelis Benjamin, Johannes M. Burgers, 55–63, Delft.

McClintock, F.A., J.B. Walsh. 1962. Frictional of Griffith cracks under pressure. In *Proceedings of the fourth U.S. national congress of applied mechanics*, 1015–1021.

Pollard, D.D., and P. Segall. 1987. Theoretical displacements and stresses near fractures in rocks: with applications to faults, joints, veins, dikes and solution surfaces. In *Fracture mechanics of rock*, ed. B.K. Atkinson, 277–349. London: Academic Press.

Sneddon, I.N., and H.A. Elliot. 1946. The opening of a Griffith crack under internal pressure. *Quarterly of Applied Mathematics* 4(3): 262–267

Chapter 5
Active and Static Fractures

Abstract In this chapter, we examine the interactions between fractures. We examine three types of fracture interaction: microfractures which become enveloped by the stress shadows of propagating macrofractures; macrofractures which propagate into the stress shadow of another parallel macrofracture and macrofractures which intersect a perpendicular or oblique macrofracture. The likelihood of a specific fracture interacting in one of these ways will increase as the overall fracture population grows, and we derive quantitative expressions for the probability of the different types of fracture interaction as a function of overall fracture density. In each of the interactions above, we assume that the fracture in question will cease propagating. We can, therefore, subdivide the fracture populations into active fractures, that are still propagating, and static fractures, that have ceased propagating. We show how to combine the expressions for the probability of fracture interaction with the cumulative density distribution functions derived in previous chapters, to obtain cumulative density distribution functions describing the volumetric density and mean linear density of active and static fractures.

Keywords Fracture modelling · Discrete fracture networks · Naturally fractured reservoir · Subcritical crack propagation · Fracture interaction

So far we have assumed that all microfractures will continue propagating until their radius reaches half the layer thickness, and that macrofractures will continue propagating indefinitely, as long as there is a driving stress. In reality, however, fractures can cease propagating for many reasons; for example, if they intersect an orthogonal fracture, or if they fall into the stress shadow of another parallel fracture. We must, therefore, modify the equations for the cumulative fracture density distribution functions to take account of this.

We can do this by subdividing the microfracture and macrofracture populations at any time into two groups: active fractures, which are still propagating, and static fractures, which have ceased propagating. For the macrofractures, however, we have already noted that the two tips propagate independently, and therefore each can become deactivated independently of the other. The easiest way to account for this is

© The Editor(s) (if applicable) and The Author(s), under exclusive license to Springer Nature Switzerland AG 2020
M. J. Welch et al., *Modelling the Evolution of Natural Fracture Networks*,
https://doi.org/10.1007/978-3-030-52414-2_5

to model half-macrofractures rather than full macrofractures. A half-macrofracture represents the portion of a macrofracture between the nucleation point and one tip; each macrofracture, therefore, comprises two conjoined half-macrofractures.

We use the label $\underline{\mathrm{MF}}$ (note the underline) to denote half-macrofractures in all subsequent equations. Thus the volumetric half-macrofracture density is denoted ${}_{\underline{\mathrm{MF}}}\mathrm{P}_{30}$ and the mean linear half-macrofracture density ${}_{\underline{\mathrm{MF}}}\mathrm{P}_{32}$. To calculate the total volumetric half-macrofracture density, we simply need to multiply the expressions for total volumetric macrofracture density derived in Chap. 4 by 2, and the total mean linear half-macrofracture density will be the same as the total mean linear half-macrofracture density. Note, however, that this will not be true of the cumulative macrofracture and half-macrofracture population functions for lengths greater than zero, as we discuss in Sect. 8.3.4.

We use the label ℓ to denote the half-macrofracture length. The half-macrofracture propagation rate $\mathrm{d}\ell/\mathrm{dt}$ is given by

$$\frac{\mathrm{d}\ell}{\mathrm{dt}} = \alpha_{\mathrm{MF}}\sigma_{\mathrm{d}}^{\mathrm{b}} \tag{5.1}$$

Hereafter we will use the prefix a to denote active microfractures and half-macrofractures and s to denote static microfractures and half-macrofractures. The volumetric density of active and static microfractures will thus be denoted ${}^{\mathrm{a}}_{\mu\mathrm{F}}\mathrm{P}_{30}$ and ${}^{\mathrm{s}}_{\mu\mathrm{F}}\mathrm{P}_{30}$, respectively, while the volumetric density of active and static half-macrofractures will thus be denoted ${}^{\mathrm{a}}_{\underline{\mathrm{MF}}}\mathrm{P}_{30}$ and ${}^{\mathrm{s}}_{\underline{\mathrm{MF}}}\mathrm{P}_{30}$, respectively. A similar notation will be used to denote the mean linear densities of active and static microfractures and half-macrofractures.

All fractures are initially considered to be active. However active fractures can cease propagating and become static fractures at any time. Static fractures cannot revert to active fractures and will, therefore, remain the same size indefinitely. It is important, however, to note that static fractures can still accumulate displacement and accommodate applied strain.

By considering the different mechanisms by which microfractures and macrofractures may cease propagating, we can derive expressions for the probability that an active fracture will be deactivated and become a static fracture. These expressions can take one of two forms: the *instantaneous probability of fracture deactivation* represents the probability that a specific fracture will cease propagating within a very short time increment dt, while the *cumulative fracture activation probability* represents the probability that an initially active fracture will still be active at the end of a longer time interval, e.g. a timestep. The best form to use will depend on the mechanism of fracture deactivation that we are considering.

Some mechanisms of fracture deactivation are random processes, in which the instantaneous probability of fracture deactivation q does not vary with time. In this the case, the cumulative fracture activation probability θ follows an exponential decay curve of the form

$$\theta(t) = e^{-qt} \tag{5.2}$$

However, other mechanisms of fracture deactivation are regular or semi-regular process, in which the instantaneous probability of fracture deactivation increases with time. In these cases, it is easier to calculate the probability that a fracture will be deactivated over a longer time interval (i.e. the cumulative activation probability), and then divide this by the time interval to obtain a pseudo-instantaneous or mean probability of fracture deactivation.

We can calculate the volumetric density of active fractures by combining the cumulative fracture activation probability with the expressions for ${}_{\mu F}P_{30}$ or ${}_{MF}P_{30}$ derived in Sect. 3.3 and 4.2, respectively. If we then multiply the resulting active fracture volumetric densities with the instantaneous fracture deactivation probability (or the mean fracture deactivation probability), we will obtain the rate at which active fractures cease propagating and become static fractures. Integrating these deactivation rates through time will, therefore, give expressions for the volumetric density of static fractures. Finally, we can repeat the previous trick of differentiating the volumetric density functions to determine the number of fractures with a given area, multiplying this by the fracture area and integrating the resultant functions across the range of fracture sizes to obtain functions for the mean linear active and static densities.

One advantage of this method is that we can separate the calculation of the fracture deactivation probabilities from the derivation of the cumulative density distibution functions for active and static fractures: the cumulative density distibution functions will include expressions for the fracture deactivation probabilities but will be independent of how these are calculated. This is important because, since fractures cease propagating as a result of interaction with other fractures, the fracture deactivation probabilities will change through time as the fracture populations grow. By calculating the fracture deactivation probabilities independently from the cumulative fracture density distibution functions, we can use the timestep approach to include this feedback: we can calculate the fracture deactivation probabilities at the start of each timestep, based on the fracture densities at the end of the previous timestep, and then use these to calculate the evolution of the cumulative fracture density distibution functions during the subsequent timestep.

However, this approach is not valid if the average lifetime (or half-life) of an active fracture is shorter than the timestep duration. Unfortunately, this is likely to be the case in the later stages of the growth of the fracture network, when the probabilities of fracture interactions are high. In this situation, we must instead derive equations that model the "residual" fracture populations as an equilibrium population, where individual fractures are short-lived but the rate of fracture nucleation is equal to the rate of fracture deactivation.

5.1 Calculating the Fracture Deactivation Probabilities

We will consider three mechanisms by which fractures may cease propagating, one of which applies to microfractures and two of which apply to half-macrofractures (remembering that each macrofracture comprises two half-macrofractures, corresponding to its two tips):

- Microfractures will cease propagating if they fall into the stress shadow of a nearby parallel propagating macrofracture.
- Half-macrofractures will cease propagating if their stress shadow interacts with the stress shadow of another parallel macrofracture.
- Half-macrofractures will also cease propagating if they intersect an orthogonal or oblique macrofracture.

We will derive expressions for the probability of fracture deactivation by each of these mechanisms.

5.1.1 Probability of Microfracture Deactivation Due to Interaction with Macrofracture Stress Shadows

To simplify the calculation of the microfracture deactivation probability, we make some assumptions about microfracture behaviour:

- The microfractures are small compared to the macrofractures and propagate at a slower rate.
- Therefore, a microfracture will cease propagating if a parallel macrofracture grows so that the centre of the microfracture lies within its stress shadow.
- The microfracture will have no effect on the macrofracture, which will continue propagating as before.
- The microfractures have insufficient size to develop a significant stress shadow themselves. We therefore ignore interactions with other microfractures.

If we assume that all microfractures are active at time $t = 0$ and that they are evenly distributed throughout the layer, then the proportion of microfractures that will have been deactivated at a later time t_N is given by the ratio of the macrofracture stress shadow volume to the total volume of the layer at time t_N. Therefore, the proportion of microfractures still active at time t_N, which we will call the cumulative microfracture activation probability θ_N, is given by:

$$\theta_N = 1 - \psi_{MF}(t_N) \tag{5.3}$$

The instantaneous probability of microfracture deactivation, q, represents the probability that a microfracture that does not currently lie within a macrofracture stress shadow will be enveloped by a macrofracture stress shadow within a very short time interval dt. This is purely a function of the rate of increase of the total macrofracture stress shadow volume ψ_{MF}:

$$q(t) = \frac{1}{1 - \psi_{MF}(t)} \frac{d\psi_{MF}(t)}{dt} \tag{5.4}$$

Note that combining (5.3) and (5.4) gives us a differential equation for θ_N that is consistent with (5.2). We can also see from (5.4) that the instantaneous probability of microfracture deactivation will increase continuously, as the total macrofracture stress shadow volume ψ_{MF} increases. However, we can also define a mean rate of microfracture deactivation $\overline{q_M}$ for a given timestep **M**, by dividing the total probability of microfracture deactivation during timestep **M** by the timestep duration:

$$\overline{q_M} = \frac{\theta_{M-1} - \theta_M}{\theta_{M-1}(t_M - t_{M-1})} = \frac{\psi_{MF}(t_M) - \psi_{MF}(t_{M-1})}{(1 - \psi_{MF}(t_{M-1}))(t_M - t_{M-1})} \tag{5.5}$$

One final issue is that (5.3) assumes that the microfractures themselves have no stress shadow, and so remain active as long as they do not lie within a macrofracture stress shadow. However, when a microfracture becomes a macrofracture, it will immediately acquire its own stress shadow. If its newly acquired stress shadow overlaps the stress shadow of an existing parallel macrofracture, the new macrofracture will be instantly deactivated (of course this sudden deactivation is an artefact of the assumption that microfractures have no stress shadows; in reality the growth of the stress shadow will be a more gradual process).

Therefore, if we want to calculate the probability that an active macrofracture will nucleate from an initial microfracture, we must modify (5.3) to include the probability that the stress shadow of the new macrofracture will overlap an existing macrofracture stress shadow. We can do this by defining an "exclusion zone" around each of the existing macrofractures, with a width twice that of the macrofracture stress shadows. If a new macrofracture nucleates within this exclusion zone, its stress shadow will immediately overlap with that of the existing macrofractures, and so it will be instantly deactivated. We can, therefore, define an effective cumulative microfracture activation probability, θ'_N, as the inverse of the total volume of all macrofracture exclusion zones.

Calculating the total volume of all macrofracture exclusion zones, χ_{MF}, is, however, more complicated than calculating the total stress shadow volume, since the exclusion zones can partially overlap. In Appendix B we show that the exact value of χ_{MF} is controlled by the spatial distribution of the macrofractures. This is described by the cumulative macrofracture spacing distribution function, S(x), which follows an exponential form. We derive an expression for the evolution of the cumulative macrofracture spacing distribution function and show that $1-\chi_{MF}$, which we call the clear zone volume, is given by

$$\theta'_N = 1 - \chi_{MF}(t_N) = \frac{\mathbb{A}}{\mathbb{B}} \tag{5.6}$$

where $\mathbb{A}$ and $\mathbb{B}$ are parameters of the cumulative macrofracture spacing distribution function. It is not possible to derive an analytical expression for $\mathbb{A}$ and $\mathbb{B}$; instead, they must be determined numerically, by recalculating their values at the end of each timestep, based on their previous values and on the increase in the total mean linear macrofracture density or the total stress shadow volume during the timestep.

Ideally, it would be useful to have an approximation for $1-\chi_{MF}$ that can be calculated analytically. Appendix B also shows how to derive such an expression, although it is only valid in the early stages of fracture development, when the total stress shadow volume is less than c.0.45. This approximation gives the effective cumulative microfracture activation probability θ'_N as

$$\theta'_N = 1 - \chi_{MF}(t_N) \approx (1 - \psi_{MF}(t_N))e^{-\frac{\psi_{MF}(t_N)}{1-\psi_{MF}(t_N)}} \tag{5.7}$$

Note that in the later stages of fracture development, when ψ exceeds c.0.45, we must revert to the numerical method of calculating θ'_N.

5.1.2 Probability of Half-Macrofracture Deactivation Due to Stress Shadow Interaction

Like microfractures, half-macrofractures may cease propagating due to stress shadow interaction. However there are several key differences:

- In order for the stress shadows around two half-macrofractures to interact, at least one of the tips must be active and must be propagating towards the tip of the other half-macrofracture.
- All half-macrofractures have stress shadows, and stress shadow interaction will occur when the stress shadows around two parallel half-macrofractures overlap.
- When this happens, both of the half-macrofractures will be deactivated.

The process of half-macrofracture stress shadow interaction is illustrated schematically in Fig. 5.1. As described in Sect. 5.1.1 and Appendix B, we can define an exclusion zone around each half-macrofracture, equal to twice the stress shadow width W, so that if another half-macrofracture tip enters that exclusion zone the two

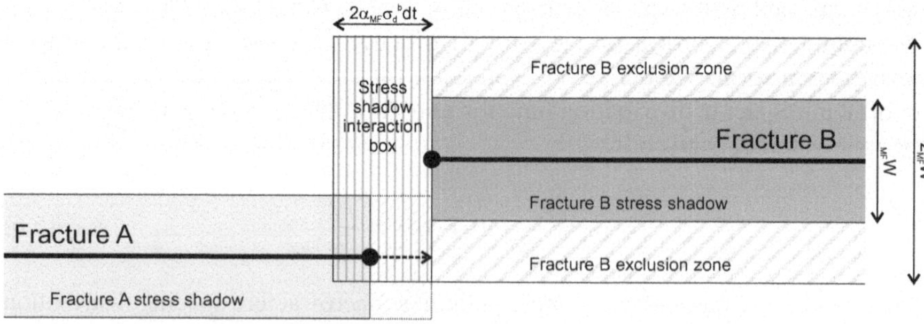

Fig. 5.1 Schematic illustration of stress shadow interaction between two layer-bound half-macrofractures, seen in plan view. Stress shadow interaction between fractures **A** and **B** will occur within the short time interval dt if the propagating tip of fracture **A** lies within the stress shadow interaction box shown around the tip of fracture **B**

stress shadows will overlap and both half-macrofractures will be deactivated. If we consider a very short time interval dt, so that the increase in the total macrofracture exclusion zone volume during dt is small, then a propagating half-macrofracture tip (e.g. the tip of fracture A in Fig. 5.1) will interact with the stress shadow around a parallel half-macrofracture (e.g. fracture B in Fig. 5.1), if tip A lies within a box projected ahead of tip B, labelled the "stress shadow interaction box". This box has width 2 W and length equal to the half-macrofracture propagation rate multiplied by dt, if tip B is static, or twice that if both the tips are propagating towards each other.

The probability of stress shadow interaction between the two half-macrofractures within the time interval dt is, therefore, given by the volume of the stress shadow interaction box as a proportion of the total clear zone volume (i.e. the volume not lying within a macrofracture exclusion zone). Furthermore, the probability that a propagating half-macrofracture tip will experience stress shadow interaction with any other half-macrofracture within time interval dt is given by the total volume of the stress shadow interaction boxes around all parallel half-macrofracture tips facing in the opposite direction, whether static or dynamic. This is proportional to the half of the total volumetric half-macrofracture density.

There is, however, a further complication when considering interaction with static half-macrofracture tips: since half-macrofractures become deactivated due to either stress shadow interaction or intersection, all static half-macrofracture tips must lie either directly adjacent to the stress shadow of another parallel half-macrofracture or else on a perpendicular-striking macrofracture. In the former case, the exclusion zone around this third half-macrofracture will overlap a part of the stress shadow interaction box of the static half-macrofracture tip B (see Fig. 5.2a), and in the latter case, the propagating half-macrofracture A will be deactivated due to intersecting

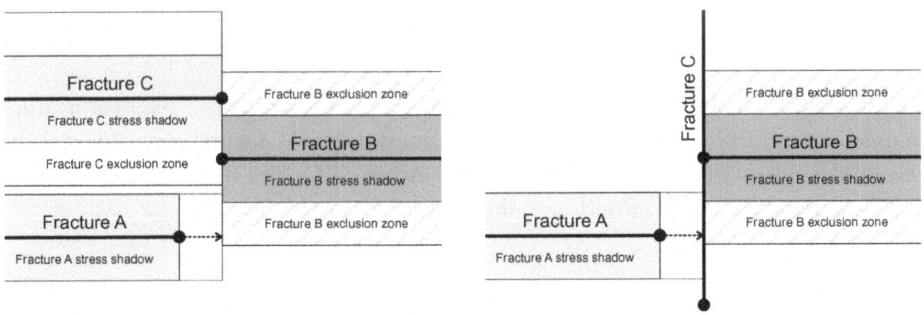

Fig. 5.2 Schematic illustration (in plan view) of the obstruction effects relating to stress shadow interaction between a propagating half-macrofracture tip (Fracture **A**) and a static half-macrofracture fracture tip (Fracture **B**), due to an obstructing fracture **C**. In the example on the left, Fracture **C** is parallel to Fracture **B** and has caused Fracture **B** to be deactivated due to stress shadow interaction. The stress shadow around Fracture **C** thus obstructs much of the stress shadow around Fracture **B**, preventing Fracture **A** from interacting with it. In the example on the right, Fracture **C** is perpendicular to Fracture **B** and has caused Fracture **B** to be deactivated due to intersection. Fracture **C** obstructs the stress shadow around Fracture **B**, preventing Fracture **A** from interacting with it

the perpendicular macrofracture C before it reaches the exclusion zone of the static half-macrofracture tip B (see Fig. 5.2b). In both cases, therefore, these "obstruction effects" will reduce the probability of stress shadow interaction between the propagating and static half-macrofractures, and reduce the effective size of the outer stress shadow interaction boxes.

We will, therefore, define a function $\xi(t)$, representing the total rate of increase of the combined length of all stress shadow interaction boxes facing in a given direction at time t, weighted to take account of obstruction effects:

$$\xi(t) = \alpha_{MF}\sigma_d^b\left({}^{a}_{\underline{MF}}P_{30}(0,t) + \frac{1}{2}{}^{I,I}\varsigma\,{}^{sII}_{\underline{MF}}P_{30}(0,t) + \frac{1}{2}{}^{I,J}\varsigma\,{}^{sIJ}_{\underline{MF}}P_{30}(0,t)\right) \quad (5.8)$$

where ${}^{sII}_{\underline{MF}}P_{30}$ and ${}^{sIJ}_{\underline{MF}}P_{30}$ represent the volumetric density of half-macrofractures that are deactivated due to stress shadow interaction and intersection, respectively, and ${}^{I,I}\varsigma$ and ${}^{I,J}\varsigma$ are the modifiers to take account of obstruction effects at the static half-macrofracture tips. In general, the mean overlap of exclusion zones of two half-macrofractures terminating due to stress shadow interaction will be 75%, so we can assume ${}^{I,I}\varsigma = 0.25$ (see Fig. 5.2a). In the case of fracture intersection, unless the intersection point lies close to one of the tips of the perpendicular macrofracture C, the propagating half-macrofracture tip A will always intersect macrofracture C before it reaches the stress shadow of the parallel half-macrofracture B. Therefore, we can approximate ${}^{I,J}\varsigma \approx 0$ (although it is possible to calculate a more precise value, if the perpendicular macrofractures are short).

We can then define a new function ${}_{IB}P_{32}$, representing the total macrofracture area per unit volume at time t plus the cross-sectional area of the stress shadow interaction boxes for the time interval dt:

$$ {}_{IB}P_{32}(t+dt) = {}_{MF}P_{32}(t) + h\xi(t)dt \quad (5.9)$$

and a second function χ_{IB}, representing the total macrofracture exclusion zone volume at time t plus the volume of the interaction boxes for the time interval dt. The total volume of stress shadow interaction boxes for the time interval dt will, therefore, be given by $d\chi_{IB}/d_{IB}P_{32} \times d_{IB}P_{32}/dt \times dt$, and hence

$$F_{I,I,N}dt = \frac{\frac{d\chi_{IB}}{d_{IB}P_{32}}\frac{d_{IB}P_{32}}{dt}dt}{1-\chi_{MF}(t)} \quad (5.10)$$

If the timestep **N** is reasonably short, we can assume that ${}_{MF}P_{32}$, $d\chi_{MF}/d_{MF}P_{32}$ and ξ will remain constant throughout the timestep. We can, therefore, and use the values calculated at the end of the previous timestep, so that $\chi_{MF}(t_n) \approx \chi_{MF,N-1}$, $\xi(t_n) \approx \xi_{N-1}$ and $d_{IB}P_{32}/dt \approx h\xi_{N-1}$. As $dt \to 0$, ${}_{IB}P_{32} \to {}_{MF}P_{32,N-1}$ and $\chi_{IB} \to \chi_{MF,N-1}$, and hence $d\chi_{IB}/d_{IB}P_{32} \approx d\chi_{MF}/d_{MF}P_{32,N-1}$. The instantaneous probability of half-macrofracture stress shadow interaction $F_{I,I,N}$ will, therefore, be given by

$$F_{I,I,N} = \frac{\frac{d\chi_{\mathrm{MF}}}{d_{MF}\mathrm{P}_{32}\,\mathbf{N-1}}}{1 - \chi_{\mathrm{MF,N-1}}} h\xi_{N-1} \tag{5.11}$$

We show how to calculate an optimal timestep duration, to minimise the errors introduced by these assumptions, in Appendix C. Expressions for $1-\chi_{\mathrm{MF,}\mathbf{N-1}}$ and $d\chi_{\mathrm{MF}}/d_{\mathrm{MF}}\mathrm{P}_{32,\mathbf{N-1}}$ are derived in Appendix B:

$$1 - \chi_{MF,N-1} = \frac{\mathbb{A}_{N-1}}{\mathbb{B}_{N-1}} \tag{5.12}$$

and

$$\frac{d\chi_{\mathrm{MF}}}{d_{MF}\mathrm{P}_{32\,\mathbf{N-1}}} = \frac{2}{\mathbb{B}_{N-1}}\left(1 - e^{-\mathbb{B}_{N-1}W}\right) \tag{5.13}$$

where $\mathbb{A}_{\mathbf{N-1}}$ and $\mathbb{B}_{\mathbf{N-1}}$ are parameters describing the spatial distribution of the macrofractures. Inserting (5.12) and (5.13) into (5.11) gives

$$F_{I,I,N} = \frac{2}{\mathbb{A}_{N-1}}\left(1 - e^{-\mathbb{B}_{N-1}W}\right)h\xi_{N-1} \tag{5.14}$$

Note that in the early stages of fracture development, when the total stress shadow volume ψ is less than c.0.45, $1-\chi_{\mathrm{MF,}\mathbf{N-1}}$ and $d\chi_{\mathrm{MF}}/d_{\mathrm{MF}}\mathrm{P}_{32,\mathbf{N-1}}$ can be approximated as

$$1 - \chi_{MF,N-1} \approx \left(1 - \psi_{MF,N-1}\right)e^{-\frac{\psi_{MF,N-1}}{1-\psi_{MF,N-1}}} \tag{5.15}$$

and

$$\frac{d\chi_{\mathrm{MF}}}{d_{\mathrm{MF}}\mathrm{P}_{32\,\mathbf{N-1}}} \approx W\frac{2 - \psi_{MF,N-1}}{1 - \psi_{MF,N-1}}e^{-\frac{\psi_{MF,N-1}}{1-\psi_{MF,N-1}}} \tag{5.16}$$

Thus the instantaneous probability of half-macrofracture stress shadow interaction $\mathrm{F_{I,I,N}}$ will be given by

$$F_{I,I,N} = \frac{W\frac{2-\psi_{MF,N-1}}{1-\psi_{MF,N-1}}e^{-\frac{\psi_{MF,N-1}}{1-\psi_{MF,N-1}}}}{\left(1 - \psi_{MF,N-1}\right)e^{-\frac{\psi_{MF,N-1}}{1-\psi_{MF,N-1}}}}h\xi_{N-1} = \frac{2 - \psi_{MF,N-1}}{\left(1 - \psi_{MF,N-1}\right)^2}Wh\xi_{N-1} \tag{5.17}$$

The cumulative half-macrofracture activation probability for stress shadow interaction $\Phi_{\mathrm{I,I,N}}$ represents the probability that an active half-macrofracture tip will not be deactivated due to stress shadow interaction during timestep **N**. Since half-macrofracture stress shadow interaction is a random process (i.e. $\mathrm{F_{I,I,N}}$ is independent of the time that the half-macrofracture has been active), $\Phi_{\mathrm{I,I,N}}$ is given by an exponential function:

$$\Phi_{I,I,N} = e^{-F_{I,I,N}(t_N - t_{N-1})} \tag{5.18}$$

5.1.3 Probability of Half-Macrofracture Deactivation Due to Intersection

Layer-bound half-macrofractures will also cease propagating if they intersect a perpendicular-striking macrofracture, as long as the frictional traction on the surface of the intersected fracture is sufficiently low to allow slip. We will assume that this is always the case, so the probability of half-macrofracture intersection is determined only by the propagation rate of the intersecting half-macrofracture and the population of perpendicular-striking macrofractures.

However, as Fig. 5.3 shows, the probability of half-macrofracture intersection is dependent not only on the density of the perpendicular-striking macrofractures, but also their spatial distribution. If the perpendicular macrofractures are randomly distributed, fracture intersection will be a random process, governed by an exponential probability function. However, if the perpendicular macrofractures are evenly

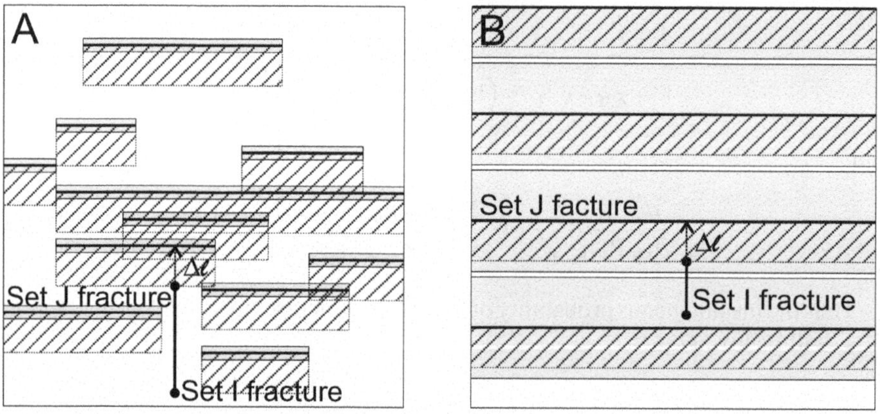

Fig. 5.3 Two examples of a northward-propagating half-macrofracture tip (the Set I fracture) intersecting a set of perpendicular east-west striking macrofractures (set J). The stress shadows around the set J fractures are shown by the grey shading. In example A, the set J fractures are surrounded by narrow stress shadows so have an approximately random distribution, whereas in example B they are surrounded by wide stress shadows and are evenly spaced; note that the total area of fracture set J is the same in both cases. The striped boxes on the lower side of fracture set J show the intersection boxes for a single timestep, during which the Set ℓ fracture will propagate a distance $\Delta\ell$. It will intersect a set J fracture during the timestep if its tip lies within one of the intersection box at the start of the timestep

spaced (for example, if their distribution is controlled by stress shadows), then fracture intersection will be a regular process. In reality, many fracture sets will be somewhere in between these two end-members: there will be a minimum separation between the fractures, determined by the stress shadow width, but there will be larger gaps that are randomly distributed. Half-macrofracture intersection then becomes a semi-regular process that is especially hard to model, as it will include elements of both regular and random behaviour.

We can derive a general expression for the probability of half-macrofracture intersection that is valid for any fracture distribution, by considering a half-macrofracture tip from set I propagating towards a set of perpendicular-striking macrofractures, set J (see Fig. 5.3). The fractures of set J are surrounded by stress shadows of width ${}^{J}W$, which cannot overlap. If the fractures in set J are randomly distributed then ${}^{J}W$ will be zero, and if they are evenly spaced then ${}^{J}W$ will be the inverse of the total mean linear density of fracture set J.

If we take a short timestep of duration Δt, the propagating half-macrofracture tip (fracture I in Fig. 5.3) will intersect a set J fracture during Δt if that fracture lies within the distance that the set I half-macrofracture tip can propagate in the time period Δt ($\Delta\ell$ in Fig. 5.3). We can, therefore, define a set of "intersection boxes", extending towards the propagating half-macrofracture tip from each of the set J fractures, with width $\Delta\ell$. The probability of fracture intersection during the timestep is equal to the total volume of all intersection boxes for fracture set J, as a proportion of the total rock volume.

The total volume of an intersection box attached to a specific set J fracture with length L is given by $Lh\Delta\ell$. If the width of the intersection boxes $\Delta\ell$ is less than the stress shadow width W then none of the intersection boxes can overlap (see Fig. 5.3b), and the total volume of all the intersection boxes is given by ${}_{MF}^{J}P_{32}\Delta\ell$. However, if the width of the intersection boxes is greater than the stress shadow width, those parts of the intersection box that lie further from the set J fracture than the stress shadow width may overlap the intersection boxes around other set J fractures (see Fig. 5.3a). In this case, we must use a "one-sided proximity zone" formula to calculate the total intersection box volume, as described in Appendix B.

A general version of the "one-sided proximity zone" formula, which gives the total intersection box volume $\chi'(\Delta\ell)$ in all circumstances, is given by (B.43) in Appendix B. Based on this formula, the cumulative half-macrofracture activation probability for fracture intersection $\Phi_{I,J,N}$, which represents the probability that an active half-macrofracture in set I will not be deactivated due to intersection during timestep **N**, is given by

$$\Phi_{I,J,N} = \begin{cases} 1 - {}_{MF}^{J}P_{32}(t_{N-1})\Delta\ell_N & if\ \Delta\ell_N \le {}^{J}W \\ \frac{{}_{MF}^{J}P_{32}(t_{N-1})}{\mathbb{B}_{2,N-1}^{J}} e^{-\mathbb{B}_{2,N-1}^{J}(\Delta\ell_N - {}^{J}W)} + \left(\frac{\mathbb{A}_{N-1}^{J}}{\mathbb{B}_{N-1}^{J}} - \frac{\mathbb{A}_{N-1}^{J}}{\mathbb{B}_{2,N-1}^{J}}\right) & if\ {}^{J}W < \Delta\ell_N \le 2^{J}W \\ \frac{\mathbb{A}_{N-1}^{J}}{\mathbb{B}_{N-1}^{J}} e^{-\mathbb{B}_{N-1}^{J}(\Delta\ell_N - 2^{J}W)} & if\ \Delta\ell_N > 2^{J}W \end{cases} \tag{5.19}$$

where ${}^{J}W$ is the macrofracture stress shadow width for set J macrofractures, ${}^{J}_{MF}P_{32}$ is the total mean linear density of set J macrofractures, and $\mathbb{A}^{J}_{N-1}$, $\mathbb{B}^{J}_{N-1}$ and $\mathbb{B}^{J}_{2,N-1}$ are parameters describing the spatial distribution of set J macrofractures at the end of the previous timestep **N−1** (see Appendix B for details). The propagation distance of a set I half-macrofracture tip during timestep **N**, $\Delta\ell_N$, is given by

$$\Delta\ell_N = \alpha_{MF}{}^{I}\sigma^{b}_{dN}\Delta t_N \tag{5.20}$$

Note that if ${}^{J}W = 0$ (i.e. there are no stress shadows) then $\mathbb{A}^{J}_{N-1} = \mathbb{B}^{J}_{N-1} = {}^{J}_{MF}P_{32}(t_{N-1})$, and (5.19) can be reduced to

$$\Phi_{I,J,N} = e^{-{}^{J}_{MF}P_{32}(t_{N-1})\Delta\ell_N} \tag{5.21}$$

Since the instantaneous probability of half-macrofracture intersection F_{IJ} is dependent on the length of the propagating half-macrofracture, strictly speaking (5.19) is valid only for half-macrofractures with zero length at the start of the timestep, and must be modified for pre-existing half-macrofractures with an initial length ℓ_0. We can do this by considering the case of a zero length half-macrofracture propagating in two stages, the first from length 0 to length ℓ_0, and the second from length ℓ_0 to length $\ell_0 + \Delta\ell$. In this case, ignoring deactivation due to other factors, there are three possible outcomes:

1. The half-macrofracture could be deactivated due to intersection during the first stage, before reaching length ℓ_0. The probability of this is given by the total intersection box volume $\chi'(\ell_0)$, which can be calculated using the appropriate "one-sided proximity zone" formula from Appendix B.
2. The half-macrofracture could be deactivated due to intersection during the second stage, after reaching length ℓ_0 but before reaching length $\ell_0 + \Delta\ell$. This represents the inverse of the cumulative half-macrofracture activation probability for fracture intersection of a half-macrofracture with an initial length ℓ_0, which we will denote $1 -{}^{\ell}\Phi_{IJ}$.
3. The half-macrofracture could reach length $\ell_0 + \Delta\ell$ without being deactivated. The probability of this is given by the inverse of the total intersection box volume, $1 - \chi'(\ell_0 + \Delta\ell)$, which can also be calculated using the appropriate "one-sided proximity zone" formula from Appendix B.

Since these are the only possible outcomes, we can say that

$$\begin{aligned} &\chi'(\ell_0) + \left(1 - {}^{\ell}\Phi_{IJ}\right) + \left(1 - \chi'(\ell_0 + \Delta\ell)\right) = 1 \\ &\Rightarrow 1 - {}^{\ell}\Phi_{IJ} = \chi'(\ell_0 + \Delta\ell) - \chi'(\ell_0) \end{aligned} \tag{5.22}$$

However, if we now consider a half-macrofracture that has already reached length ℓ_0, then the first outcome is ruled out, and the probability of the second outcome is now given by the original probability of the second outcome, divided by the inverse of the original probability of the first outcome. Therefore

$$1 - {}^{\ell}\Phi_{IJ} = \frac{\chi'(\ell_0 + \Delta\ell) - \chi'(\ell_0)}{1 - \chi'(\ell_0)} = 1 - \frac{1 - \chi'(\ell_0 + \Delta\ell)}{1 - \chi'(\ell_0)}$$
$$\Rightarrow {}^{\ell}\Phi_{IJ} = \frac{1 - \chi'(\ell_0 + \Delta\ell)}{1 - \chi'(\ell_0)} \tag{5.23}$$

We can derive a reasonable approximation for the cumulative half-macrofracture activation probability for fracture intersection of a set of half-macrofractures during a timestep **N**, by substituting the mean length of active half-macrofractures in the set, $\overline{\ell_I}$, for ℓ_0 in (5.23). Thus

$$\Phi_{I,J,N} = \frac{1 - \chi'\left(\overline{\ell_I} + \Delta\ell_N\right)}{1 - \chi'\left(\overline{\ell_I}\right)} \tag{5.24}$$

where $1-\chi'(\overline{\ell_I})$ and $1-\chi'$ $\chi'(\overline{\ell_I} + \Delta\ell_N)$ are calculated using the "one-sided proximity zone" formula from Appendix B, (B.43).

Finally, if we divide the probability that an initially active half-macrofracture will be deactivated during timestep **N** by the timestep duration, we can calculate the mean probability of half-macrofracture intersection $\overline{F}_{I,J,N}$:

$$\overline{F_{I,J,N}} = \frac{1 - \Phi_{I,J,N}}{t_N - t_{N-1}} \tag{5.25}$$

5.1.4 Combining the Half-Macrofracture Deactivation Probabilities

Since the two half-macrofracture deactivation mechanisms are independent, the probability that a half-macrofracture that is active at the start of a timestep **N** will remain active at the end of the timestep Φ_N is given by multiplying the cumulative half-macrofracture activation functions for stress shadow interaction and for intersection:

$$\Phi_N = \Phi_{I,I,N}\Phi_{I,J,N} \tag{5.26}$$

The instantaneous probability of half-macrofracture deactivation for the timestep F_N is much harder to calculate, since half-macrofracture deactivation can occur by a combination of random and regular processes. However, we can easily calculate the mean probability of half-macrofracture deactivation $\overline{F}_N$ by dividing the probability of half-macrofracture deactivation during the timestep by the timestep duration:

$$\overline{F_N} = \frac{1 - \Phi_N}{t_N - t_{N-1}} \tag{5.27}$$

As we will see in the following sections, it is also useful to define the cumulative probability that a half-macrofracture that is active at the end of a timestep **M** will remain active at the end of a later timestep **N**, $\Phi_{N,M}$:

$$\Phi_{N,M} = \Phi_{M+1} \times \Phi_{M+2} \times \cdots \times \Phi_{N} \tag{5.28}$$

5.2 Calculating Active and Static Microfracture Populations

We can combine the microfracture deactivation probabilities with the cumulative microfracture density distribution functions derived in Chap. 3 to obtain expressions for the active and static macrofracture populations.

5.2.1 *Volumetric Microfracture Density*

We can calculate the cumulative volumetric active and static microfracture densities ${}^{a}_{\mu F}P_{30}(r,t)$ and ${}^{s}_{\mu F}P_{30}(r,t)$ by modifying the cumulative microfracture density distribution function given in (3.14). The cumulative active microfracture density distribution function is obtained by multiplying the cumulative microfracture density distribution function by the cumulative microfracture activation probability:

$${}^{a}_{\mu F}P_{30}(r_N, t_N) = \theta_N B\left(r_N^{\frac{1}{\beta}} + \Gamma_N\right)^{-\beta c} \tag{5.29}$$

To calculate the cumulative volumetric static microfracture density, we must derive an expression for the rate of microfracture deactivation and integrate this through time from $t = 0$ to $t = t_N$. The rate of microfracture deactivation is equal to the active microfracture population times the instantaneous microfracture deactivation rate. However, as we can only calculate the cumulative microfracture activation probability at the end of each timestep, we must use a constant value for θ throughout each timestep, reflecting the macrofracture population at the end of the previous timestep. We must, therefore, discretise the integral into a sum of terms representing individual timesteps. We should also use the mean probability of microfracture deactivation $\overline{q_K}$ rather than the instantaneous probability of microfracture deactivation q_K when calculating the rate of microfracture deactivation during a timestep **K**. This gives

$$\frac{d\,{}^{s}_{\mu F}P_{30}(r_N, t)}{dt} = \overline{q_K}\theta_{K-1}B\left(r_N^{\frac{1}{\beta}} + \Gamma_K - \Gamma_K^{\frac{1}{\beta}}(t_K - t)\right)^{-\beta c} \tag{5.30}$$

Summing the integral of (5.30) across all timesteps up to and including timestep **N** gives

$$
\begin{aligned}
{}^{s}_{\mu F}P_{30}(r_N, t_N) &= \frac{B}{(1-\beta c)} \sum_{K=1}^{N} \left[\int_{t_{K-1}}^{t_K} \overline{q_K} \theta_{K-1} B \left(r_N^{\frac{1}{\beta}} + \Gamma_K - \Gamma_K^{\frac{1}{\beta}} (t_K - t) \right)^{-\beta c} dt \right] \\
&= \frac{B}{(1-\beta c)} \sum_{K=1}^{N} \left[\frac{\overline{q_K} \theta_{K-1}}{\Gamma_K^{\frac{1}{\beta}}} \left(\left(r_N^{\frac{1}{\beta}} + \Gamma_K \right)^{1-\beta c} - \left(r_N^{\frac{1}{\beta}} + \Gamma_{K-1} \right)^{1-\beta c} \right) \right]
\end{aligned}
\tag{5.31}
$$

Note that (5.30) and (5.31) include microfractures that have reached the layer boundaries and become layer-bound macrofractures. To calculate the population of microfractures with $r \leq h/2$ we must subtract the layer-bound macrofracture population from (5.30), and we must also include an extra term in (5.31) to account for microfractures that are deactivated as they reach $r = h/2$ because they lie in a macrofracture exclusion zone:

$$
\begin{aligned}
&{}^{s}_{\mu F}P_{30}(r_N, t_N) \\
&= B \sum_{K=1}^{N} \left[\begin{aligned} &\frac{\overline{q_K} \theta_{K-1}}{(1-\beta c)\gamma_K^{\frac{1}{\beta}}} \left(\left(r_N^{\frac{1}{\beta}} + \Gamma_K \right)^{1-\beta c} - \left(r_N^{\frac{1}{\beta}} + \Gamma_{K-1} \right)^{1-\beta c} - \left(\left(\tfrac{h}{2}\right)^{\frac{1}{\beta}} + \Gamma_K \right)^{1-\beta c} + \left(\left(\tfrac{h}{2}\right)^{\frac{1}{\beta}} + \Gamma_{K-1} \right)^{1-\beta c} \right) \\ &+ \left(\theta_{K-1} - \theta'_{K-1} \right) \left(\left(\left(\tfrac{h}{2}\right)^{\frac{1}{\beta}} + \Gamma_K \right)^{-\beta c} - \left(\left(\tfrac{h}{2}\right)^{\frac{1}{\beta}} + \Gamma_{K-1} \right)^{-\beta c} \right) \end{aligned} \right]
\end{aligned}
\tag{5.32}
$$

5.2.2 *Mean Linear Microfracture Density*

We can calculate the mean linear active and static microfracture densities ${}^{a}_{\mu F}P_{32}(r,t)$ and ${}^{s}_{\mu F}P_{32}(r,t)$ in the same way that we calculated ${}_{\mu F}P_{32}(r,t)$ in Sect. 3.4: differentiating the equations for volumetric microfracture density to determine the number of microfractures with a radius between r and r + dr, multiplying this by the area of a microfracture with radius r, and integrating the resultant function from r = 0 to r = h/2.

As we saw, there is no analytical solution to the integral equations for the mean linear microfracture density, but a reasonable approximation can be obtained by numerical integration. Combining this with the cumulative microfracture activation probability gives us an expression for the mean linear active microfracture density:

$$
\begin{aligned}
&{}^{a}_{\mu F}P_{32}(0, t_N) \\
&= \pi B \theta_N \int_0^{\frac{h}{2}} r^2 \frac{d}{dr} \left[\left(r^{\frac{1}{\beta}} + \Gamma_N \right)^{-\beta c} \right] dr
\end{aligned}
$$

$$\approx \pi \left(\frac{h}{2}\right)^2 \mathrm{B}\theta_N \sum_{i=1}^{I-1} \left[\left(\frac{i}{I}\right)^2 \left(\left(\left(\frac{ih}{2I}\right)^{\frac{1}{\beta}} + \Gamma_\mathrm{N} \right)^{-\beta c} - \left(\left(\frac{(i+1)h}{2I}\right)^{\frac{1}{\beta}} + \Gamma_\mathrm{N} \right)^{-\beta c} \right) \right] \quad (5.33)$$

Calculating the mean linear static microfracture density is more complicated as we must differentiate the summations across timesteps given in (5.31) with respect to r, multiply by fracture area, and then reintegrate numerically across the range of microfracture radii from r = 0 to r = h/2, to get:

$${}^{s}_{\mu F}\mathrm{P}_{32}(0, \mathrm{t}_N) = \frac{\pi \mathrm{B}}{(1-\beta c)} \int_0^{\frac{h}{2}} \mathrm{r}^2 \frac{d}{dr} \left[\sum_{K=1}^{N} \left[\frac{\overline{q_K}\theta_{K-1}}{\gamma_K^{\frac{1}{\beta}}} \left(\left(\mathrm{r}_N^{\frac{1}{\beta}} + \Gamma_K \right)^{1-\beta c} - \left(\mathrm{r}_N^{\frac{1}{\beta}} + \Gamma_{K-1} \right)^{1-\beta c} \right) \right] \right] \mathrm{dr}$$

$$\approx \frac{\pi \mathrm{B}}{(1-\beta c)} \left(\frac{h}{2}\right)^2 \sum_{i=1}^{I-1} \left[\left(\frac{i}{2I}\right)^2 \sum_{K=1}^{N} \left[\frac{\overline{q_K}\theta_{K-1}}{\gamma_K^{\frac{1}{\beta}}} \left(\left(\left(\frac{ih}{2I}\right)^{\frac{1}{\beta}} + \Gamma_K \right)^{1-\beta c} - \left(\left(\frac{ih}{2I}\right)^{\frac{1}{\beta}} + \Gamma_{K-1} \right)^{1-\beta c} - \left(\left(\frac{(i+1)h}{2I}\right)^{\frac{1}{\beta}} + \Gamma_K \right)^{1-\beta c} + \left(\left(\frac{(i+1)h}{2I}\right)^{\frac{1}{\beta}} + \Gamma_{K-1} \right)^{1-\beta c} \right) \right] \right] \quad (5.34)$$

5.3 Calculating Active and Static Half-Macrofracture Populations

We can obtain expressions for the active and static half-macrofracture populations in a similar manner: combining the macrofracture deactivation probabilities with the cumulative macrofracture density distribution functions derived in Chap. 4.

5.3.1 *Volumetric Half-Macrofracture Density*

We can calculate the volumetric density distribution functions for active and static half-macrofractures, ${}^{a}_{\underline{MF}}\mathrm{P}_{30}(\ell,\mathrm{t})$ and ${}^{s}_{\underline{MF}}\mathrm{P}_{30}(\ell,\mathrm{t})$, by modifying (4.10). However, since these calculations are complicated, it is helpful to represent the growth of the fractures graphically.

Figure 5.4 illustrates the growth of a set of half-macrofractures nucleating at different times, as a series of lines on a graph of half-macrofracture length against time. The solid red lines represent the growth of half-macrofractures nucleating at the end of each timestep, and the dashed red line represents the growth of half-macrofractures nucleating during timestep 7. All half-macrofractures propagate at the same rate, so within any given timestep, all the half-macrofracture growth curves are parallel. However, the propagation rate can vary between timesteps; in this example the rate decreases through time, so the growth curves flatten off to the right, but this does not have to be the case. We can trace back along the appropriate growth curve to determine the time of nucleation of a half-macrofracture with any specified length ℓ at any time t.

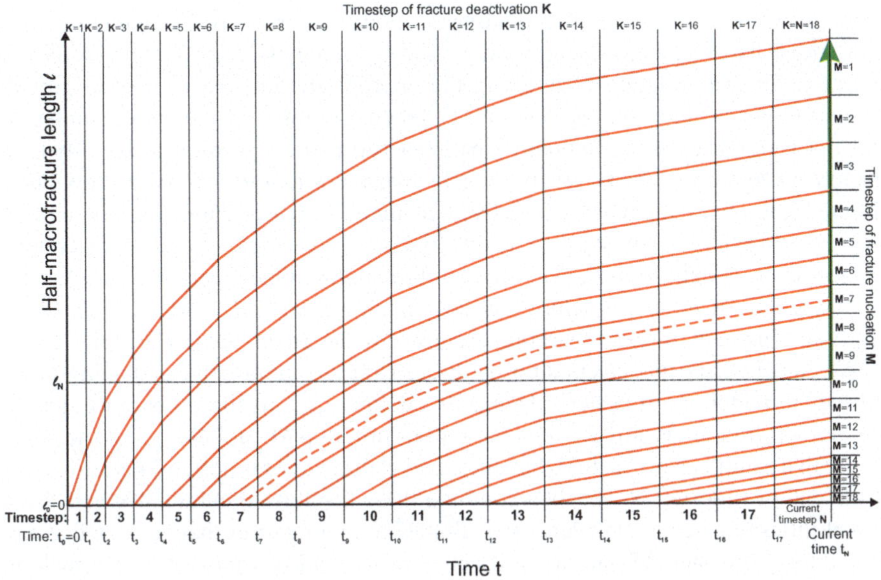

Fig. 5.4 Graph showing the growth of active half-macrofractures through time. The solid red lines represent the growth of half-macrofractures that nucleate at the end of each timestep and remain active until the end of timestep **N**. Of course half-macrofractures may nucleate at any time, and the dashed red line indicates the growth trajectory of a half-macrofracture that nucleated in the middle of timestep **7**. The vertical green line represents the population of active half-macrofractures of length ℓ_N or greater at time t_N

5.3.1.1 Volumetric Density of Active Half-Macrofractures

Calculating the volumetric density of active half-macrofractures is more complicated than calculating the volumetric density of active microfractures, since half-macrofractures of different lengths will have nucleated at different times. The probability that a half-macrofracture is still active (expressed in terms of the cumulative half-macrofracture activation probability Φ) will depend on its nucleation time t_m. We must also take into account the possibility that the seed microfracture was deactivated before it could become a macrofracture; this also depends on the time of macrofracture nucleation.

We cannot, therefore, simply multiply (4.10) by a single cumulative fracture activation probability to obtain an expression for the cumulative volumetric density of active half-macrofractures, as we did with microfractures. Instead, we must derive an expression for the number of currently active half-macrofractures as a function of half-macrofracture length (which is related to the half-macrofracture nucleation time), and then integrate this across the range of all possible lengths.

To do this, we must first modify (4.10) to apply to half-macrofractures rather than macrofractures, by substituting ℓ for L/2 and 2B for B. We then calculate the maximum density of half-macrofractures with length between ℓ and $\ell + d\ell$, if there is no fracture deactivation, by differentiating the modified version of 4.10 with respect to ℓ. We must also determine the nucleation time t_m for a half-macrofracture of current length ℓ, by extrapolating back along the appropriate half-macrofracture growth trajectory (the red lines on Fig. 5.4). We can then multiply the maximum number of half-macrofractures by the cumulative half-macrofracture activation probability for a fracture nucleating at time t_m, and by the effective microfracture activation probability at time t_m, to obtain the density of active half-macrofractures with length between ℓ and $\ell + d\ell$. Finally we must integrate this function between the specified lower half-macrofracture length limit ℓ_N and the current maximum half-macrofracture length, i.e. the length of a macrofracture that nucleated at time $t = 0$, which is given by $\alpha_{MF}\Lambda_{N,0}$. The path of this integration is shown by the green line in Fig. 5.4.

We can simplify this calculation by discretising the cumulative activation probabilities to match the timesteps. Hence, we define $\theta_{\mathbf{M-1}}$ as the probability that an initial microfracture is still active at the start of timestep **M**, and $\Phi_{\mathbf{N,M-1}}$ as the probability that a half-macrofracture that nucleated at the start of timestep **M** is still active at the end of timestep **N**. We then combine these discretised cumulative activation probabilities to calculate the density of all half-macrofractures that nucleated at any time during timestep **M** and are still active at the end of timestep **N**. By making this approximation, we can replace the integral equation described above with a sum across timesteps, so that 4.10 becomes

$$\underline{{}^{a}_{\mathrm{MF}}}\mathrm{P}_{30}(\ell_{\boldsymbol{N}}, \mathrm{t}_{\boldsymbol{N}}) = 2\mathrm{B}\left(\begin{array}{l}\sum_{\boldsymbol{M}=1}^{\boldsymbol{J}-1}\left[\left(\left(\frac{h}{2}\right)^{\frac{1}{\beta}} + \Gamma_{\boldsymbol{M}}\right)^{-\beta c}\left(\Phi_{\boldsymbol{N},\boldsymbol{M}-1}\theta'_{\boldsymbol{M}-1} - \Phi_{\boldsymbol{N},\boldsymbol{M}}\theta'_{\boldsymbol{M}}\right)\right] \\ +\Phi_{\boldsymbol{N},\boldsymbol{J}-1}\theta'_{\boldsymbol{J}-1}\left(\left(\frac{h}{2}\right)^{\frac{1}{\beta}} + \Gamma_{\boldsymbol{J}} - \frac{1}{\beta}\left(\frac{2}{h}\right)^{\frac{b}{2}}(\alpha_{\mathrm{MF}}\Lambda_{\boldsymbol{N},\boldsymbol{J}} - \ell_{\boldsymbol{N}})\right)^{-\beta c}\end{array}\right) \quad (5.35)$$

where timestep **J** is defined such that

$$\alpha_{\mathrm{MF}}\Lambda_{\boldsymbol{N},\boldsymbol{J}} \leq \ell_{\boldsymbol{N}} < \alpha_{\mathrm{MF}}\Lambda_{\boldsymbol{N},\boldsymbol{J}-1} \quad (5.36)$$

We can visualise this using Fig. 5.4. To calculate the cumulative volumetric density of all active half-macrofractures with length greater than ℓ_N at time t_N using continuous fracture activation probabilities, we would integrate the current active density of all half-macrofractures with lengths greater than ℓ_N along the vertical path marked in green. However, if we discretise the fracture activation probabilities by timestep, we convert this integral into a sum of discrete terms representing each timestep from timestep **1**, when the longest currently active half-macrofractures nucleated, to timestep **J**, when half-macrofractures with length ℓ_N nucleated. The boundaries of these terms are the points at which the solid red half-macrofracture growth trajectories cross the green integration line; between these boundaries, we assume that the cumulative fracture activation probabilities are uniform.

We can also derive a semi-continuous form of (5.35) for any time t_n within timestep **N**, in which the term for volumetric active half-macrofracture density is continuous with respect to time, although the cumulative activation probabilities remain discretised:

$$\begin{aligned}&{}^{a}_{\underline{\mathrm{MF}}}\mathrm{P}_{30}(\ell_n, \mathrm{t}_n)\\&= 2\mathrm{B}\left(\begin{array}{c}\sum\limits_{M=1}^{J-1}\left[\left(\Phi_{N-1,M-1}\theta'_{M-1} - \Phi_{N-1,M}\theta'_{M}\right)\left(\left(\frac{h}{2}\right)^{\frac{1}{\beta}} + \Gamma_M\right)^{-\beta c}\right]\\ +\Phi_{N-1,J-1}\theta'_{J-1}\left(\left(\frac{h}{2}\right)^{\frac{1}{\beta}} + \Gamma_J - \frac{1}{\beta}\left(\frac{2}{h}\right)^{\frac{b}{2}}\left(\alpha_{\mathrm{MF}}\Lambda_{N,J} - \alpha_{\mathrm{MF}}\sigma^{\mathrm{b}}_{\mathrm{dN}}(t_N - t_n) - \ell_n\right)\right)^{-\beta c}\end{array}\right)\end{aligned} \tag{5.37}$$

This will be used when calculating the volumetric density of static half-macrofractures.

5.3.1.2 Volumetric Density of Static Half-Macrofractures

We can calculate the cumulative volumetric density distribution function for static half-macrofractures at time $\mathrm{t_N}$ in the same way as we did for static microfractures: we multiply the volumetric density of active half-macrofractures, calculated using (5.37), by the mean probability of half-macrofracture deactivation $\overline{\mathrm{F}}_{\mathbf{K}}$, to obtain an expression for the rate of deactivation of half-macrofractures of length $\ell_{\mathbf{N}}$ or greater, $\mathrm{d}^{\mathrm{S}}_{\underline{\mathrm{MF}}}\mathrm{P}_{30}(\ell_{\mathbf{N}},\mathrm{t})/\mathrm{dt}$. We then integrate this through time from $\mathrm{t} = 0$ to $\mathrm{t} = \mathrm{t_N}$ to obtain an expression for the cumulative volumetric static half-macrofracture density ${}^{\mathrm{S}}_{\underline{\mathrm{MF}}}\mathrm{P}_{30}(\ell_{\mathbf{N}},\mathrm{t}_{\mathbf{N}})$. This is illustrated in Fig. 5.5.

As with microfractures, we can simplify this calculation by discretising it into a series of timesteps **K**, each with a constant mean probability of half-macrofracture deactivation $\overline{\mathrm{F}}_{\mathbf{K}}$. This gives us

$${}^{s}_{\underline{\mathrm{MF}}}\mathrm{P}_{30}(\ell_N, \mathrm{t}_N) = \int\limits_{0}^{t_N} F(t)\,{}^{a}_{\underline{\mathrm{MF}}}\mathrm{P}_{30}(\ell_N, \mathrm{t})dt \approx \sum_{K=1}^{N}\left[\overline{F_K}\int\limits_{t_{K-1}}^{t_K} {}^{a}_{\underline{\mathrm{MF}}}\mathrm{P}_{30}(\ell_N, \mathrm{t})dt\right] \tag{5.38}$$

We can now use the semi-continuous form that we derived for the volumetric density of active half-macrofractures (5.37). Inserting (5.37) into (5.38) and integrating across each timestep **K** gives

$$\begin{aligned}&{}^{s}_{\underline{\mathrm{MF}}}\mathrm{P}_{30}(\ell_N, \mathrm{t}_N)\\&\approx 2\mathrm{B}\sum_{K=1}^{N}\left[\overline{F_K}\int\limits_{t_{K-1}}^{t_K}\left(\begin{array}{c}\sum\limits_{M=1}^{J-1}\left[\left(\Phi_{K-1,M-1}\theta'_{M-1} - \Phi_{K-1,M}\theta'_{M}\right)\left(\left(\frac{h}{2}\right)^{\frac{1}{\beta}} + \Gamma_M\right)^{-\beta c}\right]\\ +\Phi_{K-1,J-1}\theta'_{J-1}\left(\left(\frac{h}{2}\right)^{\frac{1}{\beta}} + \Gamma_J - \frac{1}{\beta}\left(\frac{2}{h}\right)^{\frac{b}{2}}\left(\alpha_{\mathrm{MF}}\Lambda_{K,J} - \alpha_{\mathrm{MF}}\sigma^{\mathrm{b}}_{\mathrm{dK}}(t_K - t) - \ell_N\right)\right)^{-\beta c}\end{array}\right)dt\right]\end{aligned} \tag{5.39}$$

where timestep **J** is such that

$$\alpha_{\mathrm{MF}}\left(\Lambda_{K,J} - \sigma^{\mathrm{b}}_{\mathrm{dK}}(t_K - t)\right) \le \ell_N < \alpha_{\mathrm{MF}}\left(\Lambda_{K,J-1} - \sigma^{\mathrm{b}}_{\mathrm{dK}}(t_K - t)\right) \tag{5.40}$$

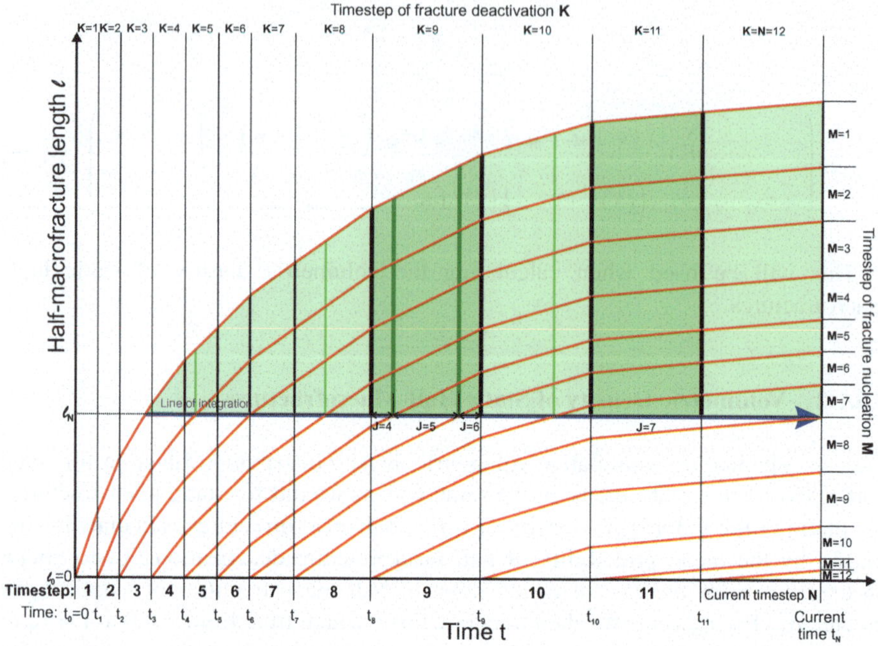

Fig. 5.5 Graphical representation of half-macrofractures represented in the integral in (5.38). To calculate the static half-macrofracture density, we integrate the active half-macrofracture density multiplied by the half-macrofracture deactivation rate with respect to time, along the blue horizontal path. However, the active half-macrofracture density is itself an integral with respect to half-macrofracture length, so overall we must integrate across the entire area shaded in green. **J** represents the timestep in which an active half-macrofracture must have nucleated in order to have length $\ell_{\mathbf{N}}$ at a specific time t. As we move along the integration path, the value of **J** will change whenever we cross a half-macrofracture growth trajectory line (shown in red). This will generally not coincide with the timestep boundaries. In this illustration, we focus on the integration across two timesteps **K**: $\mathbf{K} = \mathbf{9}$, which has multiple **J** values, and $\mathbf{K} = \mathbf{11}$, in which the **J** value is constant

However, we must remember that the active half-macrofracture volumetric density Eq. (5.37) is itself an integral with respect to half-macrofracture length, albeit discretised to a sum of terms representing the half-macrofracture nucleation timesteps. The last term in this sum, which represents the nucleation timestep of a half-macrofracture of length $\ell_{\mathbf{N}}$, varies continuously with time. This is illustrated in Fig. 5.5, where the horizontal blue line represents the path of integration through time, the vertical black lines represent the boundaries of the discretised values of the mean fracture deactivation probability $\overline{\mathrm{F}}_{\mathbf{K}}$ (dependent on the timestep of fracture deactivation **K** in (5.39)), and the vertical green lines represent the boundaries of the nucleation timestep of a half-macrofracture of length $\ell_{\mathbf{N}}$ (i.e. the timestep **J** in (5.37)). As we can see on Fig. 5.5, the two sets of boundaries do not generally coincide, so there may be several different timesteps **J** that satisfy (5.40) for different values of t between $t_{\mathbf{K-1}}$ and $t_{\mathbf{K}}$

(e.g. timestep $\mathbf{K} = \mathbf{9}$ in Fig. 5.5). This makes the definite integral across timestep $\mathbf{K}$ in (5.39) complicated.

The easiest way to solve (5.39) is to integrate each of the individual timesteps $\mathbf{M}$ in the inner sum (including the final term for $\mathbf{M} = \mathbf{J}$) separately across timestep $\mathbf{K}$. Thus we can express (5.39) as a second inner sum over timesteps from $\mathbf{M} = \mathbf{1}$ to $\mathbf{M} = \mathbf{K}$, where $\mathbf{M}$ represents the timestep of nucleation and $\mathbf{K}$ represents the timestep of deactivation of a subset of static half-macrofractures of half-length $\ell_{\mathbf{N}}$ or greater:

$$
\begin{aligned}
&{}_{\underline{\mathrm{MF}}}^{s}\mathrm{P}_{30}(\ell_N, t_N) \\
&\approx 2\mathrm{B}\sum_{K=1}^{N}\left[\overline{F_K}\sum_{M=1}^{K}\left[\int_{t_{K-1}}^{t_K}\left(< {}_{\underline{\mathrm{MF}}}^{a}\mathrm{P}_{30} >_M (\ell_N, t)\right)dt\right]\right] = 2\mathrm{B}\sum_{K=1}^{N}\left[\overline{F_K}\sum_{M=1}^{K}\left[< \int {}_{\underline{\mathrm{MF}}}^{a}\mathrm{P}_{30} >_{K,M}\right]\right]
\end{aligned}
\tag{5.41}
$$

where $< {}_{\underline{\mathrm{MF}}}^{\mathrm{a}}\mathrm{P}_{30} >_{\mathbf{M}}$ represents the terms for individual timesteps $\mathbf{M}$ in the sum in (5.39) (including the final term for $\mathbf{M} = \mathbf{J}$), and $< \int {}_{\mathrm{MF}}^{\mathrm{a}}\mathrm{P}_{30} >_{\mathbf{K,M}}$ represents the integral of these terms across timestep $\mathbf{K}$. The form of $< {}_{\underline{\mathrm{MF}}}^{\mathrm{a}}\mathrm{P}_{30} >_{\mathbf{M}}$ will vary depending on whether $\mathbf{M} < \mathbf{J}$ (when it will be independent of time t), $\mathbf{M} = \mathbf{J}$ (when it will be time dependent) or $\mathbf{M} > \mathbf{J}$ (when it will be zero). The $< \int {}_{\mathrm{MF}}^{\mathrm{a}}\mathrm{P}_{30} >_{\mathbf{K,M}}$ terms in the inner sum can, therefore, be one of six different types, depending on whether the $\mathbf{M} = \mathbf{J}$ for some, all or none of timestep $\mathbf{K}$:

Type 1 terms

For timesteps $\mathbf{M}$ where $\ell_{\mathbf{N}} \leq \alpha_{\mathrm{MF}}\Lambda_{\mathbf{K-1,M}}$ and $\ell_{\mathbf{N}} < \alpha_{\mathrm{MF}}\Lambda_{\mathbf{K,M}}$ (e.g. timesteps $\mathbf{M} = \mathbf{1}$, $\mathbf{M} = \mathbf{2}$ and $\mathbf{M} = \mathbf{3}$ for $\mathbf{K} = \mathbf{9}$ in Fig. 5.5) then $\mathbf{M} < \mathbf{J}$ for all values of t between $t_{\mathbf{K-1}}$ and $t_{\mathbf{K}}$, so only the time-independent component of the integral is relevant. These terms are, therefore, given by

$$
< \int {}_{\underline{\mathrm{MF}}}^{a}\mathrm{P}_{30} >_{K,M} = \left(\Phi_{K-1,M-1}\theta'_{M-1} - \Phi_{K-1,M}\theta'_M\right)\left(\left(\frac{h}{2}\right)^{\frac{1}{\beta}} + \Gamma_M\right)^{-\beta c}(t_K - t_{K-1})
\tag{5.42}
$$

Type 2 terms

For timesteps $\mathbf{M}$ where $\alpha_{\mathrm{MF}}\Lambda_{\mathbf{K-1,M}} < \ell_{\mathbf{N}} \leq \alpha_{\mathrm{MF}}\Lambda_{\mathbf{K-1,M-1}}$ and $\ell_{\mathbf{N}} < \alpha_{\mathrm{MF}}\Lambda_{\mathbf{K,M}}$ (e.g. timestep $\mathbf{M} = \mathbf{4}$ for $\mathbf{K} = \mathbf{9}$ in Fig. 5.5) then $\mathbf{M} = \mathbf{J}$ for values of t between $t_{\mathbf{K-1}}$ and $t_{\mathbf{K}} - (\alpha_{\mathrm{MF}}\Lambda_{\mathbf{K,M}} - \ell_{\mathbf{N}})/(\alpha_{\mathrm{MF}}\,\sigma_{\mathbf{dK}}^{\mathrm{b}})$, but $\mathbf{M} < \mathbf{J}$ for values of t between $t_{\mathbf{K}} - (\alpha_{\mathrm{MF}}\Lambda_{\mathbf{K,M}} - \ell_{\mathbf{N}})/(\alpha_{\mathrm{MF}}\,\sigma_{\mathbf{dK}}^{\mathrm{b}})$ and $t_{\mathbf{K}}$. We must, therefore, include both the time-dependent component of the integral (when $\mathbf{M} = \mathbf{J}$) and the time-independent component of the integral (when $\mathbf{M} < \mathbf{J}$), to obtain the term:

$$
\begin{aligned}
&\left\langle \int {}_{\underline{\mathrm{MF}}}^{a}\mathrm{P}_{30} \right\rangle_{K,M} \\
&= \left(\Phi_{K-1,M-1}\theta'_{M-1} - \Phi_{K-1,M}\theta'_M\right)\left(\left(\frac{h}{2}\right)^{\frac{1}{\beta}} + \Gamma_M\right)^{-\beta C}\left(\frac{\alpha_{MF}\Lambda_{K,M} - \ell_N}{\alpha_{MF}\sigma_{dK}^{b}}\right) \\
&+ \frac{-\beta\Phi_{K-1,M-1}\theta'_{M-1}}{(1-\beta C)\alpha_{MF}\sigma_{dK}^{b}}\left(\frac{h}{2}\right)^{\frac{b}{2}}\left(\left(\left(\frac{h}{2}\right)^{\frac{1}{\beta}} + \Gamma_M\right)^{1-\beta c} - \left(\left(\frac{h}{2}\right)^{\frac{1}{\beta}} + \Gamma_{M-1} - \frac{1}{\beta}\left(\frac{2}{h}\right)^{\frac{b}{2}}\left(\alpha_{MF}\Lambda_{K-1,M-1} - \ell_N\right)\right)^{1-\beta C}\right)
\end{aligned}
\tag{5.43}
$$

Type 3 terms

For timesteps **M** where $\alpha_{MF}\Lambda_{\mathbf{K-1,M-1}} < \ell_{\mathrm{N}} < \alpha_{MF}\Lambda_{\mathbf{K,M}}$ (e.g. timestep **M** = **5** for **K** = **9** in Fig. 5.5) then **M** > **J** for values of t between $t_{\mathbf{K-1}}$ and $t_{\mathbf{K}} - (\alpha_{MF}\Lambda_{\mathbf{K,M-1}} - \ell_{\mathrm{N}})/(\alpha_{MF}\ \sigma^{b}_{\mathrm{d}\mathbf{K}})$, **M** = **J** for values of t between $t_{\mathbf{K}} - (\alpha_{MF}\Lambda_{\mathbf{K,M-1}} - \ell_{\mathrm{N}})/(\alpha_{MF}\ \sigma^{b}_{\mathrm{d}\mathbf{K}})$ and $t_{\mathbf{K}} - (\alpha_{MF}\Lambda_{\mathbf{K,M}} - \ell_{\mathrm{N}})/(\alpha_{MF}\ \sigma^{b}_{\mathrm{d}\mathbf{K}})$, and **M** < **J** for values of t between $t_{\mathbf{K}} - (\alpha_{MF}\Lambda_{\mathbf{K,M}} - \ell_{\mathrm{N}})/(\alpha_{MF}\ \sigma^{b}_{\mathrm{d}\mathbf{K}})$ and $t_{\mathbf{K}}$. We can ignore the interval where **M** > **J**, but we must include both the time-dependent component of the integral (when **M** = **J**) and the time-independent component of the integral (when **M** < **J**) to obtain the term:

$$
\begin{aligned}
&< \int^{a}_{\underline{MF}} P_{30} >_{\boldsymbol{K,M}} \\
&= \left(\Phi_{\boldsymbol{K}-1,\boldsymbol{M}-1}\theta'_{\boldsymbol{M}-1} - \Phi_{\boldsymbol{K}-1,\boldsymbol{M}}\theta'_{\boldsymbol{M}}\right)\left(\left(\frac{h}{2}\right)^{\frac{1}{\beta}} + \Gamma_{\boldsymbol{M}}\right)^{-\beta c}\left(\frac{\alpha_{MF}\Lambda_{\boldsymbol{K,M}} - \ell_{N}}{\alpha_{MF}\sigma^{b}_{d\boldsymbol{K}}}\right) \\
&+ \frac{-\beta\Phi_{\boldsymbol{K}-1,\boldsymbol{M}-1}\theta'_{\boldsymbol{M}-1}}{(1-\beta c)\alpha_{MF}\sigma^{b}_{\mathrm{d}\mathbf{K}}}\left(\frac{h}{2}\right)^{\frac{b}{2}}\left(\left(\left(\frac{h}{2}\right)^{\frac{1}{\beta}} + \Gamma_{\boldsymbol{M}}\right)^{1-\beta c} - \left(\left(\frac{h}{2}\right)^{\frac{1}{\beta}} + \Gamma_{\boldsymbol{M}-1}\right)^{1-\beta c}\right)
\end{aligned}
\tag{5.44}
$$

Type 4 terms

For timesteps **M** where $\alpha_{MF}\Lambda_{\mathbf{K,M}} \leq \ell_{\mathrm{N}} < \alpha_{MF}\Lambda_{\mathbf{K,M-1}}$ and $\ell_{\mathrm{N}} > \alpha_{MF}\Lambda_{\mathbf{K-1,M-1}}$ (e.g. timestep **M** = **6** for **K** = **9** in Fig. 5.5) then **M** > **J** for values of t between $t_{\mathbf{K-1}}$ and $t_{\mathbf{K}} - (\alpha_{MF}\Lambda_{\mathbf{K,M-1}} - \ell_{\mathrm{N}})/(\alpha_{MF}\ \sigma^{b}_{\mathrm{d}\mathbf{K}})$ and **M** = **J** for values of t between $t_{\mathbf{K}} - (\alpha_{MF}\Lambda_{\mathbf{K,M-1}} - \ell_{\mathrm{N}})/(\alpha_{MF}\ \sigma^{b}_{\mathrm{d}\mathbf{K}})$ and $t_{\mathbf{K}}$. We can ignore the interval where **M** > **J**, but we must include the time-dependent component of the integral (when **M** = **J**) to obtain the term:

$$
\begin{aligned}
&< \int^{a}_{\underline{MF}} P_{30} >_{\boldsymbol{K,M}} \\
&= + \frac{-\beta\Phi_{\boldsymbol{K}-1,\boldsymbol{M}-1}\theta'_{\boldsymbol{M}-1}}{(1-\beta c)\alpha_{MF}\sigma^{b}_{\mathrm{d}\mathbf{K}}}\left(\frac{h}{2}\right)^{\frac{b}{2}}\left(\left(\left(\frac{h}{2}\right)^{\frac{1}{\beta}} + \Gamma_{\boldsymbol{M}} - \frac{1}{\beta}\left(\frac{2}{h}\right)^{\frac{b}{2}}\left(\alpha_{MF}\Lambda_{\boldsymbol{K,M}} - \ell_{\boldsymbol{N}}\right)\right)^{1-\beta c} - \left(\left(\frac{h}{2}\right)^{\frac{1}{\beta}} + \Gamma_{\boldsymbol{M}-1}\right)^{1-\beta c}\right)
\end{aligned}
\tag{5.45}
$$

Type 5 terms

It is also possible to have a timestep **M** where $\alpha_{MF}\Lambda_{\mathbf{K-1,M}} < \ell_{\mathrm{N}} \leq \alpha_{MF}\Lambda_{\mathbf{K-1,M-1}}$ but $\ell_{\mathrm{N}} \geq \alpha_{MF}\Lambda_{\mathbf{K,M}}$ (e.g. timestep **M** = **7** for **K** = **11** in Fig. 5.5). In this case, **M** = **J** for all values of t between $t_{\mathbf{K-1}}$ and $t_{\mathbf{K}}$. We, therefore, only need to include the time-dependent component of the integral, to obtain the term:

$$< \int^{a}_{\underline{MF}} P_{30} >_{\boldsymbol{K,M}}$$

$$= + \frac{-\beta \Phi_{\boldsymbol{K}-1,\boldsymbol{M}-1} \theta'_{\boldsymbol{M}-1}}{(1-\beta c)\alpha_{MF} \sigma_{d\mathbf{K}}^{b}} \left(\frac{h}{2}\right)^{\frac{b}{2}} \left(\begin{array}{c} \left(\left(\frac{h}{2}\right)^{\frac{1}{\beta}} + \Gamma_{\boldsymbol{M}} - \frac{1}{\beta}\left(\frac{2}{h}\right)^{\frac{b}{2}} \left(\alpha_{MF} \Lambda_{\boldsymbol{K},\boldsymbol{M}} - \ell_{\boldsymbol{N}}\right)\right)^{1-\beta c} \\ -\left(\left(\frac{h}{2}\right)^{\frac{1}{\beta}} + \Gamma_{\boldsymbol{M}-1} - \frac{1}{\beta}\left(\frac{2}{h}\right)^{\frac{b}{2}} \left(\alpha_{MF} \Lambda_{\boldsymbol{K}-1,\boldsymbol{M}-1} - \ell_{\boldsymbol{N}}\right)\right)^{1-\beta c} \end{array} \right) \tag{5.46}$$

Type 6 terms

Finally, for timesteps **M** where $\ell_{\mathbf{N}} \geq \alpha_{MF}\Lambda_{\mathbf{K,M-1}}$ (e.g. timestep **M** = **7** for **K** = **9** in Fig. 5.5) then **M** > **J** for all values of t between $t_{\mathbf{K-1}}$ and $t_{\mathbf{K}}$. Therefore

$$< \int {}^{a}_{\underline{MF}}P_{30} >_{\boldsymbol{K},\boldsymbol{M}} = 0 \tag{5.47}$$

and we can ignore these timesteps.

We can, therefore, obtain an expression for the volumetric density of static half-macrofractures of length $\ell_{\mathbf{N}}$ or greater at time $t_{\mathbf{N}}$ by inserting the appropriate terms 1 to 6 into the double sum in (5.41). Note that when $\ell_{\mathbf{N}} = 0$, we can use a Type 1 term for all timesteps **M** < **K**, and a Type 5 term for timestep **M** = **K**. The Type 5 term can be further simplified since $\ell_{\mathbf{N}} = \Lambda_{\mathbf{K,M}} = \Lambda_{\mathbf{K-1,M-1}} = 0$. The total volumetric density of all static half-macrofractures can thus be written as

$$\begin{aligned} & {}^{s}_{\underline{MF}}P_{30}(0, t_{\boldsymbol{N}}) \\ & \approx 2B \sum_{\boldsymbol{K}=1}^{\boldsymbol{N}} \left[\overline{F_{\boldsymbol{K}}} \left(\begin{array}{l} (t_{\boldsymbol{K}} - t_{\boldsymbol{K}-1}) \sum_{\boldsymbol{M}=1}^{\boldsymbol{K}-1} \left[\left(\Phi_{\boldsymbol{K}-1,\boldsymbol{M}-1}\theta'_{\boldsymbol{M}-1} - \Phi_{\boldsymbol{K}-1,\boldsymbol{M}}\theta'_{\boldsymbol{M}}\right) \left(\left(\frac{h}{2}\right)^{\frac{1}{\beta}} + \Gamma_{\boldsymbol{M}}\right)^{-\beta c} \right] \\ + \frac{-\beta \theta'_{\boldsymbol{K}-1}}{(1-\beta c)\alpha_{MF}\sigma_{d\mathbf{K}}^{b}} \left(\frac{h}{2}\right)^{\frac{b}{2}} \left(\left(\left(\frac{h}{2}\right)^{\frac{1}{\beta}} + \Gamma_{\boldsymbol{K}}\right)^{1-\beta c} - \left(\left(\frac{h}{2}\right)^{\frac{1}{\beta}} + \Gamma_{\boldsymbol{K}-1}\right)^{1-\beta c} \right) \end{array} \right) \right] \end{aligned} \tag{5.48}$$

5.3.2 *Mean Linear Half-Macrofracture Density*

We can calculate the mean linear density distribution functions for active and static half-macrofractures, ${}^{a}_{\underline{MF}}P_{32}(\ell,t)$ and ${}^{s}_{\underline{MF}}P_{32}(\ell,t)$, in the same way that we calculated $_{MF}P_{32}(L,t)$ in Sect. 4.3: differentiating the appropriate P_{30} functions with respect to ℓ to determine the number of half-macrofractures with half-length between ℓ and $\ell + d\ell$, multiplying by ℓh to obtain the total area of half-macrofractures with half-length between ℓ and $\ell + d\ell$, and then integrating this expression from $\ell_{\mathbf{N}}$ to the maximum possible half-macrofracture half-length ℓ_{max}.

5.3.2.1 Mean Linear Density of Active Half-Macrofractures

We can modify the differential of $_{MF}P_{30}(\ell,t)$ in (4.13) to include the cumulative activation functions $\Phi_{\mathbf{N,M-1}}$ and $\theta'_{\mathbf{M-1}}$, and also to represent half-macrofractures instead of macrofractures. This gives us

$$\frac{d_{\underline{\mathrm{MF}}}\mathrm{P}_{30}(\ell,\mathrm{t}_N)}{d\ell}=\frac{d}{d\ell}\left[2\mathrm{B}\Phi_{N,M-1}\theta'_{M-1}\left(\left(\frac{h}{2}\right)^{\frac{1}{\beta}}+\Gamma_N+\left(\frac{2}{h}\right)^{\frac{b}{2}}\frac{\ell}{\beta}\right)^{-\beta c}\right]\tag{5.49}$$

for a half-macrofracture that nucleated in timestep **M**. Since cumulative activation probabilities $\Phi_{\mathbf{N,M-1}}$ and $\theta'_{\mathbf{M-1}}$ are discretised to match the timesteps, they can be treated as constants when integrating across half-macrofractures that nucleated in the same timestep **M**. The full form of equation (4.14) derived in Sect. 4.3 is already discretised by the macrofracture nucleation timestep, and therefore requires only minor modifications to give an expression for the mean linear density of active half-macrofractures of length $\ell_{\mathbf{N}}$ or greater at time $\mathrm{t}_{\mathbf{N}}$:

$$\begin{aligned}
&{}^{a}_{\underline{\mathrm{MF}}}\mathrm{P}_{32}(\ell_N,\mathrm{t}_N)\\
&=2\mathrm{Bh}\sum_{M=1}^{K-1}\left[\Phi_{N,M-1}\theta'_{M-1}\left(\begin{array}{c}\frac{-\beta}{(1-\beta c)}\left(\frac{h}{2}\right)^{\frac{b}{2}}\left(\left(\left(\frac{h}{2}\right)^{\frac{1}{\beta}}+\Gamma_M\right)^{1-\beta c}-\left(\left(\frac{h}{2}\right)^{\frac{1}{\beta}}+\Gamma_{M-1}\right)^{1-\beta c}\right)\\+\left(\left(\left(\frac{h}{2}\right)^{\frac{1}{\beta}}+\Gamma_M\right)^{-\beta c}\alpha_{\mathrm{MF}}\Lambda_{N,M}-\left(\left(\frac{h}{2}\right)^{\frac{1}{\beta}}+\Gamma_{M-1}\right)^{-\beta c}\alpha_{\mathrm{MF}}\Lambda_{N,M-1}\right)\end{array}\right)\right]\\
&+2\mathrm{Bh}\Phi_{N,K-1}\theta'_{K-1}\left(\begin{array}{c}\frac{-\beta}{(1-\beta c)}\left(\frac{h}{2}\right)^{\frac{b}{2}}\left(\left(\left(\frac{h}{2}\right)^{\frac{1}{\beta}}+\Gamma_N+\left(\frac{2}{h}\right)^{\frac{b}{2}}\frac{\ell_N}{\beta}\right)^{1-\beta c}-\left(\left(\frac{h}{2}\right)^{\frac{1}{\beta}}+\Gamma_{K-1}\right)^{1-\beta c}\right)\\+\left(\left(\left(\frac{h}{2}\right)^{\frac{1}{\beta}}+\Gamma_N+\left(\frac{2}{h}\right)^{\frac{b}{2}}\frac{\ell_N}{\beta}\right)^{-\beta c}\ell_N-\left(\left(\frac{h}{2}\right)^{\frac{1}{\beta}}+\Gamma_{K-1}\right)^{-\beta c}\alpha_{MF}\Lambda_{N,K-1}\right)\end{array}\right)
\end{aligned}\tag{5.50}$$

where timestep **K** represents the timestep of nucleation of a half-macrofracture with current length $\ell_{\mathbf{N}}$, defined such that

$$\alpha_{\mathrm{MF}}\Lambda_{N,K}\leq\ell_N<\alpha_{\mathrm{MF}}\Lambda_{N,K-1}\tag{5.51}$$

However, this time we cannot simplify (5.50) by cancelling terms across the timesteps, since the cumulative activation probabilities will be different. We can, however, derive a simplified expression for the total mean linear density of all active half-macrofractures by setting $\ell_{\mathbf{N}}=0$:

$$\begin{aligned}
&{}^{a}_{\underline{\mathrm{MF}}}\mathrm{P}_{32}(0,\mathrm{t}_N)\\
&=2\mathrm{Bh}\sum_{M=1}^{N}\left[\Phi_{N,M-1}\theta'_{M-1}\left(\begin{array}{c}\frac{-\beta}{(1-\beta c)}\left(\frac{h}{2}\right)^{\frac{b}{2}}\left(\left(\left(\frac{h}{2}\right)^{\frac{1}{\beta}}+\Gamma_M\right)^{1-\beta c}-\left(\left(\frac{h}{2}\right)^{\frac{1}{\beta}}+\Gamma_{M-1}\right)^{1-\beta c}\right)\\+\left(\left(\left(\frac{h}{2}\right)^{\frac{1}{\beta}}+\Gamma_M\right)^{-\beta c}\alpha_{\mathrm{MF}}\Lambda_{N,M}-\left(\left(\frac{h}{2}\right)^{\frac{1}{\beta}}+\Gamma_{M-1}\right)^{-\beta c}\alpha_{\mathrm{MF}}\Lambda_{N,M-1}\right)\end{array}\right)\right]
\end{aligned}\tag{5.52}$$

5.3.2.2 Mean Linear Density of Static Half-Macrofractures

We can calculate the mean linear density for static half-macrofractures in a similar way: differentiate the volumetric density functions for static half-macrofractures with respect to ℓ to obtain an expression for the number of half-macrofractures with length between ℓ and $\ell + d\ell$, multiply this by the half-macrofracture area ℓh, and then integrate the resulting functions across a range of half-lengths, up to the maximum possible half-length ℓ_{max}. This is illustrated in Fig. 5.6.

The functions for ${}^{s}_{\underline{MF}}P_{30}$ derived in Sect. 5.3.1.2 are broken down into multiple terms based on the timestep of half-macrofracture nucleation **M** and the timestep of deactivation **K**. We can differentiate and reintegrate each of these terms independently. Thus the differential of the static half-macrofracture volumetric density with respect to length is given by

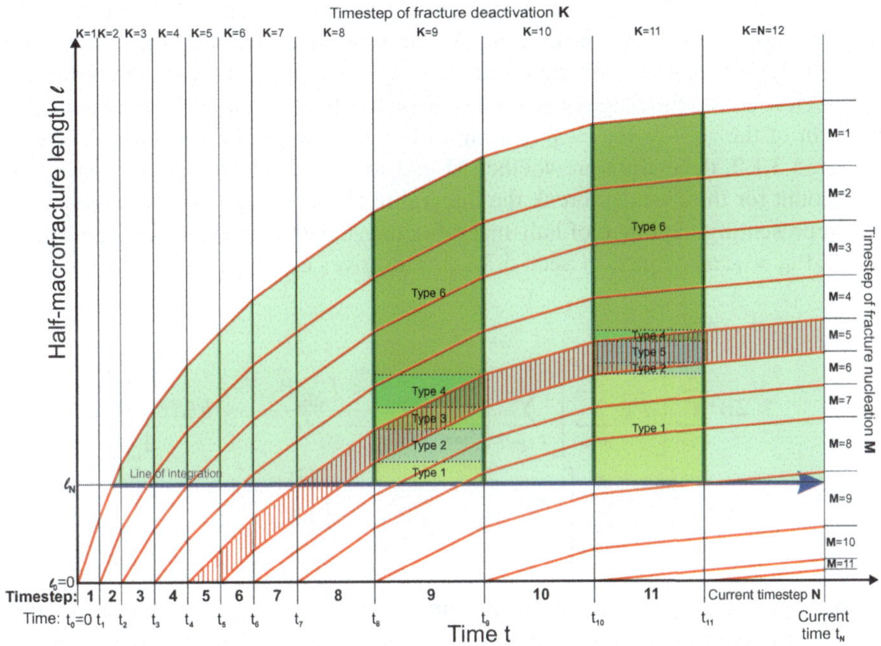

Fig. 5.6 Graphical representation of the triple summation in (5.55). Here we take the term for a specific combination of nucleation timestep **M** (in this case **M** = **5**) and deactivation timestep **K** (two examples are shown, **K** = **9** and **K** = **11**), and integrate it with respect to half-macrofracture length ℓ (i.e. in the vertical direction). The nucleation timestep **J** of an active half-macrofracture with the exact length ℓ at time t will vary with both t and ℓ. Therefore, as ℓ increases, the $< \int {}^{a}_{\underline{MF}}P_{30} >_{\mathbf{K,M}}$ term, representing the density of active half-macrofractures nucleated in timestep **M**, will vary between the six types described in Sect. 5.3.1.2, depending on whether **M** = **J** for some, all or none of timestep **K**. The green shaded areas show the range of lengths over which the different types of $< \int {}^{a}_{\underline{MF}}P_{30} >_{\mathbf{K,M}}$ term are valid, for timesteps **K** = **9** and **K** = **11**. For any timestep **K**, the total value of the integral across all lengths between ℓ_N and ℓ_{max}, and from $t_{\mathbf{K-1}}$ to $t_{\mathbf{K}}$, can be calculated by summing the integrals for each of the 6 types of $< \int {}^{a}_{\underline{MF}}P_{30} >_{\mathbf{K,M}}$ term over the intervals in which they are valid (note that some types may not be valid over any interval, for any given combination of **K** and **M**)

$$\frac{d^s_{\underline{MF}}P_{30}(\ell, t_N)}{d\ell} \approx 2B\sum_{K=1}^{N}\left[\overline{F_K}\sum_{M=1}^{K}\left[\frac{d < \int {}^a_{\underline{MF}}P_{30} >_{K,M}}{d\ell}\right]\right] \tag{5.53}$$

and the mean linear density of static half-macrofractures, ${}^s_{\underline{MF}}P_{32}(\ell_N,t_N)$, is given by

$${}^s_{\underline{MF}}P_{32}(\ell_N, t_N) \approx 2B\sum_{K=1}^{N}\left[\overline{F_K}\sum_{M=1}^{K}\left[\int_{\ell_N}^{\ell_{max}} -\frac{d < \int {}^a_{\underline{MF}}P_{30} >_{K,M}}{d\ell} h\ell d\ell\right]\right] \tag{5.54}$$

The differential of the $< \int {}^a_{\underline{MF}}P_{30} >_{K,M}$ term with respect to ℓ represents the volumetric density of active macrofractures with length between ℓ and $\ell + d\ell$, integrated in time throughout timestep **K**. However, the nucleation timestep **J** for an active half-macrofracture length ℓ at time t will change with ℓ and t. Therefore, as we integrate across the range of possible half-macrofracture lengths from ℓ_N to ℓ_{max}, the form of the $< \int {}^a_{\underline{MF}}P_{30} >_{K,M}$ term will vary between the six types described in Sect. 5.3.1.2, depending on whether **M** = **J** for some, all or none of timestep **K**. To account for this, we can break the integral in (5.54) down into six components, each representing the range of half-macrofracture lengths covered by a type T term $< \int {}^a_{\underline{MF}}P_{30} >_T$, as defined in Sect. 5.3.1.2. This gives us an extra summation:

$$\begin{aligned}
&{}^s_{\underline{MF}}P_{32}(\ell_N, t_N) \\
&\approx 2B\sum_{K=1}^{N}\left[\overline{F_K}\sum_{M=1}^{K}\left[\sum_{TypeT}\left[\int_{\ell_{min(T)}}^{\ell_{max(T)}} -\frac{d < \int {}^a_{\underline{MF}}P_{30} >_T}{d\ell} h\ell d\ell\right]\right]\right] \\
&= 2B\sum_{K=1}^{N}\left[\overline{F_K}\sum_{M=1}^{K}\left[\sum_{TypeT}\left[< \int {}^a_{\underline{MF}}P_{32} >_T\right]\right]\right]
\end{aligned} \tag{5.55}$$

where $\ell_{min(T)}$ and $\ell_{max(T)}$ represent the limits of the range of half-macrofracture lengths covered by a type T term $< \int {}^a_{\underline{MF}}P_{30} >_T$, or ℓ_N if this is greater. This is illustrated for a specific timestep $\mathbf{M} = \underline{\mathbf{5}}$ in Fig. 5.6. We can calculate the integral components for each of the six types of terms $< \int {}^a_{\underline{MF}}P_{32} >_T$ as follows:

Integral component for Type 1 range

When $\ell \leq \alpha_{MF}\Lambda_{K-1,M}$ (and $\ell < \alpha_{MF}\Lambda_{K,M}$) then (5.42) gives us

$$\begin{aligned}
&< \int {}^a_{\underline{MF}}P_{30} >_{K,M} \\
&= \left(\Phi_{K-1,M-1}\theta'_{M-1} - \Phi_{K-1,M}\theta'_M\right)\left(\left(\frac{h}{2}\right)^{\frac{1}{\beta}} + \Gamma_M\right)^{-\beta c}(t_K - t_{K-1}) \\
&\Rightarrow -\frac{d < \int {}^a_{\underline{MF}}P_{30} >_{K,M}}{d\ell} = 0 \\
&\Rightarrow < \int {}^a_{\underline{MF}}P_{32} >_{T1} = 0
\end{aligned} \tag{5.56}$$

We can, therefore, ignore this component.

Integral component for Type 2 range

When $\alpha_{MF}\Lambda_{\mathbf{K-1,M}} < \ell \le \alpha_{MF}\Lambda_{\mathbf{K-1,M-1}}$ and $\ell < \alpha_{MF}\Lambda_{\mathbf{K,M}}$ then (5.43) gives us

$$
\begin{aligned}
&< \int {}^{a}_{\underline{MF}}P_{30} >_{K,M} \\
&= \left(\Phi_{K-1,M-1}\theta'_{M-1} - \Phi_{K-1,M}\theta'_{M}\right)\left(\left(\frac{h}{2}\right)^{\frac{1}{\beta}} + \Gamma_M\right)^{-\beta c}\left(\frac{\alpha_{MF}\Lambda_{K,M} - \ell}{\alpha_{MF}\sigma_{dK}^{b}}\right) \\
&\quad - \frac{\beta\Phi_{K-1,M-1}\theta'_{M-1}}{(1-\beta c)\alpha_{MF}\sigma_{dK}^{b}}\left(\frac{h}{2}\right)^{\frac{b}{2}}\left(\left(\left(\frac{h}{2}\right)^{\frac{1}{\beta}} + \Gamma_M\right)^{1-\beta c} - \left(\left(\frac{h}{2}\right)^{\frac{1}{\beta}} + \Gamma_{M-1} - \frac{1}{\beta}\left(\frac{2}{h}\right)^{\frac{b}{2}}\left(\alpha_{MF}\Lambda_{K-1,M-1} - \ell\right)\right)^{1-\beta c}\right) \\
&\Rightarrow -\frac{d < \int {}^{a}_{\underline{MF}}P_{30} >_{K,M}}{d\ell} = \frac{\left(\Phi_{K-1,M-1}\theta'_{M-1} - \Phi_{K-1,M}\theta'_{M}\right)}{\alpha_{MF}\sigma_{dK}^{b}}\left(\left(\frac{h}{2}\right)^{\frac{1}{\beta}} + \Gamma_M\right)^{-\beta c} \\
&\quad - \frac{\Phi_{K-1,M-1}\theta'_{M-1}}{\alpha_{MF}\sigma_{dK}^{b}}\left(\left(\frac{h}{2}\right)^{\frac{1}{\beta}} + \Gamma_{M-1} - \frac{1}{\beta}\left(\frac{2}{h}\right)^{\frac{b}{2}}\left(\alpha_{MF}\Lambda_{K-1,M-1} - \ell\right)\right)^{-\beta c}
\end{aligned} \tag{5.57}
$$

$$
\begin{aligned}
&\Rightarrow \langle \int {}^{a}_{\underline{MF}}P_{32}\rangle_{T2} = \frac{\left(\Phi_{K-1,M-1}\theta'_{M-1} - \Phi_{K-1,M}\theta'_{M}\right)}{\alpha_{MF}\sigma_{dK}{}^{b}}\left(\left(\frac{h}{2}\right)^{\frac{1}{\beta}} + \Gamma_M\right)^{-\beta c}\frac{h}{2}\left(max\left\{\begin{matrix}(\ell_N)^2 \\ min\left\{\begin{matrix}(\alpha_{MF}\Lambda_{K,M})^2 \\ (\alpha_{MF}\Lambda_{K-1,M-1})^2\end{matrix}\right\}\end{matrix}\right\} - max\left\{\begin{matrix}(\ell_N)^2 \\ (\alpha_{MF}\Lambda_{K-1,M})^2\end{matrix}\right\}\right) \\
&\quad + \left(\frac{\Phi_{K-1,M-1}\theta'_{M-1}}{\alpha_{MF}\sigma_{dK}{}^{b}(1-\beta c)(2-\beta c)}h\left(\beta\left(\frac{h}{2}\right)^{\frac{b}{2}}\right)^2\right) \times \\
&\left(\left(\left(\frac{h}{2}\right)^{\frac{1}{\beta}} + \Gamma_{M-1} - \frac{1}{\beta}\left(\frac{2}{h}\right)^{\frac{b}{2}}\left(\alpha_{MF}\Lambda_{K-1,M-1} - max\left\{\begin{matrix}\ell_N \\ min\left\{\begin{matrix}\alpha_{MF}\Lambda_{K,M} \\ \alpha_{MF}\Lambda_{K-1,M-1}\end{matrix}\right\}\end{matrix}\right\}\right)\right)^{1-\beta c}\left(\left(\frac{h}{2}\right)^{\frac{1}{\beta}} + \Gamma_{M-1} - \frac{1}{\beta}\left(\frac{2}{h}\right)^{\frac{b}{2}}\left(\alpha_{MF}\Lambda_{K-1,M-1} + (1-\beta c)max\left\{\begin{matrix}\ell_N \\ min\left\{\begin{matrix}\alpha_{MF}\Lambda_{K,M} \\ \alpha_{MF}\Lambda_{K-1,M-1}\end{matrix}\right\}\end{matrix}\right\}\right)\right)\right. \\
&\left.\quad - \left(\left(\frac{h}{2}\right)^{\frac{1}{\beta}} + \Gamma_{M-1} - \frac{1}{\beta}\left(\frac{2}{h}\right)^{\frac{b}{2}}\left(\alpha_{MF}\Lambda_{K-1,M-1} - max\left\{\begin{matrix}\ell_N \\ \alpha_{MF}\Lambda_{K-1,M}\end{matrix}\right\}\right)\right)^{1-\beta c}\left(\left(\frac{h}{2}\right)^{\frac{1}{\beta}} + \Gamma_{M-1} - \frac{1}{\beta}\left(\frac{2}{h}\right)^{\frac{b}{2}}\left(\alpha_{MF}\Lambda_{K-1,M-1} + (1-\beta c)max\left\{\begin{matrix}\ell_N \\ \alpha_{MF}\Lambda_{K-1,M}\end{matrix}\right\}\right)\right)\right)
\end{aligned} \tag{5.58}
$$

Integral component for Type 3 range

When $\alpha_{MF}\Lambda_{\mathbf{K-1,M-1}} < \ell < \alpha_{MF}\Lambda_{\mathbf{K,M}}$ (i.e. $\alpha_{MF}\Lambda_{\mathbf{K,M}} > \alpha_{MF}\Lambda_{\mathbf{K-1,M-1}}$) then (5.44) gives us

$$
\begin{aligned}
&< \int {}^{a}_{\underline{MF}}P_{30} >_{K,M} \\
&= \left(\Phi_{K-1,M-1}\theta'_{M-1} - \Phi_{K-1,M}\theta'_{M}\right)\left(\left(\frac{h}{2}\right)^{\frac{1}{\beta}} + \Gamma_M\right)^{-\beta c}\left(\frac{\alpha_{MF}\Lambda_{K,M} - \ell}{\alpha_{MF}\sigma_{dK}^{b}}\right) \\
&\quad - \frac{\beta\Phi_{K-1,M-1}\theta'_{M-1}}{(1-\beta c)\alpha_{MF}\sigma_{dK}^{b}}\left(\frac{h}{2}\right)^{\frac{b}{2}}\left(\left(\left(\frac{h}{2}\right)^{\frac{1}{\beta}} + \Gamma_M\right)^{1-\beta c} - \left(\left(\frac{h}{2}\right)^{\frac{1}{\beta}} + \Gamma_{M-1}\right)^{1-\beta c}\right) \\
&\Rightarrow -\frac{d < \int {}^{a}_{\underline{MF}}P_{30} >_{K,M}}{d\ell} = \frac{\left(\Phi_{K-1,M-1}\theta'_{M-1} - \Phi_{K-1,M}\theta'_{M}\right)}{\alpha_{MF}\sigma_{dK}^{b}}\left(\left(\frac{h}{2}\right)^{\frac{1}{\beta}} + \Gamma_M\right)^{-\beta c}
\end{aligned} \tag{5.59}
$$

$$\Rightarrow < \int {}^{a}_{\underline{MF}}P_{32} >_{T3}$$

$$= \frac{\left(\Phi_{K-1,M-1}\theta'_{M-1} - \Phi_{K-1,M}\theta'_M\right)}{\alpha_{MF}\sigma_{dK}^b}\left(\left(\frac{h}{2}\right)^{\frac{1}{\beta}} + \Gamma_M\right)^{-\beta c}$$

$$\frac{h}{2}\left(max\left\{\frac{(\ell_N)^2}{(\alpha_{MF}\Lambda_{K,M})^2}\right\} - max\left\{\frac{(\ell_N)^2}{(\alpha_{MF}\Lambda_{K-1,M-1})^2}\right\}\right) \tag{5.60}$$

Integral component for Type 4 range

When $\ell > \alpha_{MF}\Lambda_{\mathbf{K-1,M-1}}$ and $\alpha_{MF}\Lambda_{\mathbf{K,M}} \leq \ell < \alpha_{MF}\Lambda_{\mathbf{K,M-1}}$ then (5.45) gives us

$$< \int {}^{a}_{\underline{MF}}P_{30} >_{K,M}$$

$$-\frac{\beta\Phi_{K-1,M-1}\theta'_{M-1}}{(1-\beta c)\alpha_{MF}\sigma_{dK}^b}\left(\frac{h}{2}\right)^{\frac{b}{2}}\left(\left(\left(\frac{h}{2}\right)^{\frac{1}{\beta}} + \Gamma_M - \frac{1}{\beta}\left(\frac{2}{h}\right)^{\frac{b}{2}}\left(\alpha_{MF}\Lambda_{K,M} - \ell\right)\right)^{1-\beta c} - \left(\left(\frac{h}{2}\right)^{\frac{1}{\beta}} + \Gamma_{M-1}\right)^{1-\beta c}\right)$$

$$\Rightarrow -\frac{d < \int {}^{a}_{\underline{MF}}P_{30} >_{K,M}}{d\ell} = \frac{\Phi_{K-1,M-1}\theta'_{M-1}}{\alpha_{MF}\sigma_{dK}^b}\left(\left(\frac{h}{2}\right)^{\frac{1}{\beta}} + \Gamma_M - \frac{1}{\beta}\left(\frac{2}{h}\right)^{\frac{b}{2}}\left(\alpha_{MF}\Lambda_{K,M} - \ell\right)\right)^{-\beta c} \tag{5.61}$$

$$\Rightarrow \langle \int {}^{a}_{\underline{MF}}P_{32} \rangle_{T4} = \left(\frac{\Phi_{K-1,M-1}\theta'_{M-1}}{\alpha_{MF}\sigma_{dK}{}^b(1-\beta c)(2-\beta c)} h\left(\beta\left(\frac{h}{2}\right)^{\frac{b}{2}}\right)^2\right) \times$$

$$\left(\begin{array}{l} -\left(\left(\frac{h}{2}\right)^{\frac{1}{\beta}} + \Gamma_M - \frac{1}{\beta}\left(\frac{2}{h}\right)^{\frac{b}{2}}\left(\alpha_{MF}\Lambda_{K,M} - max\left\{\begin{matrix}\ell_N \\ \alpha_{MF}\Lambda_{K,M-1}\end{matrix}\right\}\right)\right)^{1-\beta c}\left(\left(\frac{h}{2}\right)^{\frac{1}{\beta}} + \Gamma_M - \frac{1}{\beta}\left(\frac{2}{h}\right)^{\frac{b}{2}}\left(\alpha_{MF}\Lambda_{K,M} + (1-\beta c)max\left\{\begin{matrix}\ell_N \\ \alpha_{MF}\Lambda_{K,M-1}\end{matrix}\right\}\right)\right) \\ +\left(\left(\frac{h}{2}\right)^{\frac{1}{\beta}} + \Gamma_M - \frac{1}{\beta}\left(\frac{2}{h}\right)^{\frac{b}{2}}\left(\alpha_{MF}\Lambda_{K,M} - max\left\{\begin{matrix}\ell_N \\ \alpha_{MF}\Lambda_{K,M} \\ \alpha_{MF}\Lambda_{K-1,M-1}\end{matrix}\right\}\right)\right)^{1-\beta c}\left(\left(\frac{h}{2}\right)^{\frac{1}{\beta}} + \Gamma_M - \frac{1}{\beta}\left(\frac{2}{h}\right)^{\frac{b}{2}}\left(\alpha_{MF}\Lambda_{K,M} + (1-\beta c)max\left\{\begin{matrix}\ell_N \\ \alpha_{MF}\Lambda_{K,M} \\ \alpha_{MF}\Lambda_{K-1,M-1}\end{matrix}\right\}\right)\right) \end{array}\right) \tag{5.62}$$

Integral component for Type 5 range

When $\alpha_{MF}\Lambda_{\mathbf{K,M}} \leq \ell \leq \alpha_{MF}\Lambda_{\mathbf{K-1,M-1}}$ (i.e. $\alpha_{MF}\Lambda_{\mathbf{K,M}} \leq \alpha_{MF}\Lambda_{\mathbf{K-1,M-1}}$) then (5.46) gives us

$$\langle \int {}^{a}_{\underline{MF}}P_{30} \rangle_{K,M} = -\frac{\beta\Phi_{K-1,M-1}\theta'_{M-1}}{(1-\beta c)\alpha_{MF}\sigma_{dK}{}^b}\left(\frac{h}{2}\right)^{\frac{b}{2}}\left(\begin{array}{c} \left(\left(\frac{h}{2}\right)^{\frac{1}{\beta}} + \Gamma_M - \frac{1}{\beta}\left(\frac{2}{h}\right)^{\frac{b}{2}}(\alpha_{MF}\Lambda_{K,M} - \ell)\right)^{1-\beta c} \\ -\left(\left(\frac{h}{2}\right)^{\frac{1}{\beta}} + \Gamma_{M-1} - \frac{1}{\beta}\left(\frac{2}{h}\right)^{\frac{b}{2}}(\alpha_{MF}\Lambda_{K-1,M-1} - \ell)\right)^{1-\beta c} \end{array}\right) \tag{5.63}$$

$$\Rightarrow -\frac{d\langle \int {}^{a}_{\underline{MF}}P_{30} \rangle_{K,M}}{d\ell} = \frac{\Phi_{K-1,M-1}\theta'_{M-1}}{\alpha_{MF}\sigma_{dK}{}^b}\left(\left(\left(\frac{h}{2}\right)^{\frac{1}{\beta}} + \Gamma_M - \frac{1}{\beta}\left(\frac{2}{h}\right)^{\frac{b}{2}}(\alpha_{MF}\Lambda_{K,M} - \ell)\right)^{-\beta c} - \left(\left(\frac{h}{2}\right)^{\frac{1}{\beta}} + \Gamma_{M-1} - \frac{1}{\beta}\left(\frac{2}{h}\right)^{\frac{b}{2}}(\alpha_{MF}\Lambda_{K-1,M-1} - \ell)\right)^{-\beta c}\right)$$

$$\Rightarrow \langle \int {}^{a}_{\underline{MF}}P_{32} \rangle_{T5} = \left(\frac{\Phi_{K-1,M-1}\theta'_{M-1}}{\alpha_{MF}\sigma_{dK}{}^b(1-\beta c)(2-\beta c)} h\left(\beta\left(\frac{h}{2}\right)^{\frac{b}{2}}\right)^2\right) \times$$

$$\left(\begin{array}{c} -\left(\left(\frac{h}{2}\right)^{\frac{1}{\beta}} + \Gamma_M - \frac{1}{\beta}\left(\frac{2}{h}\right)^{\frac{b}{2}}\left(\alpha_{MF}\Lambda_{K,M} - max\left\{\begin{matrix}\ell_N \\ \alpha_{MF}\Lambda_{K-1,M-1}\end{matrix}\right\}\right)\right)^{1-\beta c}\left(\left(\frac{h}{2}\right)^{\frac{1}{\beta}} + \Gamma_M - \frac{1}{\beta}\left(\frac{2}{h}\right)^{\frac{b}{2}}\left(\alpha_{MF}\Lambda_{K,M} + (1-\beta c)max\left\{\begin{matrix}\ell_N \\ \alpha_{MF}\Lambda_{K-1,M-1}\end{matrix}\right\}\right)\right) \\ +\left(\left(\frac{h}{2}\right)^{\frac{1}{\beta}} + \Gamma_M - \frac{1}{\beta}\left(\frac{2}{h}\right)^{\frac{b}{2}}\left(\alpha_{MF}\Lambda_{K,M} - max\left\{\begin{matrix}\ell_N \\ \alpha_{MF}\Lambda_{K,M}\end{matrix}\right\}\right)\right)^{1-\beta c}\left(\left(\frac{h}{2}\right)^{\frac{1}{\beta}} + \Gamma_M - \frac{1}{\beta}\left(\frac{2}{h}\right)^{\frac{b}{2}}\left(\alpha_{MF}\Lambda_{K,M} + (1-\beta c)max\left\{\begin{matrix}\ell_N \\ \alpha_{MF}\Lambda_{K,M}\end{matrix}\right\}\right)\right) \\ +\left(\left(\frac{h}{2}\right)^{\frac{1}{\beta}} + \Gamma_{M-1} - \frac{1}{\beta}\left(\frac{2}{h}\right)^{\frac{b}{2}}\left(\alpha_{MF}\Lambda_{K-1,M-1} - max\left\{\begin{matrix}\ell_N \\ \alpha_{MF}\Lambda_{K-1,M-1}\end{matrix}\right\}\right)\right)^{1-\beta c}\left(\left(\frac{h}{2}\right)^{\frac{1}{\beta}} + \Gamma_{M-1} - \frac{1}{\beta}\left(\frac{2}{h}\right)^{\frac{b}{2}}\left(\alpha_{MF}\Lambda_{K-1,M-1} + (1-\beta c)max\left\{\begin{matrix}\ell_N \\ \alpha_{MF}\Lambda_{K-1,M-1}\end{matrix}\right\}\right)\right) \\ -\left(\left(\frac{h}{2}\right)^{\frac{1}{\beta}} + \Gamma_{M-1} - \frac{1}{\beta}\left(\frac{2}{h}\right)^{\frac{b}{2}}\left(\alpha_{MF}\Lambda_{K-1,M-1} - max\left\{\begin{matrix}\ell_N \\ \alpha_{MF}\Lambda_{K,M}\end{matrix}\right\}\right)\right)^{1-\beta c}\left(\left(\frac{h}{2}\right)^{\frac{1}{\beta}} + \Gamma_{M-1} - \frac{1}{\beta}\left(\frac{2}{h}\right)^{\frac{b}{2}}\left(\alpha_{MF}\Lambda_{K-1,M-1} + (1-\beta c)max\left\{\begin{matrix}\ell_N \\ \alpha_{MF}\Lambda_{K,M}\end{matrix}\right\}\right)\right) \end{array}\right) \tag{5.64}$$

Integral component for Type 6 range

When $\ell \geq \alpha_{MF}\Lambda_{\mathbf{K,M-1}}$ then (5.47) gives us

$$\begin{aligned} &< \textstyle\int {}^{a}_{\underline{MF}}P_{30} >_{K,M} = 0 \\ &\Rightarrow -\frac{d < \int {}^{a}_{\underline{MF}}P_{30} >_{K,M}}{d\ell} = 0 \\ &\Rightarrow < \textstyle\int {}^{a}_{\underline{MF}}P_{32} >_{T6} = 0 \end{aligned} \tag{5.65}$$

We can, therefore, ignore this component.

The full integral of the $< \int {}^{a}_{\underline{MF}}P_{30} >_{\mathbf{K,M}}$ term from $\ell_{\mathbf{N}}$ to ℓ_{max}, for half-macrofractures nucleated in timestep **M**, is given by summing the integral components for each of the six types. Note, however, that only those components where **M** = **J** for a half-macrofracture of length $\ell_{\mathbf{N}}$ during some or all of timestep **K** (i.e. the components for types 2, 3, 4 and 5) will contribute to the integral. Combining these four-term types for a specific timestep combination **M**,**K** gives us

$$\begin{aligned} \Sigma_{Type\,T}\left[\left\langle \textstyle\int {}^{a}_{\underline{MF}}P_{32}\right\rangle_T\right] = &\left(\frac{\left(\Phi_{K-1,M-1}\theta'_{M-1}-\Phi_{K-1,M}\theta'_{M}\right)}{\alpha_{MF}\sigma_{dK}{}^{b}}\left(\left(\tfrac{h}{2}\right)^{\frac{1}{\beta}}+\Gamma_M\right)^{-\beta c}\frac{h}{2}\right)\left(max\left\{\begin{matrix}(\ell_N)^2\\ (\alpha_{MF}\Lambda_{K,M})^2\end{matrix}\right\}-max\left\{\begin{matrix}(\ell_N)^2\\ (\alpha_{MF}\Lambda_{K-1,M})^2\end{matrix}\right\}\right) \\ &+\left(\frac{\Phi_{K-1,M-1}\theta'_{M-1}}{\alpha_{MF}\sigma_{dK}{}^{b}(1-\beta c)(2-\beta c)}h\left(\beta\left(\tfrac{h}{2}\right)^{\frac{b}{2}}\right)^2\right)\times \\ &\left(\begin{matrix} -\left(\left(\tfrac{h}{2}\right)^{\frac{1}{\beta}}+\Gamma_M-\frac{1}{\beta}\left(\tfrac{2}{h}\right)^{\frac{b}{2}}\left(\alpha_{MF}\Lambda_{K,M}-max\left\{\begin{matrix}\ell_N\\ \alpha_{MF}\Lambda_{K,M-1}\end{matrix}\right\}\right)\right)^{1-\beta c}\left(\left(\tfrac{h}{2}\right)^{\frac{1}{\beta}}+\Gamma_M-\frac{1}{\beta}\left(\tfrac{2}{h}\right)^{\frac{b}{2}}\left(\alpha_{MF}\Lambda_{K,M}+(1-\beta c)max\left\{\begin{matrix}\ell_N\\ \alpha_{MF}\Lambda_{K,M-1}\end{matrix}\right\}\right)\right) \\ +\left(\left(\tfrac{h}{2}\right)^{\frac{1}{\beta}}+\Gamma_M-\frac{1}{\beta}\left(\tfrac{2}{h}\right)^{\frac{b}{2}}\left(\alpha_{MF}\Lambda_{K,M}-max\left\{\begin{matrix}\ell_N\\ \alpha_{MF}\Lambda_{K,M}\end{matrix}\right\}\right)\right)^{1-\beta c}\left(\left(\tfrac{h}{2}\right)^{\frac{1}{\beta}}+\Gamma_M-\frac{1}{\beta}\left(\tfrac{2}{h}\right)^{\frac{b}{2}}\left(\alpha_{MF}\Lambda_{K,M}+(1-\beta c)max\left\{\begin{matrix}\ell_N\\ \alpha_{MF}\Lambda_{K,M}\end{matrix}\right\}\right)\right) \\ +\left(\left(\tfrac{h}{2}\right)^{\frac{1}{\beta}}+\Gamma_{M-1}-\frac{1}{\beta}\left(\tfrac{2}{h}\right)^{\frac{b}{2}}\left(\alpha_{MF}\Lambda_{K-1,M-1}-max\left\{\begin{matrix}\ell_N\\ \alpha_{MF}\Lambda_{K-1,M-1}\end{matrix}\right\}\right)\right)^{1-\beta c}\left(\left(\tfrac{h}{2}\right)^{\frac{1}{\beta}}+\Gamma_{M-1}-\frac{1}{\beta}\left(\tfrac{2}{h}\right)^{\frac{b}{2}}\left(\alpha_{MF}\Lambda_{K-1,M-1}+(1-\beta c)max\left\{\begin{matrix}\ell_N\\ \alpha_{MF}\Lambda_{K-1,M-1}\end{matrix}\right\}\right)\right) \\ -\left(\left(\tfrac{h}{2}\right)^{\frac{1}{\beta}}+\Gamma_{M-1}-\frac{1}{\beta}\left(\tfrac{2}{h}\right)^{\frac{b}{2}}\left(\alpha_{MF}\Lambda_{K-1,M-1}-max\left\{\begin{matrix}\ell_N\\ \alpha_{MF}\Lambda_{K-1,M}\end{matrix}\right\}\right)\right)^{1-\beta c}\left(\left(\tfrac{h}{2}\right)^{\frac{1}{\beta}}+\Gamma_{M-1}-\frac{1}{\beta}\left(\tfrac{2}{h}\right)^{\frac{b}{2}}\left(\alpha_{MF}\Lambda_{K-1,M-1}+(1-\beta c)max\left\{\begin{matrix}\ell_N\\ \alpha_{MF}\Lambda_{K-1,M}\end{matrix}\right\}\right)\right) \end{matrix}\right) \end{aligned} \tag{5.66}$$

(5.66) can be simplified if we know the value of $\ell_{\mathbf{N}}$ relative to $\Lambda_{\mathbf{K,M-1}}$, $\Lambda_{\mathbf{K,M}}$, $\Lambda_{\mathbf{K-1,M-1}}$ and $\Lambda_{\mathbf{K-1,M}}$. Again we can define six different types of terms, in this case reflecting the value of $\ell_{\mathbf{N}}$ relative to the timesteps **M** and **K** (i.e. the lowest term component in the summation, represented by position of the blue line relative to the "Type" boxes on Fig. 5.6):

Type 1 terms

If $\ell_{\mathbf{N}} \leq \alpha_{MF}\Lambda_{\mathbf{K-1,M}}$ and $\ell_{\mathbf{N}} < \alpha_{MF}\Lambda_{\mathbf{K,M}}$ (i.e. $\alpha_{MF}\Lambda_{\mathbf{K,M}} > 0$) then (5.66) can be reduced to

$$\begin{aligned} &\sum_{Type\,T}\left[< \textstyle\int {}^{a}_{\underline{MF}}P_{32} >_T\right] \\ &= \frac{\left(\Phi_{K-1,M-1}\theta'_{M-1}-\Phi_{K-1,M}\theta'_{M}\right)}{\alpha_{MF}\sigma^{b}_{dK}}\frac{h}{2}\left(\left(\frac{h}{2}\right)^{\frac{1}{\beta}}+\Gamma_M\right)^{-\beta c}\left(\left(\alpha_{MF}\Lambda_{K,M}\right)^2-\left(\alpha_{MF}\Lambda_{K-1,M}\right)^2\right) \end{aligned}$$

$$+\Phi_{\boldsymbol{K}-1,\boldsymbol{M}-1}\theta'_{\boldsymbol{M}-1}\frac{-\beta}{(1-\beta c)}h\left(\frac{h}{2}\right)^{\frac{b}{2}}(t_{\boldsymbol{K}}-t_{\boldsymbol{K}-1})\left(\left(\left(\frac{h}{2}\right)^{\frac{1}{\beta}}+\Gamma_{\boldsymbol{M}}\right)^{1-\beta c}-\left(\left(\frac{h}{2}\right)^{\frac{1}{\beta}}+\Gamma_{\boldsymbol{M}-1}\right)^{1-\beta c}\right) \quad (5.67)$$

Type 2 terms

If $\alpha_{\mathrm{MF}}\Lambda_{\mathbf{K-1,M}} < \ell_{\mathbf{N}} \leq \alpha_{\mathrm{MF}}\Lambda_{\mathbf{K-1,M-1}}$ and $\ell_{\mathbf{N}} < \alpha_{\mathrm{MF}}\Lambda_{\mathbf{K,M}}$ then (5.66) can be reduced to

$$\sum_{TypeT}\left[<\int_{\underline{\mathrm{MF}}}^{a}\mathrm{P}_{32}>_T\right]$$

$$=\frac{\left(\Phi_{\boldsymbol{K}-1,\boldsymbol{M}-1}\theta'_{\boldsymbol{M}-1}-\Phi_{\boldsymbol{K}-1,\boldsymbol{M}}\theta'_{\boldsymbol{M}}\right)}{\alpha_{MF}\sigma_{d\boldsymbol{K}}^{b}}\frac{h}{2}\left(\left(\frac{h}{2}\right)^{\frac{1}{\beta}}+\Gamma_{\boldsymbol{M}}\right)^{-\beta c}\left(\left(\alpha_{MF}\Lambda_{\boldsymbol{K},\boldsymbol{M}}\right)^2-(\ell_{\boldsymbol{N}})^2\right)$$

$$-\Phi_{\boldsymbol{K}-1,\boldsymbol{M}-1}\theta'_{\boldsymbol{M}-1}\frac{-\beta}{(1-\beta c)}h\left(\frac{h}{2}\right)^{\frac{b}{2}}(t_{\boldsymbol{K}}-t_{\boldsymbol{K}-1})\left(\left(\frac{h}{2}\right)^{\frac{1}{\beta}}+\Gamma_{\boldsymbol{M}-1}\right)^{1-\beta c}$$

$$+\left(\frac{\Phi_{\boldsymbol{K}-1,\boldsymbol{M}-1}\theta'_{\boldsymbol{M}-1}\beta^2 h}{\alpha_{\mathrm{MF}}\sigma_{\mathbf{dK}}^{\mathrm{b}}(1-\beta c)(2-\beta c)}\left(\frac{h}{2}\right)^{b}\right)\left(\begin{array}{c}\left(\left(\frac{h}{2}\right)^{\frac{1}{\beta}}+\Gamma_{\boldsymbol{M}}\right)^{2-\beta c}+\left(\left(\frac{h}{2}\right)^{\frac{1}{\beta}}+\Gamma_{\boldsymbol{M}}\right)^{1-\beta c}\frac{(2-\beta c)}{-\beta}\left(\frac{2}{h}\right)^{\frac{b}{2}}\alpha_{MF}\Lambda_{\boldsymbol{K},\boldsymbol{M}}\\ -\left(\left(\frac{h}{2}\right)^{\frac{1}{\beta}}+\Gamma_{\boldsymbol{M}-1}-\frac{1}{\beta}\left(\frac{2}{h}\right)^{\frac{b}{2}}\left(\alpha_{\mathrm{MF}}\Lambda_{\boldsymbol{K}-1,\boldsymbol{M}-1}-\ell_{\boldsymbol{N}}\right)\right)^{2-\beta c}\\ -\left(\left(\frac{h}{2}\right)^{\frac{1}{\beta}}+\Gamma_{\boldsymbol{M}-1}-\frac{1}{\beta}\left(\frac{2}{h}\right)^{\frac{b}{2}}\left(\alpha_{\mathrm{MF}}\Lambda_{\boldsymbol{K}-1,\boldsymbol{M}-1}-\ell_{\boldsymbol{N}}\right)\right)^{1-\beta c}\frac{(2-\beta c)}{-\beta}\left(\frac{2}{h}\right)^{\frac{b}{2}}\ell_{\boldsymbol{N}}\end{array}\right) \quad (5.68)$$

Type 3 terms

If $\alpha_{\mathrm{MF}}\Lambda_{\mathbf{K-1,M-1}} < \ell_{\mathbf{N}} < \alpha_{\mathrm{MF}}\Lambda_{\mathbf{K,M}}$ (i.e. $\alpha_{\mathrm{MF}}\Lambda_{\mathbf{K,M}} > \alpha_{\mathrm{MF}}\Lambda_{\mathbf{K-1,M-1}}$) then (5.66) can be reduced to

$$\sum_{TypeT}\left[<\int_{\underline{\mathrm{MF}}}^{a}\mathrm{P}_{32}>_T\right]$$

$$=\frac{\left(\Phi_{\boldsymbol{K}-1,\boldsymbol{M}-1}\theta'_{\boldsymbol{M}-1}-\Phi_{\boldsymbol{K}-1,\boldsymbol{M}}\theta'_{\boldsymbol{M}}\right)}{\alpha_{MF}\sigma_{d\boldsymbol{K}}^{b}}\left(\left(\frac{h}{2}\right)^{\frac{1}{\beta}}+\Gamma_{\boldsymbol{M}}\right)^{-\beta c}\frac{h}{2}\left(\left(\alpha_{MF}\Lambda_{\boldsymbol{K},\boldsymbol{M}}\right)^2-(\ell_{\boldsymbol{N}})^2\right)$$

$$+\left(\frac{\Phi_{\boldsymbol{K}-1,\boldsymbol{M}-1}\theta'_{\boldsymbol{M}-1}\beta^2 h}{\alpha_{\mathrm{MF}}\sigma_{\mathbf{dK}}^{\mathrm{b}}(1-\beta c)(2-\beta c)}\left(\frac{h}{2}\right)^{b}\right)\left(\begin{array}{c}\left(\left(\frac{h}{2}\right)^{\frac{1}{\beta}}+\Gamma_{\boldsymbol{M}}\right)^{2-\beta c}+\left(\left(\frac{h}{2}\right)^{\frac{1}{\beta}}+\Gamma_{\boldsymbol{M}}\right)^{1-\beta c}\frac{(2-\beta c)}{-\beta}\left(\frac{2}{h}\right)^{\frac{b}{2}}\alpha_{MF}\Lambda_{\boldsymbol{K},\boldsymbol{M}}\\ -\left(\left(\frac{h}{2}\right)^{\frac{1}{\beta}}+\Gamma_{\boldsymbol{M}-1}\right)^{2-\beta c}-\left(\left(\frac{h}{2}\right)^{\frac{1}{\beta}}+\Gamma_{\boldsymbol{M}-1}\right)^{1-\beta c}\frac{(2-\beta c)}{-\beta}\left(\frac{2}{h}\right)^{\frac{b}{2}}\alpha_{MF}\Lambda_{\boldsymbol{K},\boldsymbol{M}-1}\end{array}\right) \quad (5.69)$$

Type 4 terms

If $\ell_{\mathbf{N}} > \alpha_{\mathrm{MF}}\Lambda_{\mathbf{K-1,M-1}}$ and $\alpha_{\mathrm{MF}}\Lambda_{\mathbf{K,M}} \leq \ell_{\mathbf{N}} < \alpha_{\mathrm{MF}}\Lambda_{\mathbf{K,M-1}}$ then (5.66) can be reduced to

$$\sum_{TypeT}\left[<\int_{\underline{\mathrm{MF}}}^{a}\mathrm{P}_{32}>_T\right]$$

$$=\left(\frac{\Phi_{\boldsymbol{K}-1,\boldsymbol{M}-1}\theta'_{\boldsymbol{M}-1}\beta^2 h}{\alpha_{\mathrm{MF}}\sigma_{\mathbf{dK}}^{\mathrm{b}}(1-\beta c)(2-\beta c)}\left(\frac{h}{2}\right)^{b}\right)$$

$$\times\left(\begin{array}{c}\left(\left(\frac{h}{2}\right)^{\frac{1}{\beta}}+\Gamma_{M}-\frac{1}{\beta}\left(\frac{2}{h}\right)^{\frac{b}{2}}\left(\alpha_{\mathrm{MF}}\Lambda_{K,M}-\ell_{N}\right)\right)^{2-\beta c}\\+\left(\left(\frac{h}{2}\right)^{\frac{1}{\beta}}+\Gamma_{M}-\frac{1}{\beta}\left(\frac{2}{h}\right)^{\frac{b}{2}}\left(\alpha_{\mathrm{MF}}\Lambda_{K,M}-\ell_{N}\right)\right)^{1-\beta c}\frac{(2-\beta c)}{-\beta}\left(\frac{2}{h}\right)^{\frac{b}{2}}\ell_{N}\\-\left(\left(\frac{h}{2}\right)^{\frac{1}{\beta}}+\Gamma_{M-1}\right)^{2-\beta c}-\left(\left(\frac{h}{2}\right)^{\frac{1}{\beta}}+\Gamma_{M-1}\right)^{1-\beta c}\frac{(2-\beta c)}{-\beta}\left(\frac{2}{h}\right)^{\frac{b}{2}}\alpha_{MF}\Lambda_{K,M-1}\end{array}\right) \tag{5.70}$$

Type 5 terms

If $\alpha_{\mathrm{MF}}\Lambda_{\mathbf{K,M}}\leq\ell_{\mathbf{N}}\leq\alpha_{\mathrm{MF}}\Lambda_{\mathbf{K-1,M-1}}$ (i.e. $\alpha_{\mathrm{MF}}\Lambda_{\mathbf{K,M}}\leq\alpha_{\mathrm{MF}}\Lambda_{\mathbf{K-1,M-1}}$) then (5.66) can be reduced to

$$\sum_{TypeT}\left[<\int{}^{a}_{\underline{\mathrm{MF}}}\mathrm{P}_{32}>_{T}\right]$$

$$=-\Phi_{K-1,M-1}\theta'_{M-1}\frac{-\beta}{(1-\beta c)}h\left(\frac{h}{2}\right)^{\frac{b}{2}}(t_{K}-t_{K-1})\left(\left(\frac{h}{2}\right)^{\frac{1}{\beta}}+\Gamma_{M-1}\right)^{1-\beta c}$$

$$+\left(\frac{\Phi_{K-1,M-1}\theta'_{M-1}\beta^{2}h}{\alpha_{\mathrm{MF}}\sigma_{\mathbf{dK}}^{\mathrm{b}}(1-\beta c)(2-\beta c)}\left(\frac{h}{2}\right)^{b}\right)$$

$$\times\left(\begin{array}{c}\left(\left(\frac{h}{2}\right)^{\frac{1}{\beta}}+\Gamma_{M}-\frac{1}{\beta}\left(\frac{2}{h}\right)^{\frac{b}{2}}\left(\alpha_{\mathrm{MF}}\Lambda_{K,M}-\ell_{N}\right)\right)^{2-\beta c}\\+\left(\left(\frac{h}{2}\right)^{\frac{1}{\beta}}+\Gamma_{M}-\frac{1}{\beta}\left(\frac{2}{h}\right)^{\frac{b}{2}}\left(\alpha_{\mathrm{MF}}\Lambda_{K,M}-\ell_{N}\right)\right)^{1-\beta c}\frac{(2-\beta c)}{-\beta}\left(\frac{2}{h}\right)^{\frac{b}{2}}\ell_{N}\\-\left(\left(\frac{h}{2}\right)^{\frac{1}{\beta}}+\Gamma_{M-1}-\frac{1}{\beta}\left(\frac{2}{h}\right)^{\frac{b}{2}}\left(\alpha_{\mathrm{MF}}\Lambda_{K-1,M-1}-\ell_{N}\right)\right)^{2-\beta c}\\-\left(\left(\frac{h}{2}\right)^{\frac{1}{\beta}}+\Gamma_{M-1}-\frac{1}{\beta}\left(\frac{2}{h}\right)^{\frac{b}{2}}\left(\alpha_{\mathrm{MF}}\Lambda_{K-1,M-1}-\ell_{N}\right)\right)^{1-\beta c}\frac{(2-\beta c)}{-\beta}\left(\frac{2}{h}\right)^{\frac{b}{2}}\ell_{N}\end{array}\right) \tag{5.71}$$

Type 6 terms

If $\ell_{\mathbf{N}}\geq\alpha_{\mathrm{MF}}\Lambda_{\mathbf{K,M-1}}$ then (5.66) can be reduced to

$$\sum_{TypeT}\left[<\int{}^{a}_{\underline{\mathrm{MF}}}\mathrm{P}_{32}>_{T}\right]=0 \tag{5.72}$$

The cumulative mean linear density of static half-macrofractures of length $\ell_{\mathbf{N}}$, ${}^{\mathrm{s}}_{\mathrm{MF}}\mathrm{P}_{32}(\ell_{\mathbf{N}},\mathrm{t}_{\mathbf{N}})$, can thus be obtained by inserting (5.67), (5.68), (5.68), (5.70), (5.71) or (5.72), as appropriate, into (5.55). If $\ell_{\mathbf{N}}=0$ then it can be further simplified as we can use a Type 1 term when $\mathbf{M}<\mathbf{K}$ and a Type 5 term when $\mathbf{M}=\mathbf{K}$. Thus the total mean linear static half-macrofracture density is given by

$$
{}_{\underline{MF}}^{s}P_{32}(0,t_N) \approx 2B\sum_{K=1}^{N}\left[\overline{F_K}\left(\begin{array}{c}\sum_{M=1}^{K-1}\left[\begin{array}{c}\frac{(\Phi_{K-1,M-1}\theta'_{M-1}-\Phi_{K-1,M}\theta'_M)}{\alpha_{MF}\sigma_{dK}{}^b}\frac{h}{2}\left(\left(\frac{h}{2}\right)^{\frac{1}{\beta}}+\Gamma_M\right)^{-\beta c}\left((\alpha_{MF}\Lambda_{K,M})^2-(\alpha_{MF}\Lambda_{K-1,M})^2\right)\\ +\Phi_{K-1,M-1}\theta'_{M-1}\frac{-\beta}{(1-\beta c)}h\left(\frac{h}{2}\right)^{\frac{b}{2}}(t_K-t_{K-1})\left(\left(\left(\frac{h}{2}\right)^{\frac{1}{\beta}}+\Gamma_M\right)^{1-\beta c}-\left(\left(\frac{h}{2}\right)^{\frac{1}{\beta}}+\Gamma_{M-1}\right)^{1-\beta c}\right)\end{array}\right]\\ +\theta'_{K-1}\left(\begin{array}{c}\frac{\beta^2}{(1-\beta c)(2-\beta c)\alpha_{MF}\sigma_{dK}{}^b}h\left(\frac{h}{2}\right)^b\left(\left(\left(\frac{h}{2}\right)^{\frac{1}{\beta}}+\Gamma_K\right)^{2-\beta c}-\left(\left(\frac{h}{2}\right)^{\frac{1}{\beta}}+\Gamma_{K-1}\right)^{2-\beta c}\right)\\ -\frac{-\beta}{(1-\beta c)}h\left(\frac{h}{2}\right)^{\frac{b}{2}}(t_K-t_{K-1})\left(\left(\frac{h}{2}\right)^{\frac{1}{\beta}}+\Gamma_{K-1}\right)^{1-\beta c}\end{array}\right)\end{array}\right)\right] \quad (5.73)
$$

5.4 Calculating Residual Macrofracture Populations

The equations for active and static half-macrofracture populations derived in Sect. 5.3 work well during most of the period in which the macrofracture network is growing. However, they begin to break down as the fracture network approaches saturation, when the probability of fracture deactivation is high, and the mean lifetime of each macrofracture becomes significantly less that the duration of the timestep (i.e. as $\Phi_N \rightarrow 0$). In this situation, the assumption that fracture deactivation rates are constant for the duration of the timestep is no longer valid.

Therefore, if we want to simulate the growth of the fracture network through to saturation, we must find an alternative method of calculating the half-macrofracture populations in these final timesteps. We can do this by treating the active fracture population as an equilibrium (or "residual") population, where the rate of fracture nucleation exactly balances the rate of fracture deactivation. We can thus calculate the equilibrium rate of half-macrofracture nucleation and use this to determine the increase in the static half-macrofracture density during the timestep. We can also calculate the size distribution of the equilibrium half-macrofracture population and use this to determine the increase in mean linear half-macrofracture density.

This technique is particularly useful when simulating the growth of a fully anisotropic fracture network, where the secondary fracture set develops only after the primary fracture set has already stopped growing (see Sect. 8.5). In this scenario, the deactivation rate of the secondary fracture set is high throughout its growth, so the residual growth technique is the only technique that can accurately calculate its cumulative density distribution functions.

5.4.1 *Calculating the Volumetric Density of Residual Active Half-Macrofractures*

The residual active half-macrofracture population is defined as the population where the rate of half-macrofracture deactivation equals the rate of half-macrofracture nucleation, so that the total number of active half-macrofractures remains constant. Therefore

$$\underline{\text{MF}}\dot{\text{P}}_{30}(\text{t}_n) = F_N{}^a_{r\underline{\text{MF}}}\text{P}_{30}(\text{t}_n) \tag{5.74}$$

where ${}^a_{r\underline{\text{MF}}}\text{P}_{30}(\text{t}_n)$ is the total volumetric density of residual active half-macrofractures, and $\underline{\text{MF}}\dot{\text{P}}_{30}(\text{t}_n)$ is the half-macrofracture nucleation rate, at any time t_n during a timestep **N**. Note that we must use the instantaneous probability of macrofracture deactivation $\text{F}_\mathbf{N}$, rather than the mean probability of macrofracture deactivation $\overline{\text{F}}_\mathbf{N}$, as the mean macrofracture lifetime is short compared with the timestep duration.

The half-macrofracture nucleation rate, $\underline{\text{MF}}\dot{\text{P}}_{30}$, can be calculated by differentiating 4.10 with respect to time, then multiplying the resulting equation by the effective cumulative microfracture activation probability θ', and also by a factor of 2 to convert from macrofractures to half-macrofractures. This gives us:

$$\begin{aligned}\underline{\text{MF}}\dot{\text{P}}_{30}(\text{t}_n) &= \theta'(\text{t}_n)\frac{d_{\underline{\text{MF}}}\text{P}_{30}}{dt} \\ = -2\text{B}\theta'(\text{t}_n)\beta c\gamma_N^{\frac{1}{\beta}} &\left(\left(\frac{h}{2}\right)^{\frac{1}{\beta}} + \Gamma_{N-1} + \gamma_N^{\frac{1}{\beta}}(t_n - t_{N-1})\right)^{-(\beta c+1)}\end{aligned} \tag{5.75}$$

In deriving (5.75), we have made the assumption that the rate of change of the effective cumulative microfracture activation probability, θ', is negligible compared with the rate of macrofracture nucleation, and hence $\theta'(\text{t}_n)$ can be treated as a constant when differentiating $\underline{\text{MF}}\text{P}_{30}$. However, this does not mean we can assume that the change in the effective cumulative microfracture activation probability over the course of the entire timestep is negligible compared with the change in the rate of macrofracture nucleation during the timestep. In other words, we can assume that

$$\frac{d\theta'}{dt}\underline{\text{MF}}\text{P}_{30} \ll \theta'(\text{t}_n)\frac{d_{\underline{\text{MF}}}\text{P}_{30}}{dt} \tag{5.76}$$

but we cannot assume that

$$\frac{\theta'_{N-1} - \theta'_N}{\theta'_{N-1}} \ll \frac{\left|\frac{d_{\underline{\text{MF}}}\text{P}_{30}}{dt}_{N-1} - \frac{d_{\underline{\text{MF}}}\text{P}_{30}}{dt}_N\right|}{\frac{d_{\underline{\text{MF}}}\text{P}_{30}}{dt}_{N-1}} \tag{5.77}$$

As the macrofracture network approaches saturation and $\theta' \to 0$, we must, therefore, take into account the variation in $\theta'(\text{t}_n)$ when calculating the rate of macrofracture nucleation at any time t_n during timestep **N**. We can calculate the rate of change of θ' from the number of propagating macrofracture tips and the rate of tip propagation:

$$\frac{d\theta'}{dt} = \frac{d\theta'}{d_{\text{MF}}\text{P}_{32}}\frac{d_{\text{MF}}\text{P}_{32}}{dt} = -\frac{d\chi_{MF}}{d_{\text{MF}}\text{P}_{32}}{}^a_{r\underline{\text{MF}}}\text{P}_{30}(t_n)\alpha_{MF}\sigma^b_{dN}h \tag{5.78}$$

An expression for $d\chi_{\text{MF}}/d_{\text{MF}}\text{P}_{32}$ is derived in Appendix B:

$$\frac{d\chi_{MF}}{d_{\mathrm{MF}}\mathrm{P}_{32}} = \frac{2}{\mathbb{B}}\left(1 - e^{-\mathbb{B}W}\right) \tag{5.79}$$

where $\mathbb{B}$ is one of the parameters describing the macrofracture spatial distribution. Note that, strictly speaking, $\mathbb{B}$ is a function of $_{\mathrm{MF}}\mathrm{P}_{32}$ and will thus evolve through time; however, we can simplify the calculation by using a constant value $\mathbb{B}_{\mathrm{N-1}}$, calculated at the end of the previous timestep. Note also that since we are in the later stages of fracture growth, where the fracture network is approaching saturation, we cannot use the approximation for $\chi_{\mathrm{MF}}/d\psi_{\mathrm{MF}}$ given by (B.33).

Inserting (5.79) into (5.78) and integrating gives

$$\begin{aligned}
&\theta'(\mathrm{t}_n) \\
&= \theta'_{N-1} - \int_{\mathrm{t}_{N-1}}^{\mathrm{t}_n} \frac{2}{\mathbb{B}_{N-1}}\left(1 - e^{-\mathbb{B}_{N-1}W}\right)_{r}^{a}\underline{_{\mathrm{MF}}}\mathrm{P}_{30}(t)\alpha_{MF}\sigma_{dN}^{b}h\,dt \\
&= \theta'_{N-1} - \frac{2\alpha_{MF}\sigma_{dN}^{b}h}{F_N\mathbb{B}_{N-1}}\left(1 - e^{-\mathbb{B}_{N-1}W}\right)\int_{\mathrm{t}_{N-1}}^{\mathrm{t}_n} \underline{_{\mathrm{MF}}}\dot{\mathrm{P}}_{30}(\mathrm{t})dt
\end{aligned} \tag{5.80}$$

We can now insert (5.80) into (5.75) to derive a differential equation for the half-macrofracture nucleation rate:

$$\begin{aligned}
&\underline{_{\mathrm{MF}}}\dot{\mathrm{P}}_{30}(\mathrm{t}_n) \\
&= -2\mathrm{B}\left(\theta'_{N-1} - \frac{2\alpha_{MF}\sigma_{dN}^{b}h}{F_N\mathbb{B}_{N-1}}\left(1 - e^{-\mathbb{B}_{N-1}W}\right)\int_{\mathrm{t}_{N-1}}^{\mathrm{t}_n} \underline{_{\mathrm{MF}}}\dot{\mathrm{P}}_{30}(\mathrm{t})dt\right) \\
&\times \beta c\gamma_N^{\frac{1}{\beta}}\left(\left(\frac{h}{2}\right)^{\frac{1}{\beta}} + \Gamma_{N-1} + \gamma_N^{\frac{1}{\beta}}(t_n - t_{N-1})\right)^{-(\beta c+1)}
\end{aligned} \tag{5.81}$$

Equation (5.81) is difficult to solve because it contains two terms that are dependent on $\mathrm{t_n}$. However, we can simplify the problem by considering two end-member states: one in which the half-macrofracture nucleation rate is controlled predominantly by changes in the microfracture growth rate, and the other in which the half-macrofracture nucleation rate is controlled predominantly by changes in the exclusion zone volume (which controls the effective cumulative microfracture activation probability). In each case, we can derive a simplified version of (5.81) with only one $\mathrm{t_n}$ term. We can determine which of these end-member states to use in any given situation by comparing the residual half-macrofracture volumetric densities calculated for the two end-member states, and taking the lowest.

5.4.1.1 Residual Half-Macrofracture Population Controlled by Microfracture Growth Rate

In this scenario we assume that the proportional change in the exclusion zone volume during the timestep is negligible compared with the change in the rate at which microfractures can grow to radius h/2, i.e.

$$\left(\theta'_{N-1}-\theta'_{N}\right)\left(\left(\frac{h}{2}\right)^{\frac{1}{\beta}}+\Gamma_{N}\right)^{-(\beta c+1)} \ll \theta'_{N-1}\left|\left(\left(\frac{h}{2}\right)^{\frac{1}{\beta}}+\Gamma_{N-1}\right)^{-(\beta c+1)}-\left(\left(\frac{h}{2}\right)^{\frac{1}{\beta}}+\Gamma_{N}\right)^{-(\beta c+1)}\right| \quad (5.82)$$

In this case, we can simply use (5.75) to calculate the half-macrofracture nucleation rate throughout the timestep. Combining this with (5.74) gives us an expression for the total volumetric density of residual active half-macrofractures at any time t_n during timestep **N**:

$$ {}^{a}_{r\underline{\mathrm{MF}}}\mathrm{P}_{30}(0,\mathrm{t}_n) = -2\mathrm{B}\theta'_{N-1}\frac{\beta c\gamma_N^{\frac{1}{\beta}}}{F_N}\left(\left(\frac{h}{2}\right)^{\frac{1}{\beta}}+\Gamma_{N-1}+\gamma_N^{\frac{1}{\beta}}(t_n-t_{N-1})\right)^{-(\beta c+1)} \quad (5.83)$$

Therefore, the total volumetric density of residual active half-macrofractures at the end of timestep **N** is given by:

$$ {}^{a}_{r\underline{\mathrm{MF}}}\mathrm{P}_{30}(0,\mathrm{t}_N) = -2\mathrm{B}\theta'_{N-1}\frac{\beta c\gamma_N^{\frac{1}{\beta}}}{F_N}\left(\left(\frac{h}{2}\right)^{\frac{1}{\beta}}+\Gamma_{N}\right)^{-(\beta c+1)} \quad (5.84)$$

Ideally we should use $\theta'_{\mathbf{N}}$ rather than $\theta'_{\mathbf{N-1}}$ in (5.84). However in practice, this is difficult because $\theta'_{\mathbf{N}}$ is dependent on $_{\underline{\mathrm{MF}}}\mathrm{P}_{30}(0,\mathrm{t}_{\mathbf{N}})$, so we must first use (5.84) to calculate $_{\underline{\mathrm{MF}}}\mathrm{P}_{30}(0,\mathrm{t}_{\mathbf{N}})$, and then use $_{\underline{\mathrm{MF}}}\mathrm{P}_{30}(0,\mathrm{t}_{\mathbf{N}})$ to update $\theta'_{\mathbf{N}}$.

If we assume that macrofracture deactivation is a random process governed by an exponential decay curve, we can obtain the cumulative volumetric density distribution function for residual active half-macrofractures by inserting the half-macrofracture length, propagation rate and instantaneous deactivation probability into the exponential fracture activation probability function (5.2). Thus the volumetric density of residual active half-macrofractures of length $\ell_{\mathbf{N}}$ or greater is given by

$$ {}^{a}_{r\underline{\mathrm{MF}}}\mathrm{P}_{30}(\ell_N,t_N) = {}^{a}_{r\underline{\mathrm{MF}}}\mathrm{P}_{30}(0,t_N)e^{-\frac{F_N\ell_N}{\alpha_{MF}\sigma_{dN}^{b}}} \quad (5.85)$$

5.4.1.2 Residual Half-Macrofracture Population Controlled by Exclusion Zone Volume

In this scenario, we assume that the change in the rate at which microfractures can grow to radius h/2 is negligible compared with the change in the exclusion zone volume, and hence in the cumulative microfracture activation probability, during the timestep, i.e.

$$\theta'_{N-1}\left|\left(\left(\frac{h}{2}\right)^{\frac{1}{\beta}}+\Gamma_{N-1}\right)^{-(\beta c+1)}-\left(\left(\frac{h}{2}\right)^{\frac{1}{\beta}}+\Gamma_{N}\right)^{-(\beta c+1)}\right| \ll \left(\theta'_{N-1}-\theta'_{N}\right)\left(\left(\frac{h}{2}\right)^{\frac{1}{\beta}}+\Gamma_{N}\right)^{-(\beta c+1)} \quad (5.86)$$

We can, therefore, assume that the rate at which microfractures grow to radius h/2 is constant throughout the timestep, i.e.

$$\left(\left(\frac{h}{2}\right)^{\frac{1}{\beta}}+\Gamma_{N-1}+\gamma_N^{\frac{1}{\beta}}(t_n-t_{N-1})\right)^{-(\beta c+1)} \approx \left(\left(\frac{h}{2}\right)^{\frac{1}{\beta}}+\Gamma_{N-1}\right)^{-(\beta c+1)} \quad (5.87)$$

(5.81) can, therefore, be approximated as

$$\underline{_{MF}}\dot{P}_{30}(t_n) \approx -2B\left(\theta'_{N-1}-\frac{2\alpha_{MF}\sigma^b_{dN}h}{F_N\mathbb{B}_{N-1}}\left(1-e^{-\mathbb{B}_{N-1}W}\right)\int_{t_{N-1}}^{t_N}\underline{_{MF}}\dot{P}_{30}(t)dt\right)\beta c\gamma_N^{\frac{1}{\beta}}\left(\left(\frac{h}{2}\right)^{\frac{1}{\beta}}+\Gamma_{N-1}\right)^{-(\beta c+1)} \quad (5.88)$$

The differential Eq. (5.88) can be restructured into the form

$$x(t_n) = -A\left(B-C\int_{t_{N-1}}^{t_n}x(t)dt\right) \quad (5.89)$$

which has the solution

$$\frac{dx}{dt_n} = ACx(t_n)$$
$$x(t_n) = -ABe^{AC(t_n-t_{N-1})} \quad (5.90)$$

Thus (5.88) can be solved to give

$$\underline{_{MF}}\dot{P}_{30}(t_n) = -2B\beta c\gamma_N^{\frac{1}{\beta}}\left(\left(\frac{h}{2}\right)^{\frac{1}{\beta}}+\Gamma_{N-1}\right)^{-(\beta c+1)}$$
$$\theta'_{N-1}e^{2B\beta c\gamma_N^{\frac{1}{\beta}}\left(\left(\frac{h}{2}\right)^{\frac{1}{\beta}}+\Gamma_{N-1}\right)^{-(\beta c+1)}\frac{2\alpha_{MF}\sigma^b_{dN}h}{F_N\mathbb{B}_{N-1}}\left(1-e^{-\mathbb{B}_{N-1}W}\right)(t_n-t_{N-1})} \quad (5.91)$$

and the total volumetric density of residual active half-macrofractures at the end of timestep **N** is given by:

$$
{}^{a}_{r\underline{\mathrm{MF}}}\mathrm{P}_{30}(0, \mathrm{t}_N) = -2\mathrm{B}\frac{\beta c\gamma_N^{\frac{1}{\beta}}}{F_N}\left(\left(\frac{h}{2}\right)^{\frac{1}{\beta}} + \Gamma_N\right)^{-(\beta c+1)} \times \theta'_{N-1} e^{\frac{4\mathrm{B}\beta c\gamma_N^{\frac{1}{\beta}} \alpha_{MF}\sigma_{dN}^{b} h}{F_N \mathbb{B}_{N-1}}\left(1-e^{-\mathbb{B}_{N-1}W}\right)\left(\left(\frac{h}{2}\right)^{\frac{1}{\beta}}+\Gamma_N\right)^{-(\beta c+1)}(t_n-t_{N-1})} \tag{5.92}
$$

Note that in this case, we can substitute $\Gamma_\mathbf{N}$ for $\Gamma_\mathbf{N-1}$ in (5.92), since $\Gamma_\mathbf{N}$ is not dependent on ${}_{\underline{\mathrm{MF}}}\mathrm{P}_{30}(0,\mathrm{t}_\mathbf{N})$.

The cumulative volumetric density distribution function for residual active half-macrofractures can be obtained by inserting (5.92) into (5.85).

5.4.2 Calculating the Volumetric Density of Residual Static Half-Macrofractures

By definition, the population of active half-macrofractures will be small relative to the increase in the static half-macrofracture density during the residual growth stage. We can, therefore, generally ignore the active half-macrofractures and just calculate the increase in the static half-macrofracture density, $\Delta^{s}_{r\underline{\mathrm{MF}}}\mathrm{P}_{30}(\mathbf{N})$, for each timestep **N**.

As in Sect. 5.3.1.2, to calculate the increase in static half-macrofracture density during a timestep, we multiply the active half-macrofracture density by the probability of macrofracture deactivation and integrate the result throughout the timestep. However, this calculation is easier for the residual fracture population, since we assume that all the active half-macrofractures formed during the current timestep, so we do not need to iterate through previous timesteps of fracture nucleation.

As with active residual half-macrofractures, there will be two forms of the equations for residual volumetric density of static half-macrofractures: one form describing residual half-macrofracture populations controlled by the microfracture growth rate, and one form describing residual half-macrofracture populations controlled by the exclusion zone volume. The form that gives the smallest increase in volumetric half-macrofracture density will indicate which of these two factors is actually controlling the rate of half-macrofracture nucleation at any time.

5.4.2.1 Residual Half-Macrofracture Population Controlled by Microfracture Growth Rate

If the macrofracture nucleation rate is controlled predominantly by changes in the microfracture growth rate, we can assume that the cumulative half-macrofracture activation probability, θ', remains constant throughout the timestep. We can then calculate the increase in total volumetric static half-macrofracture density during a timestep **N** by inserting (5.74) and (5.75) into (5.38):

$$\begin{aligned}
&\Delta^{s}_{r\underline{\mathrm{MF}}}\mathrm{P}_{30}(0,\mathbf{N}) \\
&= \int_{t_{N-1}}^{t_N} F_N\, {}^{a}_{r\underline{\mathrm{MF}}}\mathrm{P}_{30}(0,t)dt = \int_{t_{N-1}}^{t_N} {}_{\underline{\mathrm{MF}}}\dot{\mathrm{P}}_{30}(t)dt = \int_{t_{N-1}}^{t_N} \theta'_{N-1}\frac{d_{\underline{\mathrm{MF}}}\mathrm{P}_{30}}{dt}dt \\
&= \theta'_{N-1}\left[{}_{\underline{\mathrm{MF}}}\mathrm{P}_{30}(t)\right]_{t_{N-1}}^{t_N} \\
&= 2\mathrm{B}\theta'_{N-1}\left(\left(\left(\frac{h}{2}\right)^{\frac{1}{\beta}}+\Gamma_N\right)^{-\beta c}-\left(\left(\frac{h}{2}\right)^{\frac{1}{\beta}}+\Gamma_{N-1}\right)^{-\beta c}\right)
\end{aligned} \tag{5.93}$$

We can multiply this by the cumulative volumetric density distribution function for residual active half-macrofractures given by (5.85) to derive an expression for the increase in the volumetric density of residual static half-macrofractures of length $\ell_{\mathbf{N}}$ or greater during a timestep $\mathbf{N}$:

$$\Delta^{s}_{r\underline{\mathrm{MF}}}\mathrm{P}_{30}(\ell_N,\mathbf{N}) = 2\mathrm{B}\theta'_{N-1}\left(\left(\left(\frac{h}{2}\right)^{\frac{1}{\beta}}+\Gamma_N\right)^{-\beta c}-\left(\left(\frac{h}{2}\right)^{\frac{1}{\beta}}+\Gamma_{N-1}\right)^{-\beta c}\right)e^{-\frac{F_N\ell_N}{\alpha_{MF}\sigma^b_{dN}}} \tag{5.94}$$

5.4.2.2 Residual Half-Macrofracture Population Controlled by Exclusion Zone Volume

If the macrofracture nucleation rate is controlled predominantly by changes in the exclusion zone volume, and hence the cumulative microfracture activation probability, then the integral in (5.93) becomes more complicated as we must take into account the change in the cumulative half-macrofracture activation probability, θ′, throughout the timestep. However, we know that the incremental increase in the total volumetric density of residual static half-macrofractures, $\Delta^{s}_{r\underline{\mathrm{MF}}}\mathrm{P}_{30}(0,\mathbf{N})$, is given by

$$\Delta^{s}_{r\underline{\mathrm{MF}}}\mathrm{P}_{30}(0,\mathbf{N}) = \int_{t_{N-1}}^{t_N} {}_{\underline{\mathrm{MF}}}\dot{\mathrm{P}}_{30}(t)dt \tag{5.95}$$

We can, therefore, insert (5.91) into (5.95) and integrate to obtain:

$$\begin{aligned}
&{}^{s}_{r\underline{\mathrm{MF}}}\mathrm{P}_{30}(0,\mathbf{N}) = -2\mathrm{B}\beta c\gamma_N^{\frac{1}{\beta}}\left(\left(\frac{h}{2}\right)^{\frac{1}{\beta}}+\Gamma_{N-1}\right)^{-(\beta c+1)} \\
&\theta'_{N-1}\int_{t_{N-1}}^{t_N} e^{2\mathrm{B}\beta c\gamma_N^{\frac{1}{\beta}}\left(\left(\frac{h}{2}\right)^{\frac{1}{\beta}}+\Gamma_{N-1}\right)^{-(\beta c+1)}\frac{2\alpha_{MF}\sigma^b_{dN}h}{F_N\mathbb{B}_{N-1}}\left(1-e^{-\mathbb{B}_{N-1}W}\right)(t-t_{N-1})}dt \\
&= \frac{\theta'_{N-1}F_N\mathbb{B}_{N-1}}{2\alpha_{MF}\sigma^b_{dN}h\left(1-e^{-\mathbb{B}_{N-1}W}\right)}\left(1-e^{\frac{4\mathrm{B}\beta c\gamma_N^{\frac{1}{\beta}}\sigma^b_{dN}h}{F_N\mathbb{B}_{N-1}}\left(1-e^{-\mathbb{B}_{N-1}W}\right)\left(\left(\frac{h}{2}\right)^{\frac{1}{\beta}}+\Gamma_{N-1}\right)^{-(\beta c+1)}(t_N-t_{N-1})}\right)
\end{aligned} \tag{5.96}$$

Again we can multiply this by the cumulative volumetric density distribution function for residual active half-macrofractures given by (5.85) to derive an expression for the increase in the volumetric density of residual static half-macrofractures of length $\ell_{\mathbf{N}}$ or greater during a timestep $\mathbf{N}$:

$$\Delta^{s}_{r\underline{\mathrm{MF}}}\mathrm{P}_{30}(\ell_N, \mathbf{N}) = \frac{\theta'_{N-1} F_N \mathbb{B}_{N-1}}{2\alpha_{MF}\sigma_{dN}^{b} h\left(1 - e^{-\mathbb{B}_{N-1}\bar{W}}\right)} \left(1 - e^{\frac{4\mathrm{B}\beta c \gamma_N^{\frac{1}{\beta}} \sigma_{dN}^{b} h}{F_N \mathbb{B}_{N-1}}\left(1 - e^{-\mathbb{B}_{N-1}W}\right)\left(\left(\frac{h}{2}\right)^{\frac{1}{\beta}} + \Gamma_{N-1}\right)^{-(\beta c+1)}(t_N - t_{N-1})}\right) e^{-\frac{F_N \ell_N}{\alpha_{MF}\sigma_{dN}^{b}}} \tag{5.97}$$

5.4.3 *Calculating the Mean Linear Density of Residual Active Half-Macrofractures*

We can calculate the mean linear density of residual half-macrofractures in the same way as in Sects. 4.3 and 5.3.2: by differentiating the appropriate volumetric half-macrofracture density distribution functions with respect to ℓ to determine the number of half-macrofractures with length between ℓ and $\ell + d\ell$, multiplying this by ℓh to obtain their total area, and then integrating the resulting expression with respect to length. However, this is much simpler for residual half-macrofractures, because the cumulative density distribution functions for residual half-macrofractures all follow the form of (5.85):

$$_{r\underline{\mathrm{MF}}}\mathrm{P}_{30}(\ell_N, \mathrm{t}_N) = {}_{r\underline{\mathrm{MF}}}\mathrm{P}_{30}(0, \mathrm{t}_N) e^{-\frac{F_N \ell_N}{\alpha_{MF}\sigma_{dN}^{b}}} \tag{5.98}$$

We can, therefore, carry out the differentiation and integration on the length-dependent exponential term independently of the expression for total residual half-macrofracture volumetric density. Note that this will also give us an expression for the mean length that a residual half-macrofracture will reach before being deactivated, $\overline{\ell_N}$:

$$\begin{aligned} \overline{\ell_N} &= \int_0^\infty -\frac{d\Phi}{d\ell}\ell d\ell \\ &= \int_0^\infty -\frac{d}{d\ell}\left(e^{-\frac{F_N \ell}{\alpha_{MF}\sigma_{dN}^{b}}}\right)\ell d\ell \\ &= \left[-e^{-\frac{F_N \ell}{\alpha_{MF}\sigma_{dN}^{b}}}\left(\frac{\alpha_{MF}\sigma_{dN}^{b}}{F_N} + \ell\right)\right]_0^\infty = \frac{\alpha_{MF}\sigma_{dN}^{b}}{F_N} \end{aligned} \tag{5.99}$$

Thus the total mean linear density of all residual active half-macrofractures is given by the total volumetric density of residual half-macrofractures times the mean residual half-macrofracture length:

$$ {}^{a}_{r\underline{MF}}P_{32}(0, t_N) = {}^{a}_{r\underline{MF}}P_{30}(0, t_N)\overline{\ell_N} = {}^{a}_{r\underline{MF}}P_{30}(0, t_N)\frac{\alpha_{MF}\sigma^{b}_{dN}}{F_N} \quad (5.100) $$

where ${}^{a}_{r\underline{MF}}P_{30}(0,t_N)$ is given by (5.84) or (5.92) as appropriate.

We can use a similar procedure to calculate the mean linear density of residual active half-macrofractures of length ℓ_N or greater:

$$ \int_{\ell_N}^{\infty} -\frac{d}{d\ell}\left(e^{-\frac{F_N\ell}{\alpha_{MF}\sigma^{b}_{dN}}}\right)\ell d\ell $$

$$ = \int_{\ell_N}^{\infty} \frac{F_N}{\alpha_{MF}\sigma^{b}_{dN}} e^{-\frac{F_N\ell}{\alpha_{MF}\sigma^{b}_{dN}}}\ell d\ell = \left[-e^{-\frac{F_N\ell}{\alpha_{MF}\sigma^{b}_{dN}}}\left(\frac{\alpha_{MF}\sigma^{b}_{dN}}{F_N} + \ell\right)\right]_{\ell_N}^{\infty} $$

$$ = e^{-\frac{F_N\ell_N}{\alpha_{MF}\sigma^{b}_{dN}}}\left(\frac{\alpha_{MF}\sigma^{b}_{dN}}{F_N} + \ell_N\right) \quad (5.101) $$

and therefore

$$ {}^{a}_{r\underline{MF}}P_{32}(\ell_N, t_N) = {}^{a}_{r\underline{MF}}P_{30}(0, t_N)e^{-\frac{F_N\ell_N}{\alpha_{MF}\sigma^{b}_{dN}}}\left(\frac{\alpha_{MF}\sigma^{b}_{dN}}{F_N} + \ell_N\right) \quad (5.102) $$

Strictly speaking, (5.100) and (5.102) will only give approximations of the mean linear residual active half-macrofracture density, since (5.99) assumes that half-macrofracture deactivation is a purely random process and that F_N remains constant throughout the timestep. However as we saw in Sect. 5.1.3, half-macrofracture interaction is in fact a semi-regular process, and thus (5.99) will only give an approximation of the true mean residual half-macrofracture length. We can calculate more precise values for both the true mean residual half-macrofracture length and the mean linear residual active half-macrofracture density by splitting Φ into two components Φ_{II} and Φ_{IJ}, and inserting the semi-regular version of Φ_{IJ} given in (5.19) into (5.99).

5.4.4 Calculating the Mean Linear Density of Residual Static Half-Macrofractures

We can use the same method to calculate the incremental increase in the mean linear density of residual static half-macrofractures during a timestep **N**. Thus the incremental increase in the total mean linear density of all residual static half-macrofractures during timestep **N**, $\Delta^{s}_{r\underline{MF}}P_{32}(0,\mathbf{N})$ is given by

$$\Delta^{s}_{r\underline{MF}}P_{32}(0,\mathbf{N}) = \Delta^{s}_{r\underline{MF}}P_{30}(0,\mathbf{N})\frac{\alpha_{MF}\sigma^{b}_{dN}}{F_N} \tag{5.103}$$

and the incremental increase in the mean linear density of residual static half-macrofractures of length ℓ_N or greater during timestep **N** is given by

$$\Delta^{s}_{r\underline{MF}}P_{32}(\ell_N,\mathbf{N}) = \Delta^{s}_{r\underline{MF}}P_{30}(0,\mathbf{N})e^{-\frac{F_N\ell_N}{\alpha_{MF}\sigma^{b}_{dN}}}\left(\frac{\alpha_{MF}\sigma^{b}_{dN}}{F_N} + \ell_N\right) \tag{5.104}$$

where $\Delta^{s}_{r\underline{MF}}P_{30}(0,\mathbf{N})$ is given by (5.93) or (5.96) as appropriate.

Note again that the incremental increase in total mean linear density of residual static half-macrofractures can be expressed as the incremental increase in total volumetric density of residual half-macrofractures times the mean residual half-macrofracture length. However, as before, (5.103) and (5.104) will only give approximations of the mean linear residual static half-macrofracture density, because (5.99) assumes that half-macrofracture deactivation is a purely random process. We can calculate more precise values by splitting Φ into two components Φ_{II} and Φ_{IJ}, and inserting the semi-regular version of Φ_{IJ} given in (5.19) into (5.99).

Chapter 6
Elastic Moduli, Stress and Fracture Growth

Abstract In this chapter, we examine the impact of the growing fracture populations on the in situ stress in the fractured layer. We describe two potential scenarios. The first is the stress shadow scenario, in which displacement on a fracture accommodates all the horizontal elastic strain in a zone surrounding the fracture, known as the stress shadow. The second is the evenly distributed stress scenario, in which some of the horizontal strain is accommodated by displacement on the fractures, while the remainder is accommodated by elastic strain throughout the host rock, and there are no stress shadows. We show that in systems with stress shadows, the in situ stress outside them is the same as in an equivalent unfractured layer, while in the absence of stress shadows, we can derive modifiers for the bulk rock elastic moduli to take into account the effects of the fractures, based on the fracture densities calculated in previous chapters.

Keywords Fracture modelling · Discrete fracture networks · Naturally fractured reservoir · Mechanical stratigraphy · Stress shadows

In this chapter we consider the effect that the growing fracture populations will have on the elastic moduli and the in situ stress within the fractured layer. This is important because the in situ stress controls the fracture driving stress, and therefore the rate of fracture growth. However, the growth of the fracture network will also affect the in situ stress, because fracture displacement accommodates some of the horizontal strain that would otherwise be accommodated elastically in the rockmass, either locally (in the form of a stress shadow) or distributed throughout the layer. This creates a feedback mechanism that is likely to be an important control on the final fracture density and distribution.

The boundary conditions for tectonic deformation can typically be defined in terms of a constant effective vertical stress $\sigma_v{}'$, and a biaxial horizontal strain applied at specified rates $\dot{\varepsilon}_{h\min}$ and $\dot{\varepsilon}_{h\max}$, in a specified orientation and for a specified duration. In a perfectly elastic host rock, this information, combined with the bulk rock elastic moduli for the fractured layer, is sufficient to calculate the in situ stress and the driving stress on each fracture set, throughout the deformation episode. If the horizontal strain

© The Editor(s) (if applicable) and The Author(s), under exclusive license to Springer Nature Switzerland AG 2020

M. J. Welch et al., *Modelling the Evolution of Natural Fracture Networks*,
https://doi.org/10.1007/978-3-030-52414-2_6

is applied at a constant rate, the fracture driving stress will increase linearly with time, at least until the fracture network starts to grow.

However in practice, a rock subject to strain over geological timescales will often experience viscoelastic (time dependent) deformation processes such as creep, grain sliding and pressure solution, which gradually convert elastic strain into inelastic strain. In the initial stages of deformation, these "strain relaxation" processes will be slow and thus the elastic strain in the rock will increase in proportion to the applied strain rate. However as the elastic strain increases, the strain relaxation processes will accelerate until an equilibrium is reached between the applied strain rate and the rate of strain relaxation. Thus the deformation episode will be characterised by an early stage of increasing elastic strain and fracture driving stress, followed by a late stage of constant elastic strain and fracture driving stress. The ultimate fracture geometry will depend on the rate of fracture growth relative to the rate of strain relaxation: fractures networks may grow rapidly in the early stage, with later strain accommodated by increasing displacement on the existing fractures, or they may grow slowly in the late stage. The impact of strain evolution on the final fracture geometry is explored further in Welch et al (2019).

For timesteps in which the fracture driving stress is increasing, it is useful to define a "weighted mean" driving stress, $\bar{\sigma}_{dN}$:

$$\overline{\sigma_{dN}} = \left(\frac{\left(\sigma_{dN-1} + \frac{d\sigma_d}{dt}(t_N - t_{N-1})\right)^{b+1} - \sigma_{dN-1}^{b+1}}{(t_N - t_{N-1})(b+1)\frac{d\sigma_d}{dt}} \right)^{\frac{1}{b}} \tag{6.1}$$

where σ_{dN-1} is the fracture driving stress at the end of the previous timestep $N-1$, and $d\sigma_d/dt$ is the rate of change of fracture driving stress during the current timestep N. Since the rate of fracture propagation is proportional to the fracture driving stress to the power of the subcritical fracture propagation index b, the weighted mean driving stress can be substituted for σ_{dN}, in all the equations derived in Chaps. 3, 4 and 5 to calculate the volumetric fracture density and mean linear fracture density.

The bulk rock elastic moduli for a fractured layer take account of the elastic strain accommodated on the fractures as well as the elastic strain in the host rock. Therefore they may also change through time as the fracture network grows. The exact form of the bulk rock elastic moduli will depend on the distribution of stress throughout the layer: whether there are fracture stress shadows or the stress is evenly distributed. We will examine both scenarios.

6.1 Elastic Moduli in the Stress Shadow Scenario

In the stress shadow scenario, elastic displacement on a fracture accommodates all the horizontal elastic strain in a zone immediately surrounding the fracture, known as the stress shadow (see Sect. 4.4). This generates a local stress anomaly within the stress shadow. Outside of the stress shadow, the in situ stress in the host rock is unaffected by the fracture and is equal to the in situ stress in an equivalent unfractured layer. The fracture driving stress outside the stress shadows will, therefore, be independent of the fracture population, while the fracture driving stress inside the stress shadows will be zero.

The horizontal effective stresses in an unfractured layer, and thus the horizontal effective stresses outside the stress shadows in a fractured layer, can be calculated from the effective vertical stress, σ_v', the horizontal elastic strains, $\varepsilon_{h\min}$ and $\varepsilon_{h\max}$, the elastic properties of the intact rock, E_r (Young's Modulus) and ν_r (Poisson's ratio), and the initial compaction factor, υ:

$$\sigma'_{h\min}(t) = \frac{E_r}{1-\nu_r^2}\left(\varepsilon_{h\min}(t) + \nu_r\varepsilon_{h\max}(t)\right) + \frac{\upsilon(1-\nu_r) + (1-\upsilon)\nu_r}{1-\nu_r}\sigma'_v$$
$$\sigma'_{h\max}(t) = \frac{E_r}{1-\nu_r^2}\left(\nu_r\varepsilon_{h\min}(t) + \varepsilon_{h\max}(t)\right) + \frac{\upsilon(1-\nu_r) + (1-\upsilon)\nu_r}{1-\nu_r}\sigma'_v \quad (6.2)$$

where $\varepsilon_{h\min}(t)$ and $\varepsilon_{h\max}(t)$ represent the horizontal elastic strains at time t, which will be a function of the applied strain rates and the strain relaxation rate as well as time.

The initial compaction factor υ accounts for compaction-related inelastic strain. In a purely elastic layer, the initial horizontal stress with no horizontal strain, σ_{h0}', is given by

$$\sigma'_{h0} = \frac{\nu_r}{1-\nu_r}\sigma'_v \quad (6.3)$$

This implies that $\sigma_{h0}' < \sigma_v'$, and there will, hence, be a differential stress in the layer even before a horizontal strain is applied. This differential stress will often be sufficient to drive inelastic deformation, thus causing compaction of the layer. This compaction may take the form of distributed ductile processes that reduce the overall porosity of the host rock, e.g. mechanical processes such as grain realignment or chemical processes such as pressure solution. However, if the differential stress is sufficient to drive Mode 2 shear failure, it could generate the compaction-related deformation bands observed in some sandstones, chalks and other lithologies (e.g. Maltman 1988; Knipe 1989; Fossen 2010; Wennberg et al. 2013). It is likely that these early fractures will heal over geological time and will not be reactivated during tectonic deformation. They will, however, lead to a permanent reduction in the differential stress in the layer, and hence an increase in the horizontal effective stress σ_{h0}'.

In Eq. (6.2), we account for this compaction through the initial compaction factor, υ. The initial compaction factor can vary from $\upsilon = 0$, which represents an elastic equilibrium with no compaction where σ_{h0}' is given by (6.3), to $\upsilon = 1$, which represents a viscoelastic equilibrium where $\sigma_{h0}' = \sigma_v'$. A common initial horizontal stress state is the critical stress state, in which optimally oriented Mode 2 shear fractures lie on the failure envelope (Zoback and Townend 2001). This will often be the situation following fault-controlled compaction. In the critical stress state, the initial horizontal stress σ_{h0}' is given by

$$\sigma_{h0}' = \frac{\left(\sin\omega\cos\omega - \mu_{fr}\cos^2\omega\right)}{\left(\sin\omega\cos\omega + \mu_{fr}\sin^2\omega\right)}\sigma_v' \tag{6.4}$$

and hence the initial compaction factor υ is given by

$$\upsilon = \frac{(1-\nu_r)\left(\sin\omega\cos\omega - \mu_{fr}\cos^2\omega\right) - \nu_r}{1-2\nu_r} \tag{6.5}$$

where ω is the dip of the Mode 2 fractures and μ_{fr} is the fracture friction coefficient.

6.2 Elastic Moduli in the Evenly Distributed Stress Scenario

In the evenly distributed stress scenario, elastic displacement on the fractures accommodates some of the applied horizontal strain, while the remainder is accommodated by elastic strain distributed evenly throughout the host rock. No local stress anomalies develop around the fractures.

In this scenario, the fractures will introduce an anisotropy in the elastic response of the bulk rock to an applied external strain. If we assume that the fracture sets are orthogonal to the principal horizontal stress directions, we can use the orthotropic form of Hooke's Law to relate the bulk rock elastic strain (i.e. the total elastic strain accommodated by both the fractures and the host rock) to the in situ stress:

$$\begin{pmatrix} \varepsilon_{h\min} \\ \varepsilon_{h\max} \\ \varepsilon_v \end{pmatrix} = \begin{pmatrix} \frac{1}{E_{h\min}} & -\frac{\nu_{h\max,h\min}}{E_{h\max}} & -\frac{\nu_{v,h\min}}{E_v} \\ -\frac{\nu_{h\min,h\max}}{E_{h\min}} & \frac{1}{E_{h\max}} & -\frac{\nu_{v,h\max}}{E_v} \\ -\frac{\nu_{h\min,v}}{E_{h\min}} & -\frac{\nu_{h\max,v}}{E_{h\max}} & \frac{1}{E_v} \end{pmatrix} \begin{pmatrix} \sigma_{h\min}' \\ \sigma_{h\max}' \\ \sigma_v' \end{pmatrix} \tag{6.6}$$

Based on the fracture strain expressions (4.22), we can derive the expressions for the bulk rock elastic moduli in (6.6):

$$E_{h\min} = \frac{E_r}{1 + \left(1-\nu_r^2\right)M_{hh}\frac{\pi}{2}\,{}^{h\min}_{\mathrm{MF}}\mathrm{P}_{32}(0, \mathrm{t}_{N-1})}$$

$$E_{h\max} = \frac{E_r}{1 + \left(1-\nu_r^2\right)M_{hh}\frac{\pi}{2}\,{}^{h\max}_{\mathrm{MF}}\mathrm{P}_{32}(0, \mathrm{t}_{N-1})}$$

$$E_v = \frac{E_r}{1 + \left(1-\nu_r^2\right)M_{vv}\frac{\pi}{2}\left({}^{h\min}_{\mathrm{MF}}\mathrm{P}_{32}(0, \mathrm{t}_{N-1}) + {}^{h\max}_{\mathrm{MF}}\mathrm{P}_{32}(0, \mathrm{t}_{N-1})\right)} \tag{6.7}$$

and

$$\begin{aligned}
\nu_{v,h\min} &= \frac{\nu_r - \left(1-\nu_r^2\right)M_{vh}\frac{\pi}{2}{}^{h\min}_{\mathrm{MF}}\mathrm{P}_{32}(0,\mathrm{t}_{N-1})}{1+\left(1-\nu_r^2\right)M_{vv}\frac{\pi}{2}\left({}^{h\min}_{\mathrm{MF}}\mathrm{P}_{32}(0,\mathrm{t}_{N-1})+{}^{h\max}_{\mathrm{MF}}\mathrm{P}_{32}(0,\mathrm{t}_{N-1})\right)}\\
\nu_{h\min,v} &= \frac{\nu_r - \left(1-\nu_r^2\right)M_{hv}\frac{\pi}{2}{}^{h\min}_{\mathrm{MF}}\mathrm{P}_{32}(0,\mathrm{t}_{N-1})}{1+\left(1-\nu_r^2\right)M_{hh}\frac{\pi}{2}{}^{h\min}_{\mathrm{MF}}\mathrm{P}_{32}(0,\mathrm{t}_{N-1})}\\
\nu_{v,h\max} &= \frac{\nu_r - \left(1-\nu_r^2\right)M_{vh}\frac{\pi}{2}{}^{h\max}_{\mathrm{MF}}\mathrm{P}_{32}(0,\mathrm{t}_{N-1})}{1+\left(1-\nu_r^2\right)M_{vv}\frac{\pi}{2}\left({}^{h\min}_{\mathrm{MF}}\mathrm{P}_{32}(0,\mathrm{t}_{N-1})+{}^{h\max}_{\mathrm{MF}}\mathrm{P}_{32}(0,\mathrm{t}_{N-1})\right)}\\
\nu_{h\max,v} &= \frac{\nu_r - \left(1-\nu_r^2\right)M_{hv}\frac{\pi}{2}{}^{h\max}_{\mathrm{MF}}\mathrm{P}_{32}(0,\mathrm{t}_{N-1})}{1+\left(1-\nu_r^2\right)M_{hh}\frac{\pi}{2}{}^{h\max}_{\mathrm{MF}}\mathrm{P}_{32}(0,\mathrm{t}_{N-1})}\\
\nu_{h\min,h\max} &= \frac{\nu_r}{1+\left(1-\nu_r^2\right)M_{hh}\frac{\pi}{2}{}^{h\min}_{\mathrm{MF}}\mathrm{P}_{32}(0,\mathrm{t}_{N-1})}\\
\nu_{h\max,h\min} &= \frac{\nu_r}{1+\left(1-\nu_r^2\right)M_{hh}\frac{\pi}{2}{}^{h\max}_{\mathrm{MF}}\mathrm{P}_{32}(0,\mathrm{t}_{N-1})}
\end{aligned} \tag{6.8}$$

where ${}^{h\min}_{\mathrm{MF}}\mathrm{P}_{32}$ represents the total mean linear density of all macrofractures (active and static) striking perpendicular to the $\varepsilon_{h\min}$ orientation, ${}^{h\max}_{\mathrm{MF}}\mathrm{P}_{32}$ represents the total mean linear density of all macrofractures striking perpendicular to the $\varepsilon_{h\max}$ orientation, and M_{hh}, M_{vh}, M_{hv} and M_{vv} are a set of Fracture Mode factors defined as follows:

- For vertical Mode 1 dilatant fractures striking perpendicular to $\varepsilon_{h\min}$

$$\begin{aligned}
M_{hh} &= \begin{cases} 1 \text{ for } \sigma'_{h\min} < 0 \\ 0 \text{ for } \sigma'_{h\min} \geq 0 \end{cases}\\
M_{vh} &= M_{hv} = M_{vv} = 0
\end{aligned} \tag{6.9}$$

- For vertical Mode 1 dilatant fractures striking perpendicular to $\varepsilon_{h\max}$

$$\begin{aligned}
M_{hh} &= \begin{cases} 1 \text{ for } \sigma'_{h\max} < 0 \\ 0 \text{ for } \sigma'_{h\max} \geq 0 \end{cases}\\
M_{vh} &= M_{hv} = M_{vv} = 0
\end{aligned} \tag{6.10}$$

- For normal Mode 2 shear fractures with dip ω and friction coefficient μ_{fr}

$$\begin{aligned}
M_{hh} &= \cos\omega\left(\sin\omega\cos\omega + \mu_{fr}\sin^2\omega\right)\\
M_{vh} &= -\cos\omega\left(\sin\omega\cos\omega - \mu_{fr}\cos^2\omega\right)\\
M_{hv} &= -\cos\omega\left(\sin\omega\cos\omega + \mu_{fr}\sin^2\omega\right)\\
M_{vv} &= \cos\omega\left(\sin\omega\cos\omega - \mu_{fr}\cos^2\omega\right)
\end{aligned} \tag{6.11}$$

We can simplify the constitutive relationships considerably by defining fracture multipliers for Young's Modulus and Poisson's ratio, Ê and $\hat{\nu}$, respectively:

$$\begin{aligned}
\hat{E}_{hmin} &= 1 + \left(1 - \nu_r^2\right) M_{hh} \frac{\pi}{2} {}^{hmin}_{MF}\mathrm{P}_{32}(0, \mathrm{t}_{N-1}) \\
\hat{E}_{hmax} &= 1 + \left(1 - \nu_r^2\right) M_{hh} \frac{\pi}{2} {}^{hmax}_{MF}\mathrm{P}_{32}(0, \mathrm{t}_{N-1}) \\
\hat{E}_{v} &= 1 + \left(1 - \nu_r^2\right) M_{vv} \frac{\pi}{2} \left({}^{hmin}_{MF}\mathrm{P}_{32}(0, \mathrm{t}_{N-1}) + {}^{hmax}_{MF}\mathrm{P}_{32}(0, \mathrm{t}_{N-1})\right)
\end{aligned} \tag{6.12}$$

and

$$\begin{aligned}
\hat{\nu}_{v,hmin} &= \nu_r - \left(1 - \nu_r^2\right) M_{vh} \frac{\pi}{2} {}^{hmin}_{MF}\mathrm{P}_{32}(0, \mathrm{t}_{N-1}) \\
\hat{\nu}_{hmin,v} &= \nu_r - \left(1 - \nu_r^2\right) M_{hv} \frac{\pi}{2} {}^{hmin}_{MF}\mathrm{P}_{32}(0, \mathrm{t}_{N-1}) \\
\hat{\nu}_{v,hmax} &= \nu_r - \left(1 - \nu_r^2\right) M_{vh} \frac{\pi}{2} {}^{hmax}_{MF}\mathrm{P}_{32}(0, \mathrm{t}_{N-1}) \\
\hat{\nu}_{hmax,v} &= \nu_r - \left(1 - \nu_r^2\right) M_{hv} \frac{\pi}{2} {}^{hmax}_{MF}\mathrm{P}_{32}(0, \mathrm{t}_{N-1})
\end{aligned} \tag{6.13}$$

We can now write the bulk rock constitutive relationship as

$$\begin{pmatrix} \varepsilon_{hmin} \\ \varepsilon_{hmax} \\ \varepsilon_{v} \end{pmatrix} = \frac{1}{E_r} \begin{pmatrix} \hat{E}_{hmin} & -\nu_r & -\hat{\nu}_{v,hmin} \\ -\nu_r & \hat{E}_{hmax} & -\hat{\nu}_{v,hmax} \\ -\hat{\nu}_{hmin,v} & -\hat{\nu}_{hmax,v} & \hat{E}_{v} \end{pmatrix} \begin{pmatrix} \sigma'_{hmin} \\ \sigma'_{hmax} \\ \sigma'_{v} \end{pmatrix} \tag{6.14}$$

and (6.2) becomes

$$\begin{aligned}
\sigma'_{hmin}(t) &= \frac{E_r}{\hat{E}_{hmin}\hat{E}_{hmax} - \nu_r^2} \left(\hat{E}_{hmax}\varepsilon_{hmin}(t) + \nu_r \varepsilon_{hmax}(t)\right) \\
&\quad + \left(\upsilon \frac{\left(\hat{E}_{hmax} + \nu_r\right)(1 - 2\nu_r)}{\hat{E}_{hmin}\hat{E}_{hmax} - \nu_r^2} + \frac{\hat{E}_{hmax}\hat{\nu}_{v,hmin} + \nu_r \hat{\nu}_{v,hmax}}{\hat{E}_{hmin}\hat{E}_{hmax} - \nu_r^2}\right)\sigma'_{v} \\
\sigma'_{hmax}(t) &= \frac{E_r}{\hat{E}_{hmin}\hat{E}_{hmax} - \nu_r^2} \left(\hat{E}_{hmin}\varepsilon_{hmax}(t) + \nu_r \varepsilon_{hmin}(t)\right) \\
&\quad + \left(\upsilon \frac{\left(\hat{E}_{hmin} + \nu_r\right)(1 - 2\nu_r)}{\hat{E}_{hmin}\hat{E}_{hmax} - \nu_r^2} + \frac{\hat{E}_{hmin}\hat{\nu}_{v,hmax} + \nu_r \hat{\nu}_{v,hmin}}{\hat{E}_{hmin}\hat{E}_{hmax} - \nu_r^2}\right)\sigma'_{v}
\end{aligned} \tag{6.15}$$

Note that the bulk moduli defined in (6.7) and (6.8), and the fracture multipliers defined in (6.12) and (6.13), do not take account of the microfractures, which are assumed to accommodate a negligible amount of strain in comparison to the layer-bound macrofractures. However, in situations where the microfractures are significant, it is possible to expand (6.7), (6.8), (6.12) and (6.13) to include additional terms representing elastic strain accommodated on the microfractures. These terms will be based on the total fracture porosity for Mode 1 microfractures or the total volumetric heave for Mode 2 microfractures (see Sect. 4.4), which must be calculated numerically using a similar procedure to that used to calculate the mean linear microfracture density ${}_{\mu F}P_{32}$ (Sect. 3.4).

References

Fossen, H. 2010. Deformation bands formed during soft-sediment deformation: Observations from SE Utah. *Marine and Petroleum Geology* 27: 215–222.

Knipe, R.J. 1989. Deformation mechanisms—recognition from natural tectonites. *Journal of Structural Geology* 11: 127–146.

Maltman, A.J. 1988. The importance of shear zones in naturally deformed wet sediments. *Tectonophysics* 145: 163–175.

Welch, M.J., Mikael Lüthje, and A.C. Glad. 2019. Influence of fracture nucleation and propagation rates on fracture geometry: insights from geomechanical modelling. *Petroleum Geoscience* 25: 470–489.

Wennberg, O.P., G. Casini, A. Jahanpanah, F. Lapponi, J. Ineson, B.G. Wall, and P. Gillespie. 2013. Deformation bands in chalk, examples from the Shetland Group of the Oseberg Field, North Sea, Norway. *Journal of Structural Geology* 56: 103–117.

Zoback, M.D., and J. Townend. 2001. Implications of hydrostatic pore pressures and high crustal strength for the deformation of intraplate lithosphere. *Tectonophysics* 336: 19–30.

Chapter 7
Applying the Method to Geological Formations

Abstract In this chapter, we describe a workflow for building implicit fracture models and explicit DFNs of actual geological structures. This workflow combines the cumulative fracture density distribution functions and the stress–strain relationships derived in previous chapters to calculate the key properties describing the fracture populations in different locations across the structure, based on the mechanical properties and deformation experienced at each location. It then integrates this data to simulate the growth of individual fractures across the structure, resulting in an explicit DFN. We discuss the requirements for the static geomodel used in the workflow. Finally, we show how to validate the results by comparing the implicit fracture model with the explicit DFN and we discuss the potential sources of inaccuracy in these outputs.

Keywords Fracture modellin · Discrete fracture networks · Naturally fractured reservoir · Reservoir modelling

To apply this method to fractured layers in real geological formations, we must describe a workflow that includes all the elements of the calculation in an appropriate order. This will allow us to predict the fracture populations across real geological structures, as a function of the mechanical properties and deformation history (effective vertical stress, horizontal strain rate, strain orientation and duration of deformation). We can also use the data generated by the implicit component of the method (the fracture nucleation and propagation rates) to generate an explicit DFN where individual fractures are represented as geometric objects.

7.1 Geometry of the Static Geomodel

The calculations we have described so far assume that the layer is planar and flat and that the mechanical properties and deformational history are homogeneous and extend laterally for an infinite distance. By contrast, real geological layers tend to have

© The Editor(s) (if applicable) and The Author(s), under exclusive license to Springer Nature Switzerland AG 2020

M. J. Welch et al., *Modelling the Evolution of Natural Fracture Networks*,
https://doi.org/10.1007/978-3-030-52414-2_7

complex geometries and laterally variable mechanical properties and deformation histories—consider, for example, the variation in total strain between a fold hinge and a fold limb or the complex fracture patterns that can develop in a fault damage zone. For this reason, static geomodels typically discretise the three-dimensional geometry into a grid of discrete gridblocks, which are usually hexahedral as this simplifies the gridding. The mechanical properties and in situ stress are assumed to be uniform within each gridblock but can vary between gridblocks.

We can incorporate this complexity and variability into our fracture models by running the implicit components of the method independently for each gridblock, using the mechanical properties and local in situ stress specified for that gridblock. This will allow us to easily compare the predicted fracture density and geometry in different parts of the geological structure—e.g. we can compare the predicted mean fracture spacing and length between a fold hinge and a fold limb, or between the footwall and hanging wall of a fault. However, to obtain accurate results, there are some constraints on the gridblock geometry:

- The gridblocks should extend vertically through the entire fractured layer—i.e. the layer should be one gridblock thick.
- The length and width of the gridblocks should be greater than the typical maximum length of the macrofractures, so that the majority of macrofractures that nucleate within a gridblock will remain entirely within that gridblock.
- The thickness and depth of the layer should not vary by too much within each gridblock—i.e. they should be relatively flat-lying (or at least should have been at the time of deformation) and should not have a significant taper.

It may be necessary to upscale the gridblocks in the original static geomodel to meet these criteria.

Finally, we can combine the implicit fracture data for each gridblock to generate a single, integrated explicit DFN spanning the entire geomodel. This allows us to generate large, geomechanically consistent DFNs that reflect the variability in geometry, mechanical properties and in situ stress across a large geological structure.

7.2 Generating Implicit Fracture Data

Figure 7.1 illustrates a workflow to generate implicit fracture population data for each gridblock in the static geomodel, in the form of piecewise cumulative fracture density distribution functions ${}_{\mu F}P_{30}(r)$, ${}_{\mu F}P_{32}(r)$, ${}_{\underline{MF}}P_{30}(l)$ and ${}_{\underline{MF}}P_{32}(l)$, for a series of specified index values $r_1, r_2, r_3\ldots$ and $l_1, l_2, l_3\ldots$. Note that we can also use this workflow to generate other useful data described in the previous sections, such as the fracture aperture and porosity, the mean fracture length, the fracture connectivity or the total strain accommodated on the fractures. This data can be broken down by fracture set if required.

The workflow starts by calculating the initial stress state of each gridblock prior to deformation based on the depth of burial, fluid pressure and compaction-related

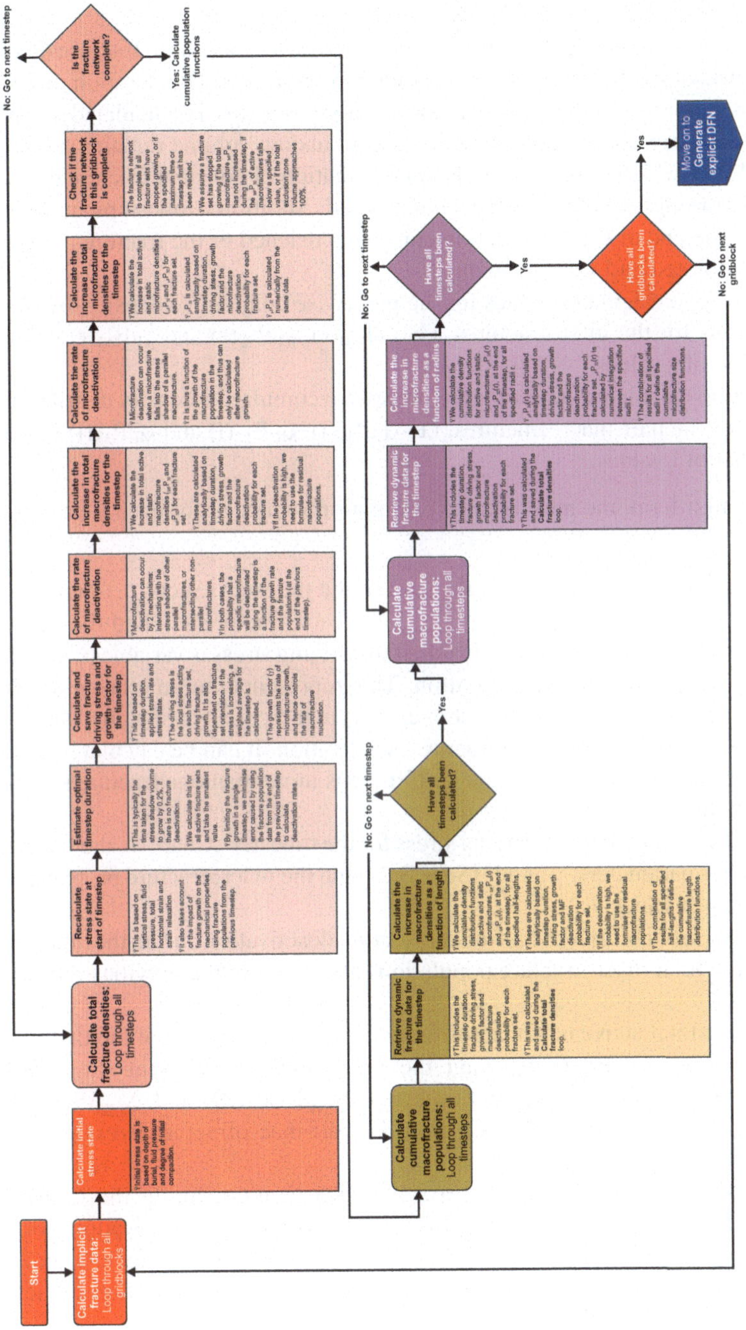

Fig. 7.1 Flow diagram for a workflow that combines all the different elements of the method described here to calculate an implicit fracture model for a fractured layer discretised into gridblocks

strain. The deformation episode is then broken down into a series of timesteps. The workflow loops through these three times:

- The first loop calculates the evolution of the total active and static fracture densities ${}_{\mu F}P_{32}(0)$, ${}_{\underline{MF}}P_{30}(0)$ and ${}_{\underline{MF}}P_{32}(0)$, for each fracture set. The duration of each timestep is determined dynamically during this loop; this is calculated so that the maximum increase in macrofracture stress shadow volume remains below a specified limit, which ensures that the approximation of constant fracture deactivation rates throughout the timestep remains valid. Additional timesteps are added until either the end of the deformation episode is reached or the fracture network becomes saturated and stops growing.
- The second loop calculates values for the piecewise cumulative density distribution functions for the microfractures, i.e. ${}_{\mu F}P_{30}(r)$, ${}_{\mu F}P_{32}(r)$ and ${}_{\mu F}P_{33}(r)$ for the specified list of r values.
- The third loop calculates values for the piecewise cumulative density distribution functions for the half-macrofractures, i.e. ${}_{\underline{MF}}P_{30}(l)$, ${}_{\underline{MF}}P_{32}(l)$ and ${}_{\underline{MF}}P_{33}(l)$ for the specified list of l values.

Within the first loop, the approximate calculation sequence for each timestep is as follows:

1. Calculate the in situ stress state at the start of the timestep (Sects. 6.1 and 6.2).
2. Calculate the initial driving stress, $\sigma_{d\mathbf{N}-1}$, for each fracture set, and the rate of increase in fracture driving stress, $d\sigma_d/dt$, if the driving stress is variable (Chap. 6).
3. Calculate the optimal timestep duration. This represents the time taken for the half-macrofracture stress shadow volume, $\psi_{\underline{MF}}$, to increase by a specified amount (typically c.0.2%), assuming no fracture deactivation. It can be calculated using the algorithms derived in Appendix C. If there is more than one fracture set, use the lowest value.
4. Calculate the weighted mean driving stress for the timestep (Chap. 6), the fracture growth factor parameters, γ and Γ (Sect. 3.2), and the macrofracture propagation distance, $\alpha_{MF}\Lambda$ (Sect. 4.2).
5. Calculate the probabilities of half-macrofracture deactivation during the timestep, based on the total macrofracture populations at the end of the previous timestep (Sect. 5.1).
6. Calculate the total active half-macrofracture densities at the end of the timestep, ${}^{a}_{\underline{MF}}P_{30}(0, t_{\mathbf{N}})$ and ${}^{a}_{\underline{MF}}P_{32}(0, t_{\mathbf{N}})$, and the increments in the total static half-macrofracture densities during the timestep, $\Delta^{s}_{\underline{MF}}P_{30}(0, \mathbf{N})$ and $\Delta^{s}_{\underline{MF}}P_{32}(0, \mathbf{N})$ (Sects. 5.3 and 5.4). Also calculate the total half-macrofracture stress shadow volume at the end of the timestep, $\psi_{\underline{MF},\mathbf{N}}$.
7. Calculate the probabilities of microfracture deactivation during the timestep (Sect. 5.1). Note that the probability of microfracture deactivation for a specific fracture set is given by the change in the half-macrofracture exclusion zone volume during the timestep, so cannot be calculated before the total half-macrofracture density data is updated.

8. Calculate the total active microfracture densities ${}^{a}_{\mu F}P_{30}(r_{min}, t_N)$ and ${}^{a}_{\mu F}P_{32}(0, t_N)$ at the end of the timestep, and the increments in static half-macrofracture populations $\Delta^{s}_{\mu F}P_{30}(r_{min}, t_N)$ and $\Delta^{s}_{\mu F}P_{32}(0, t_N)$ during the timestep (Sect. 5.2).
9. Check if the end of the deformation episode has been reached, or if the fracture network has reached saturation. If not, continue to the next timestep.

Within the second and third loops, most of these steps can be skipped by reusing data calculated in the first loop. If the timestep duration, the weighted mean driving stress, the fracture growth factor parameters, the macrofracture propagation distance, and the fracture deactivation probabilities are cached during the first loop, then only step 8 need be calculated in the second loop and only step 6 need be calculated in the third loop.

7.3 Generating the Explicit DFN

Figure 7.2 presents a workflow for generating an explicit DFN model for a fractured layer across an entire geomodel, using the implicit fracture data calculated for each of the gridblocks in the geomodel. Note that we can either populate the DFN with both microfractures and macrofractures, or with macrofractures only. In the latter case, we add seed macrofractures to the DFN according to the macrofracture nucleation rates calculated with the implicit fracture data. This will reduce the total number of fractures in the DFN while retaining all the long fractures that are likely to act as major fluid flow pathways, thus making it more suitable as input for fluid flow modelling.

The workflow for DFN generation uses the timesteps defined for each gridblock when calculating the implicit fracture data. It loops through all of the timesteps in all of the gridblocks in the geomodel, in chronological order. It can, thus, use the driving stress, nucleation rate and propagation rate data calculated for each fracture set in each timestep in the implicit calculation phase to control the nucleation and growth of the fractures in the explicit DFN.

For each timestep in each gridblock, the exact sequence of steps is as follows

1. **If the DFN includes microfractures, add new microfractures**: Since the total microfracture density is infinite for a power-law distribution, a minimum radius must be specified for the microfractures in the DFN. Use (3.14) to determine how many microfractures will reach the minimum radius during timestep. Place the new microfractures at random locations within the gridblock, and check to determine if they lie within the stress shadow of an existing macrofracture; any that do must be discarded.
2. **If the DFN includes microfractures, grow the microfractures**: Increase the radius of all active microfractures in the gridblock by an amount determined by (3.8) and the fracture driving stress for the timestep. First, however, check if the microfracture has fallen into the stress shadow of a macrofracture since the previous timestep; any that have must be deactivated.

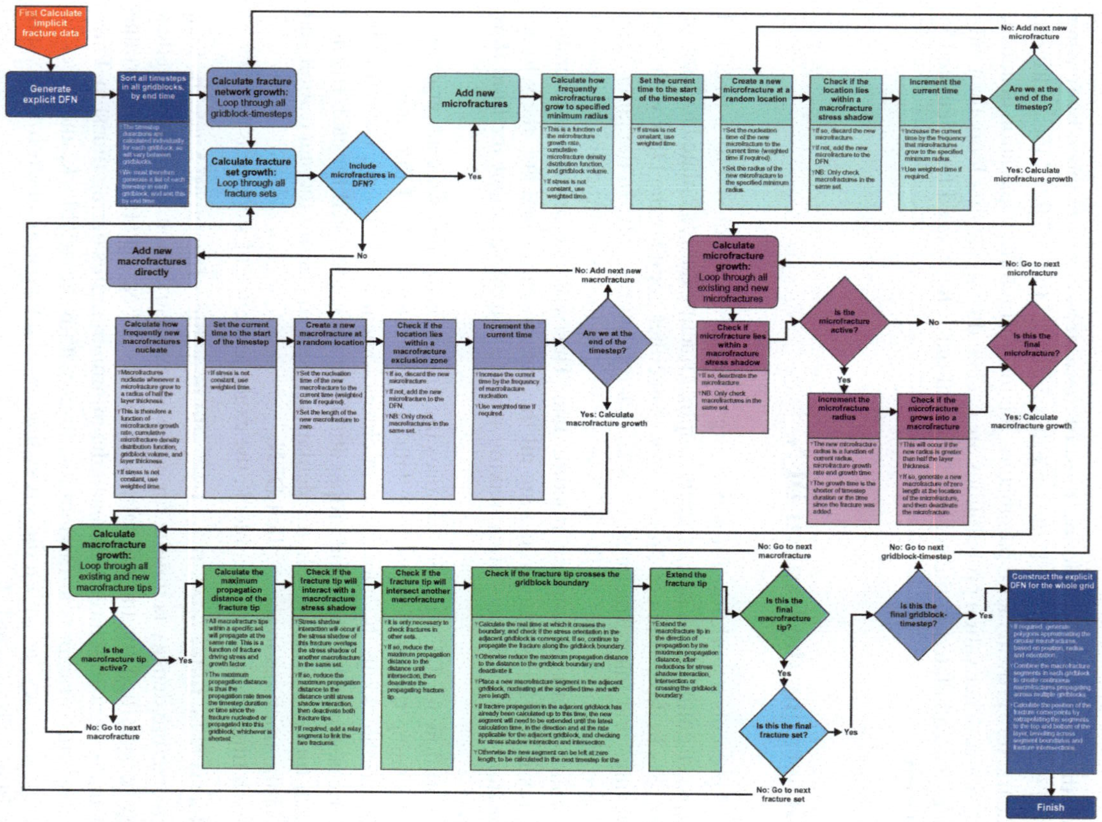

Fig. 7.2 Flow diagram for a workflow to generate an explicit DFN model for a fractured layer discretised into gridblocks

3. **Nucleate macrofractures**: If the DFN includes microfractures, then new macrofractures will nucleate whenever a microfracture grows to radius h/2. If the DFN does not include microfractures, new macrofractures must be added directly at a rate determined by (4.10). The new macrofractures should have an initial length of zero. Place them at random locations within the gridblock, and check to determine if they lie within the exclusion zone of an existing macrofracture; any that do must be discarded.
4. **Macrofracture propagation and interaction**: Since all macrofracture tips propagate at the same rate, the maximum propagation distance for a macrofracture tip during the current timestep can be calculated using (5.1). However, the actual propagation distance of any specific macrofracture tip will be less than the maximum if it intersects another macrofracture, if it interacts with the stress shadow of a parallel macrofracture or if it crosses the gridblock boundary. In the first two cases, extend the macrofracture tip to the point of intersection or stress shadow interaction, and then deactivate it (in the case of stress shadow interaction, also deactivate the other macrofracture tip). If it crosses the gridblock boundary, create a new seed segment in the adjacent gridblock, connected to the original segment, and continue extending the new segment. Note that the orientation and propagation rate of macrofractures, and even the timestep durations, may be different in the adjacent gridblock.

The process of checking for fracture interaction can be simplified by converting from global xyz coordinates, with north-south, east-west and vertical axes, to a local coordinate system with axes orthogonal to the two fracture sets. If we define an ijk coordinate system with an i axis parallel to the strike of the primary fracture set, a j axis parallel to the strike of the secondary fracture set and a vertical k axis, then a macrofracture from the primary set will intersect a macrofracture from the secondary set if its j coordinate lies between the j coordinates of the tips of the secondary fracture, and the i coordinate of the secondary fracture lies between the i coordinates of the two tips of the primary fracture. A point will lie within the stress shadow of a macrofracture from the primary set if its i coordinate lies between the i coordinates of the two tips of the macrofracture and its j coordinate lies within the range of the j coordinate of the macrofracture plus or minus the stress shadow width W; this can be used to calculate stress shadow interactions of both propagating macrofracture tips and microfractures. It is slightly more complicated to calculate whether a propagating macrofracture will intersect the gridblock boundary, as this may not be orthogonal to the ijk axes, but we can do this using standard geometric algorithms that give the intersection point of two oblique coplanar lines.

Note that the macrofractures generated by this workflow are segmented across the gridblock boundaries. This allows the DFN to replicate the complex macrofracture geometries that can arise when the layer height or strain orientation varies laterally. In Chap. 9, we will see how this method is able to replicate the curved macrofractures that are often observed around larger faults in the outcrop. Once we have finished simulating fracture propagation in all the gridblocks, we must, therefore, recombine the connected planar segments into single, generally non-planar entities representing entire macrofractures. Since the ijk coordinate systems are valid within only one gridblock, we must convert the fracture coordinates back to global xyz coordinates; this can be done using standard geometric algorithms for coordinate transformation. To ensure smooth alignment of connected segments, the corner points of these segments should also be bevelled.

7.4 Comparing the Implicit and Explicit Output

We can test whether the equations derived in Chaps. 3, 4 and 5 accurately represent the processes and assumptions described in Chap. 2, by comparing the implicit cumulative fracture density distribution functions calculated using these equations with the cumulative density distributions calculated empirically for the fractures in the explicit DFNs. In particular, since fracture interaction and deactivation is determined empirically in the explicit DFN, a close match between the implicit and explicit cumulative fracture population functions would give independent confirmation that the fracture deactivation probabilities derived in Sect. 5.1, and the cumulative density distribution functions for active and static fractures derived in Sects. 5.2 and 5.3, are valid for the assumptions given.

Figure 7.3 shows a comparison between the implicit and explicit cumulative density distribution functions ${}_{\mu F}P_{30}(r)$, ${}_{\mu F}P_{32}(r)$, ${}_{\underline{MF}}P_{30}(l)$ and ${}_{\underline{MF}}P_{32}(l)$, calculated for six different fracture sets from five different models. The models all comprise planar, homogeneous layers with uniform mechanical properties and applied strain boundary conditions (full details of the input parameters for each model are given in Table 7.1). The size of the models was adjusted to ensure it was significantly greater than the maximum macrofracture length, and the cumulative population functions for the explicit DFN were calculated only for fractures in the centre of the model, to exclude edge effects.

We can see that a good match is obtained for all the fracture sets, which include microfractures and macrofractures, fractures generated with evenly distributed stress and with stress shadows, fractures generated by uniaxial, isotropic and anisotropic applied strains and, for the latter case, both primary and secondary fracture sets. This shows that the equation derived in Chaps. 3, 4 and 5 are a valid representation of the conceptual model for layer-bound fracture growth described in Chap. 2. However, some minor discrepancies are observed, which can be explained by two factors:

- **Random variabilities**: Since the fractures in the explicit DFN are nucleated at random locations, there will always be an element of random variability in the final geometry of the fracture network, and no two models will be identical even with identical starting parameters. By contrast, the implicit cumulative fracture population functions are deterministic, so identical starting parameters will always generate identical results. Note that in some circumstances, positive feedback processes can cause small initial variations in the explicit DFN to become amplified as the fracture network grows. This is particularly the case in models with

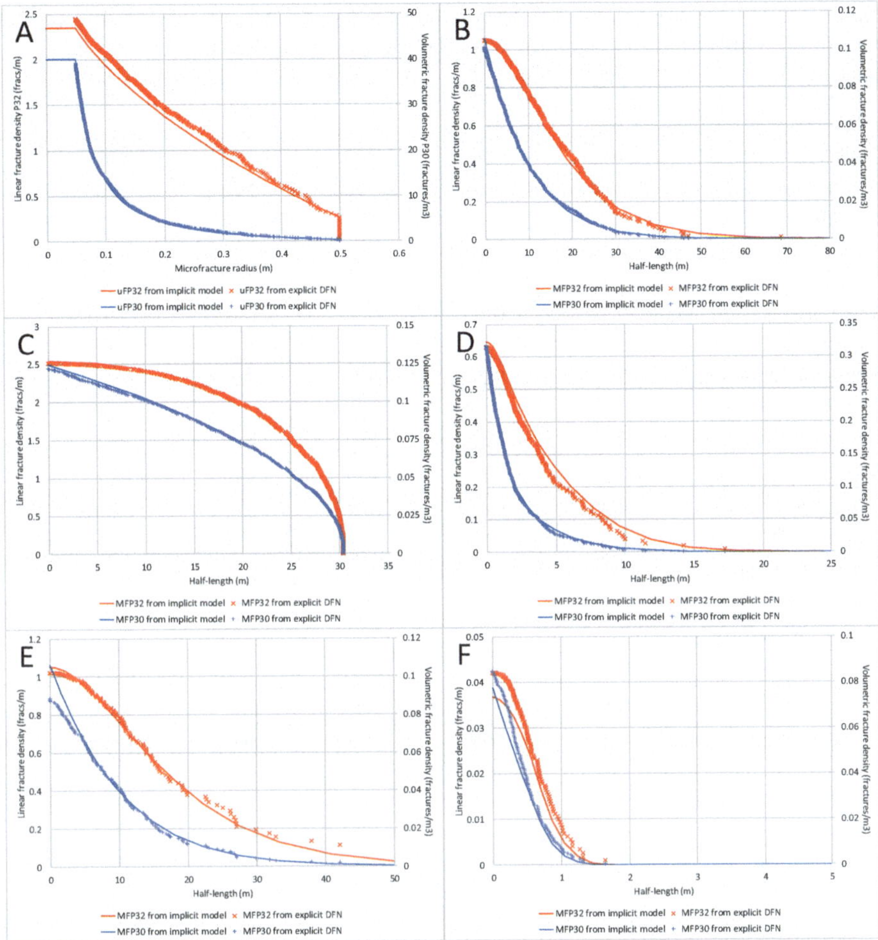

Fig. 7.3 Comparison of implicit and explicit cumulative fracture density distribution functions for different scenarios: **a** single microfracture set with uniaxial horizontal strain and stress shadows, **b** single macrofracture set with uniaxial horizontal strain and stress shadows, **c** single macrofracture set with uniaxial horizontal strain and evenly distributed stress, **d** primary macrofracture set with isotropic horizontal strain ($\dot{\varepsilon}_{hmin} = \dot{\varepsilon}_{hmax}$) and stress shadows, **e** and **f** primary and secondary macrofracture sets, respectively, from a model with anisotropic horizontal strain ($\dot{\varepsilon}_{hmin} = 0.5\dot{\varepsilon}_{hmax}$) and stress shadows. ${}_{\mu F}P_{30}(r)$ and ${}_{\underline{MF}}P_{30}(l)$ is shown in blue, ${}_{\mu F}P_{32}(r)$ and ${}_{\underline{MF}}P_{32}(l)$ is shown in red; solid lines indicate implicit cumulative fracture density distribution functions, crosses indicate the cumulative fracture density distributions calculated empirically for the explicit DFN

multiple fracture sets, where the fractures from one set can block the growth of fractures from other sets. We will discuss this further in Sect. 8.5.

- **Systematic biases**: These result from approximations or discretisation in the calculations. These may affect the cumulative fracture population functions for both the implicit and the explicit model. Causes of systematic biases may include.

Table 7.1 Input parameters for the models illustrated in Fig. 7.3. Note that an effective vertical stress of 29.43 MPa represents deformation at 2000 m depth with hydrostatic fluid pressure

Property	Figure 7.3a	Figure 7.3b	Figure 7.3c	Figure 7.3d	Figure 7.3e, f
Thickness h (m)	1	1	1	1	1
Model size (m)	10	300	300	150	150
Effective vertical stress σ_v' (MPa)	29.43	29.43	29.43	29.43	29.43
Min horizontal strain rate $\dot{\varepsilon}_{hmin}$ (/ma)	−0.01	−0.01	−0.01	−0.01	−0.01
Max horizontal strain rate $\dot{\varepsilon}_{hmax}$ (/ma)	0	0	0	−0.01	−0.008
Strain relaxation time constant (ma)	0.2	0.2	0.2	0.2	0.2
Stress shadows?	Yes	Yes	No	Yes	Yes
Subcritical index b	2	10	10	10	10
Initial μF density coefficient B	0.001	0.001	0.001	0.001	0.001
Initial μF distribution coefficient c	1.5	2	2	2	2
Young's Mod E (GPa)	10	10	10	10	10
Poisson's ratio ν	0.25	0.25	0.25	0.25	0.25
Friction coefficient μ_{fr}	0.5	0.5	0.5	0.5	0.5
Crack surface energy (J/m^2)	1000	1000	1000	1000	1000
Initial compaction factor υ	1	1	1	1	1
Fracture mode	2	2	2	2	2

- Numerical calculation of area and volume of microfractures, when calculating the implicit mean linear microfracture density, ${}_{\mu F}P_{32}$.
- The assumption that macrofracture density, and hence fracture deactivation rates, are constant throughout each timestep when calculating implicit fracture populations.
- Variability in the macrofracture nucleation rate during a timestep, when calculating implicit residual fracture populations.
- Mismatch in the timestep durations between adjacent gridblocks, leading to timing discrepancies when an explicit macrofracture crosses a gridblock boundary.

Note also that the discrepancies in the mean linear half-macrofracture density are typically greater than the discrepancies in the volumetric half-macrofracture density. This is because the cumulative mean linear half-macrofracture density is calculated by integrating the volumetric half-macrofracture density multiplied by the fracture length, so small discrepancies in the volumetric density of long macrofractures can have a significant effect on the cumulative mean linear half-macrofracture density for all macrofracture lengths.

In general, the systematic discrepancies can be minimised by reducing the timestep duration, so that the variation in macrofracture density is minimised, or by increasing the gridblock volume, so that the gridblocks are large compared with the size of individual fractures.

We should of course bear in mind that these comparisons of the implicit and explicit cumulative fracture density distribution functions can only confirm that the equations for the implicit fracture density distribution are correctly representing the conceptual model for layer-bound fracture growth described in Chap. 2. In order to demonstrate that this is an accurate representation of the growth of natural fracture networks in real geological layers, we must compare the results of these models with fracture patterns mapped in outcrop or in the subsurface. This is addressed in Chaps. 9 and 10. However, we will first examine some of the key controls on the model output, and show how these can be adjusted to produce accurate results.

Chapter 8
Controls on Fracture Evolution

Abstract In this chapter, we use the new method to study some of the controls on fracture network evolution, with a particular emphasis on the properties that can be used to calibrate the models against outcrop or subsurface fracture networks. Thus, we show that the main controls on mean linear fracture density (P_{32}) are the layer thickness and the stress shadows, but not the duration of deformation, since fracture networks tend to grow very rapidly (tens of thousands of years) and reach saturation early in the deformation episode. We show that lateral variation in the mechanical properties or the horizontal strain rate can control the fracture distribution and can result in a sharp "deformation front" sweeping across the structure, dividing it into unfractured and fully fractured zones. We show that the principal controls on fracture length and volumetric fracture density are the subcritical fracture propagation index (which controls the fracture propagation rate) and the initial microfracture density. However, because the fracture lengths do not scale linearly with layer thickness, this will affect the aspect ratio of the fractures. The subcritical fracture propagation index can also control the fracture mode (Mode 1 dilatant fractures or Mode 2 shear fractures), as well as the fluid overpressure and depth of burial. We then examine methods of quantifying the connectivity and anisotropy of the fracture network and look at the controls on these attributes. We discuss the controls on fracture orientation and different methods of calculating the horizontal strain depending on the cause of the deformation. Finally, we discuss how the data from these models can be used to calculate fracture porosity and permeability.

Keywords Fracture modelling · Discrete fracture networks · Naturally fractured reservoir · Fracture growth · Fracture density · Fracture length · Fracture anisotropy · Fracture connectivity

So far, we have presented a new, efficient method for simulating the growth of large fracture networks. In subsequent chapters, we will demonstrate the application of this method at a range of scales. This provides an opportunity to study the various factors controlling the evolution of natural fracture networks, as well as to assess

© The Editor(s) (if applicable) and The Author(s), under exclusive license to Springer Nature Switzerland AG 2020

M. J. Welch et al., *Modelling the Evolution of Natural Fracture Networks*,
https://doi.org/10.1007/978-3-030-52414-2_8

the capabilities of our method. We will then investigate potential applications for the method and develop workflows for modelling real fracture networks.

Before we can have confidence in its predictions, a method of this complexity must first be calibrated against observed fracture networks in outcrop or in the subsurface. This can be done qualitatively, by comparing the modelled and observed fracture geometries and distributions visually, or quantitatively, by comparing measurable indices such as fracture density or connectivity.

In this chapter, we investigate how we can calibrate the fracture models generated with the method by adjusting the input parameters. We will focus on parameters that can be measured quantitatively in outcrop or subsurface fracture networks: mean linear fracture density, volumetric fracture density, fracture length, fracture distribution, fracture mode, fracture orientation, fracture connectivity and fracture anisotropy. We demonstrate the effects of varying input parameters using a series of simple models comprising of planar, homogeneous layers with uniform mechanical properties and applied strain boundary conditions; full details of the input parameters for each model are given in Table 8.2.

8.1 Controls on Mean Linear Fracture Density

The mean linear fracture density (P_{32}), described in Sect. 1.1.1, is an ideal parameter with which to calibrate fracture models for two reasons:

- It is easy to measure, both in outcrop (using scanlines or similar techniques) and in the subsurface (by measuring fracture densities along boreholes, on core or borehole images).
- It is also a key parameter controlling the porosity, permeability and elastic properties of the fracture network, so it is often particularly important to be able to predict accurate P_{32} values.

8.1.1 Stress Shadows

The stress shadows around the layer-bound macrofractures provide the most important control on the mean linear fracture density since they determine the minimum spacing of the macrofractures and the zones in which microfractures cannot grow. Models generated without stress shadows (the evenly distributed stress scenario described in Sect. 6.2) will generally have a much higher mean linear fracture density than equivalent models with stress shadows (compare Figs. 8.5 and 8.6). Expressions for the stress shadow width, *W*, around Mode 1 and Mode 2 fractures were derived in Sect. 8.4:

$$W = \frac{\pi}{2}h \tag{8.1}$$

for Mode 1 macrofractures, and

$$W = \frac{\pi}{2} h \cos\omega\left(\sin\omega\cos\omega + \mu_{fr}\sin^2\omega\right) \tag{8.2}$$

for Mode 2 shear fractures.

These expressions show that the stress shadow width is dependent on the layer thickness *h*, the fracture mode and (for Mode 2 fractures) the friction coefficient μ_{fr}. However, it is independent of the in situ stress and the elastic moduli of the layer.

8.1.2 Layer Thickness

As (8.1) and (8.2) show, the width of the stress shadows around both Mode 1 and Mode 2 fractures is directly proportional to the height of the fractures, and hence the layer thickness *h*. The fracture spacing, therefore, increases in proportion to the layer thickness and the mean linear half-macrofracture density $_{MF}P_{32}$ decreases in inverse proportion to the layer thickness. Figure 8.1 shows this effect for layers from 0.1 to 100 m thickness.

This is consistent with the findings of a number of previous outcrop and modelling studies such as Pollard and Segall 1987, Bai and Pollard 2000, Bai et al. 2000a, b, and can be observed directly in many outcrops where we see layer-bound fractures in thin brittle layers of varying thicknesses (see Fig. 8.2).

The mean length of the layer-bound macrofractures is also higher in thicker layers; however, while the fracture spacing increases in direct proportion to the layer thickness, the mean fracture length increases at a lower rate. As a result, the fracture network geometry does not scale directly with layer thickness, as Fig. 8.1 shows. This is discussed further in Sect. 8.3.3.

8.1.3 Fracture Mode and Friction Coefficient

If we compare (8.1) and (8.2), we can see that Mode 1 and Mode 2 macrofractures have different stress shadow widths and that the width of the stress shadows around Mode 2 macrofractures is dependent on the fracture friction coefficient μ_{fr} and the fracture dip ω (which is itself a function of μ_{fr}). As Fig. 8.3 shows, the stress shadow around Mode 1 fractures is typically 2–3 times wider than that around Mode 2 fractures; this reflects the fact that Mode 1 fractures accommodate more horizontal strain than Mode 2 fractures because their displacement is purely horizontal rather than inclined. Note that the stress shadows around Mode 2 fractures grow slightly wider as the friction coefficient μ_{fr} increases. This may seem counter-intuitive, but it is due to the fact that when the fracture friction coefficient is high, a given reduction in the horizontal stress will cause a greater reduction in the frictional traction on the

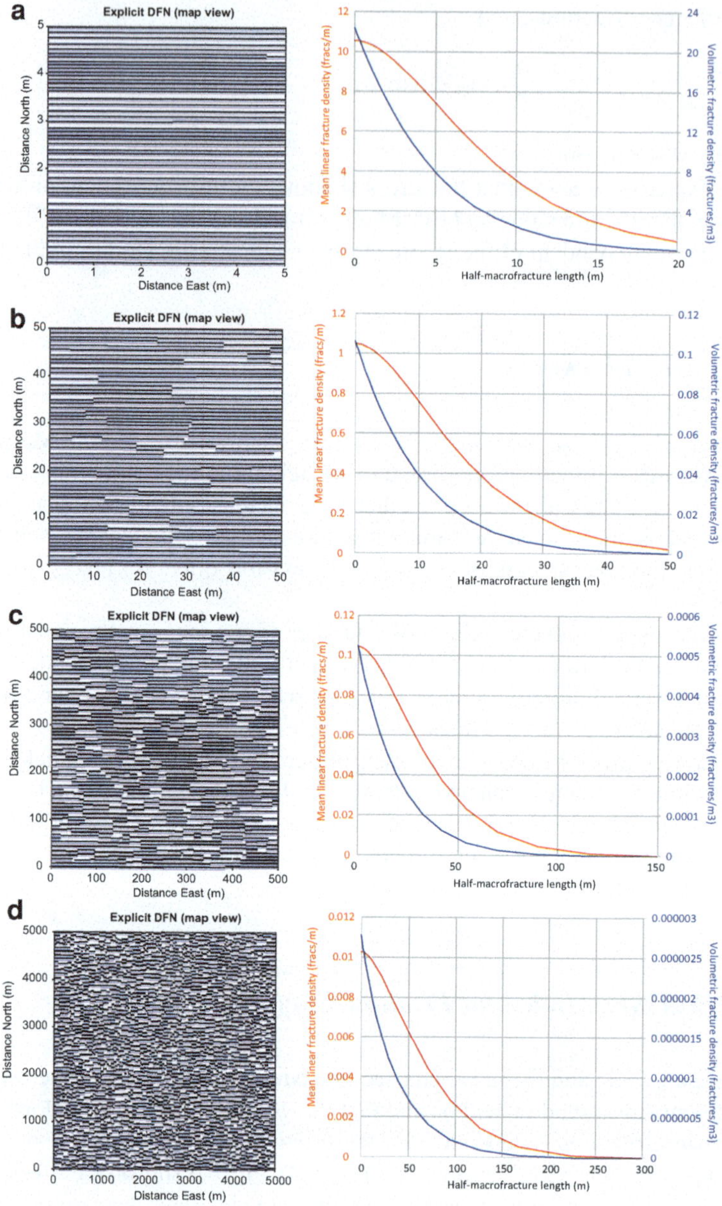

Fig. 8.1 Explicit DFN (left) and implicit cumulative half-macrofracture density distribution (right) calculated for a saturated fracture network with stress shadows, in a layer with thickness 0.1 m (**a**), 1 m (**b**), 10 m (**c**) and 100 m (**d**); note that the horizontal scale increases in proportion to the layer thickness. The fractures are Mode 2 shear fractures; the DFNs are shown in plan view, with the black lines representing the centrelines of the fractures and the blue boxes representing the stress shadow around the fractures. Other input parameters are as listed in Table 8.2

Fig. 8.2 Various outcrops showing layer-bound fractures in thin brittle layers of varying thicknesses, in interbedded sandstones and shales in Skull Creek, Colorado (**a**), interbedded limestones and shales in La Ciotat, southern France (**b**) and interbedded limestones and shales in Kilve, southwest England (**c**). Note in each case the fracture spacing varies in proportion to the bed thickness

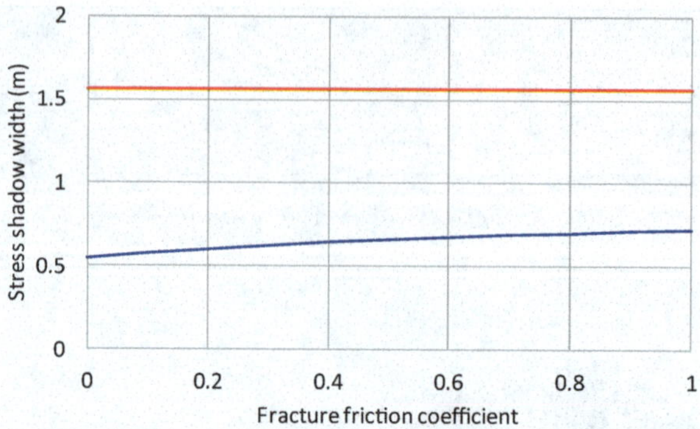

Fig. 8.3 Stress shadow widths around Mode 1 macrofractures (red line) and around Mode 2 macrofractures (blue line), as a function of the fracture friction coefficient μ_{fr}, calculated from (8.1) and (8.2) for a 1 m thick layer. Note that the Mode 1 fracture stress shadows only form when the normal stress on the fracture is tensile, whereas the calculated Mode 2 fracture stress shadow widths assume compressive stress acting on the fracture

fracture surface, and hence a greater increase in displacement. Note that this only applies when the effective normal stress acting on the fracture $\sigma_n{}'$ is compressive; when $\sigma_n{}'$ is tensile there is no frictional traction and the displacement (and hence stress shadow widths) on Mode 2 fractures will be greater. Figure 8.4 compares the fracture networks and mean linear fracture densities generated for Mode 1 fractures and for Mode 2 fractures with varying friction coefficients.

8.1.4 Duration of Deformation

The total applied horizontal strain is another major control on the mean linear fracture density of a fractured layer. When horizontal strain is applied at a constant rate, this is proportional to the total duration of the deformation episode.

Since we are modelling the growth of the fracture networks through time, it seems obvious that if the deformation stops early, before the fracture network has reached saturation, the mean linear fracture density will be less. As Figs. 8.5 and 8.6 show, this is indeed the case; early termination of the deformation leads to a proportionally smaller mean linear macrofracture density. However, there are some important caveats that mean that the duration of deformation is less significant in controlling the final mean linear fracture density than might be supposed:

- Fracture networks often develop very rapidly and reach full saturation early in the deformation episode (after tens or hundreds of thousands of years rather than millions of years). All subsequent horizontal strain is accommodated purely by accumulating displacement on existing fractures, rather than by the nucleation or propagation of new fractures. This is especially likely if the growth of the fracture

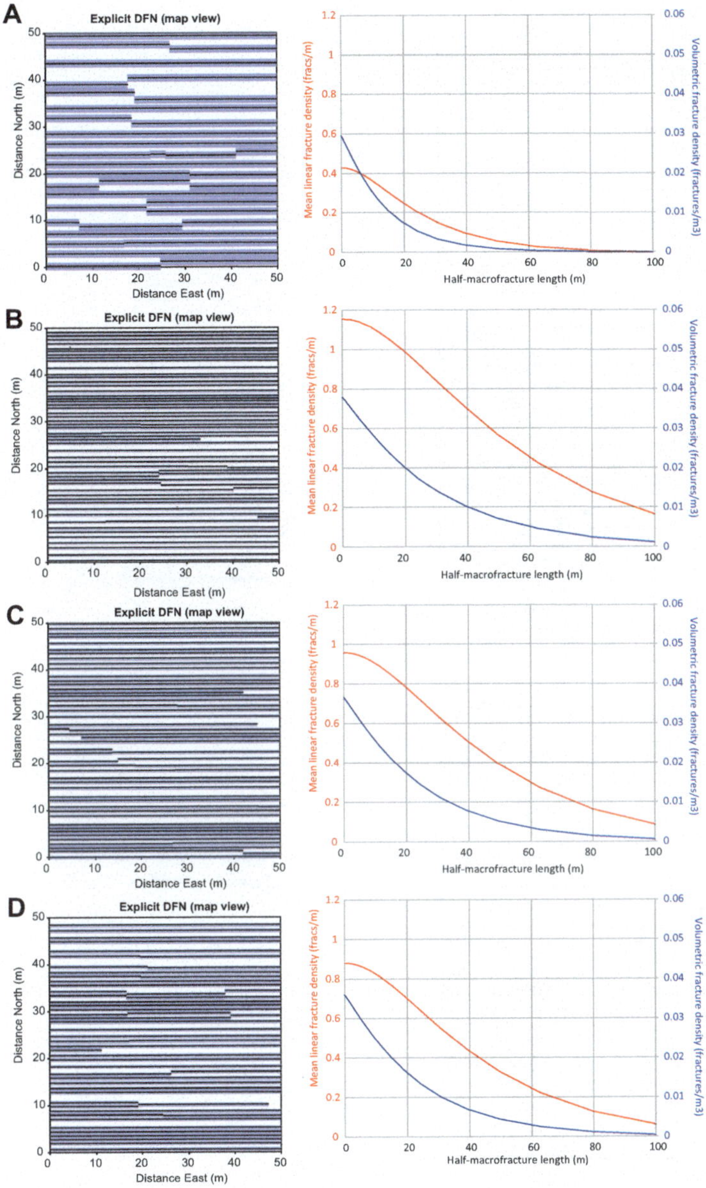

Fig. 8.4 Explicit DFN (left) and implicit cumulative density distribution for half-macrofractures (right) calculated for Mode 1 fractures (**a**), and for Mode 2 fractures with friction coefficient $\mu_{fr} = 0$ (**b**), $\mu_{fr} = 0.5$ (**c**) and $\mu_{fr} = 1$ (**d**). The DFNs are shown in plan view, with the black lines representing the centrelines of the fractures and the blue boxes representing the stress shadow around the fractures. Other input parameters are as listed in Table 8.2

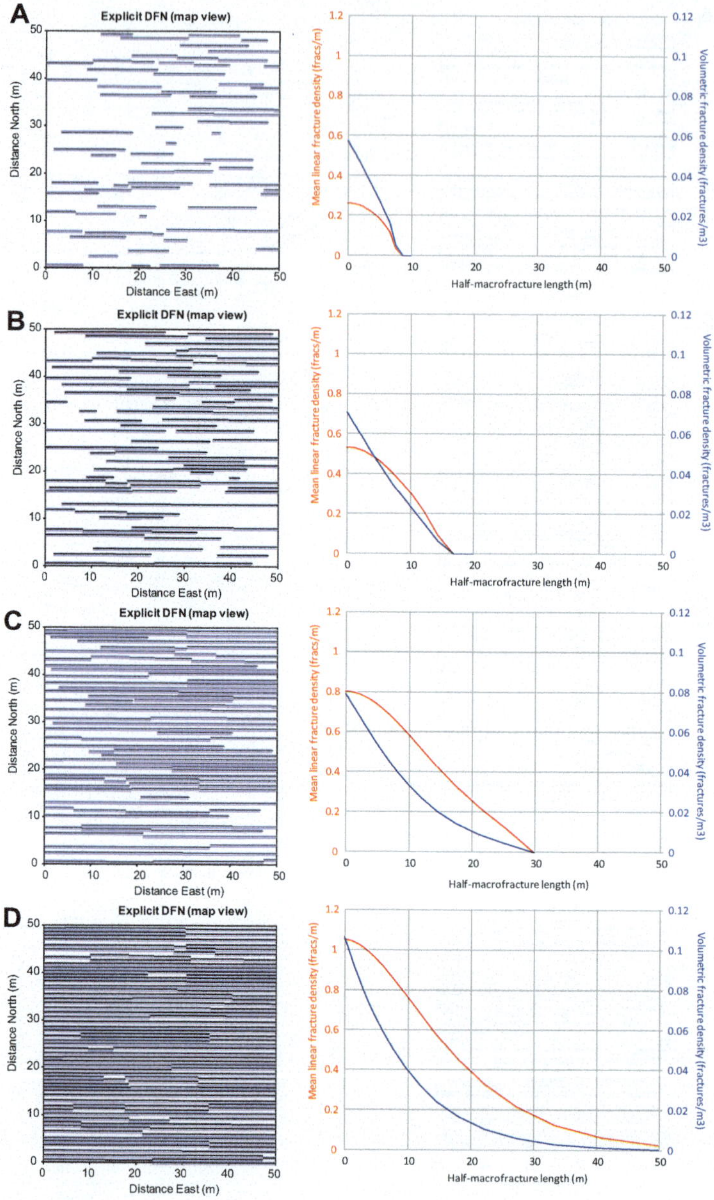

Fig. 8.5 Explicit DFN (left) and implicit cumulative density distribution for half-macrofractures (right) for a propagating fracture network in a homogeneous layer with stress shadows, after 0.1449 ma (**a**), after 0.1451 ma (**b**), after 0.1452 ma (**c**) and after 0.2 ma (full saturation; **d**). The fractures are Mode 2 shear fractures; the DFNs are shown in plan view, with the black lines representing the centrelines of the fractures and the blue boxes representing the stress shadow around the fractures. Other input parameters are as listed in Table 8.2

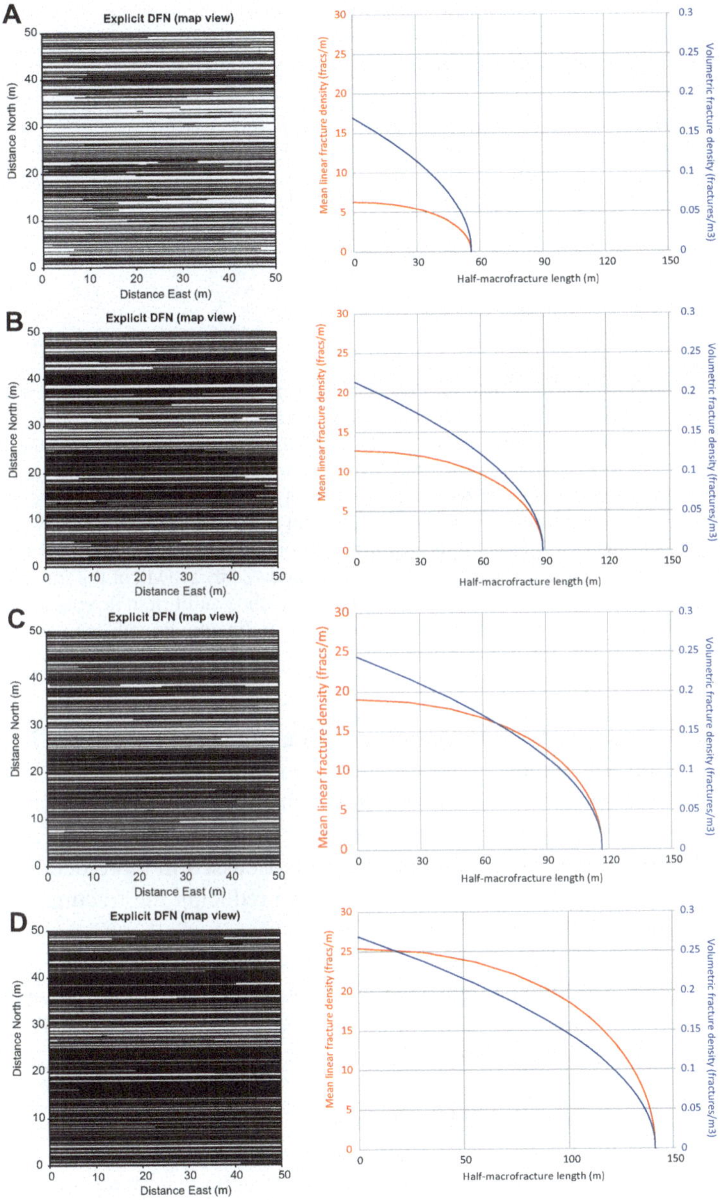

Fig. 8.6 Explicit DFN (left) and implicit cumulative density distribution for half-macrofractures (right) for a propagating fracture network in a homogeneous layer with no stress shadows, after 0.154 ma (**a**), after 0.163 ma (**b**), after 0.172 ma (**c**) and after 0.181 ma (**d**). The fractures are Mode 2 shear fractures; the DFNs are shown in plan view, with the black lines representing the centrelines of the fractures. Other input parameters are as listed in Table 8.2

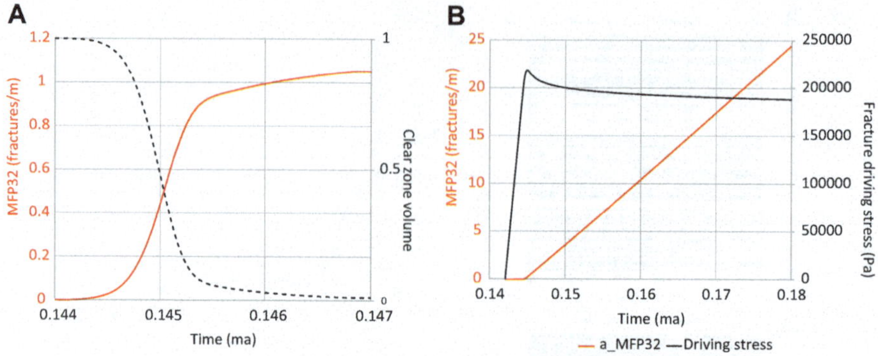

Fig. 8.7 Charts showing the growth of a fracture network with stress shadows (**a**) and with no stress shadows (**b**). The red lines on both charts show the total mean linear macrofracture density. In both cases, the fracture networks contain only one fracture set perpendicular to the minimum horizontal strain. In model A, the dashed black line shows the clear zone volume (i.e. the space available for new macrofractures to nucleate); growth of the fracture network ceases when this drops to zero and the network reaches saturation. In model B, the solid black line shows the driving stress. This initially increases, causing the fracture network to start growing. Equilibrium is then reached where the applied horizontal strain is all accommodated by growth of the fracture network and the driving stress is approximately constant. Over time, the equilibrium driving stress will fall slightly as the number of active fractures increases; however, fracture growth can continue almost indefinitely in this scenario. Other input parameters are as listed in Table 8.2

network is limited by fracture interaction, for example, stress shadow interaction or intersection with another fracture set, as shown in Fig. 8.7a.

- Processes such as grain-sliding, creep and pressure solution will cause time-dependent strain relaxation in the host rock, which will limit the total elastic horizontal stress that can accumulate. This is discussed in more detail in Welch et al. (2019) but can have a significant effect in reducing the fracture nucleation and propagation rate.
- Lateral variations in the horizontal strain rate or the mechanical properties (especially Young's Modulus) often have a far greater effect on the growth of the fracture network than the total duration of the deformation episode. This will be discussed in more detail in Sect. 8.2.

The duration of deformation will have a stronger influence on the final mean linear fracture density if the likelihood of macrofracture deactivation is low, for example, when the fracture network comprises a single set of fractures with no stress shadows, as shown on Fig. 8.6. This is because, in this scenario, the amount of horizontal strain accommodated by displacement on the fractures is proportional to the mean linear fracture density (see Sect. 6.2). Initially, therefore, when the fracture density is low, most of the applied horizontal strain is accommodated elastically within the host rock. This causes a rapid increase in the fracture driving stress, which in turn causes acceleration in the growth of the fracture network. As the fracture network grows, it can accommodate more of the applied horizontal strain, so the rate of increase in the fracture driving stress starts to drop. Equilibrium is soon reached, where the applied horizontal strain rate equals the rate at which strain can be accommodated by growth of the fracture network. At this point, the fracture driving stress will be approximately constant. Over time, as the number of active fractures increases, there will be a gradual increase in the ratio of growth in mean linear fracture density to fracture driving stress. Therefore, the equilibrium fracture driving stress will drop slowly. However, fractures will continue to grow, and hence the mean linear fracture density will continue to increase, almost indefinitely. This is shown in Fig. 8.7b.

8.2 Controls on Fracture Distribution

The fracture distribution (i.e. the lateral variation in fracture density) is also relatively easy to measure, by mapping outcrops or by comparing the mean linear fracture density encountered in different boreholes for subsurface fracture networks. If our fracture models can replicate this observed variation in fracture density, then we can use them to predict fracture distribution away from exposures or well control.

The main causes of this lateral variation in fracture density are variations in the applied horizontal strain, or in the mechanical properties of the fractured layer. Higher strain rates will cause fracture networks to develop earlier and more quickly and so will higher Young's Modulus (since the stress induced by a given amount of horizontal strain will be greater where Young's Modulus is higher). Conversely, higher friction coefficient and crack surface energy will delay the growth of the fracture network since they make it harder for the fractures to propagate.

As we saw in Sect. 8.1.4, when the strain rate and mechanical properties are homogeneous, fracture networks often develop very rapidly, and hence the fracture network at any specific location will reach saturation very soon after it first starts to grow. This means that even a slight lateral gradient in the strain rate or the Young's Modulus can cause the fracture network to sweep across the model as a "deformation front". At any specific point in time, we will see a sharp boundary between unfractured and fully saturated zones, rather than a gradual increase in fracture density across the model. Figure 8.8 illustrates this phenomenon for a model with a very slight lateral gradient in Young's Modulus.

The mechanical strength properties (friction coefficient and crack surface energy) have a much milder effect on the rate of growth of the fracture network than the Young's Modulus. The lateral variation in these properties will, therefore, generate a much more diffuse deformation front, with an "intermediate zone" characterised by a gradient in fracture density separating the unfractured and the fully saturated parts of the model (Fig. 8.9). It should, however, be noted that, since the mechanical strength is correlated with Young's Modulus in many rock types (see e.g. Chang et al. 2006), the effect of mechanical properties may be overshadowed by the effect of Young's Modulus variation on fracture propagation.

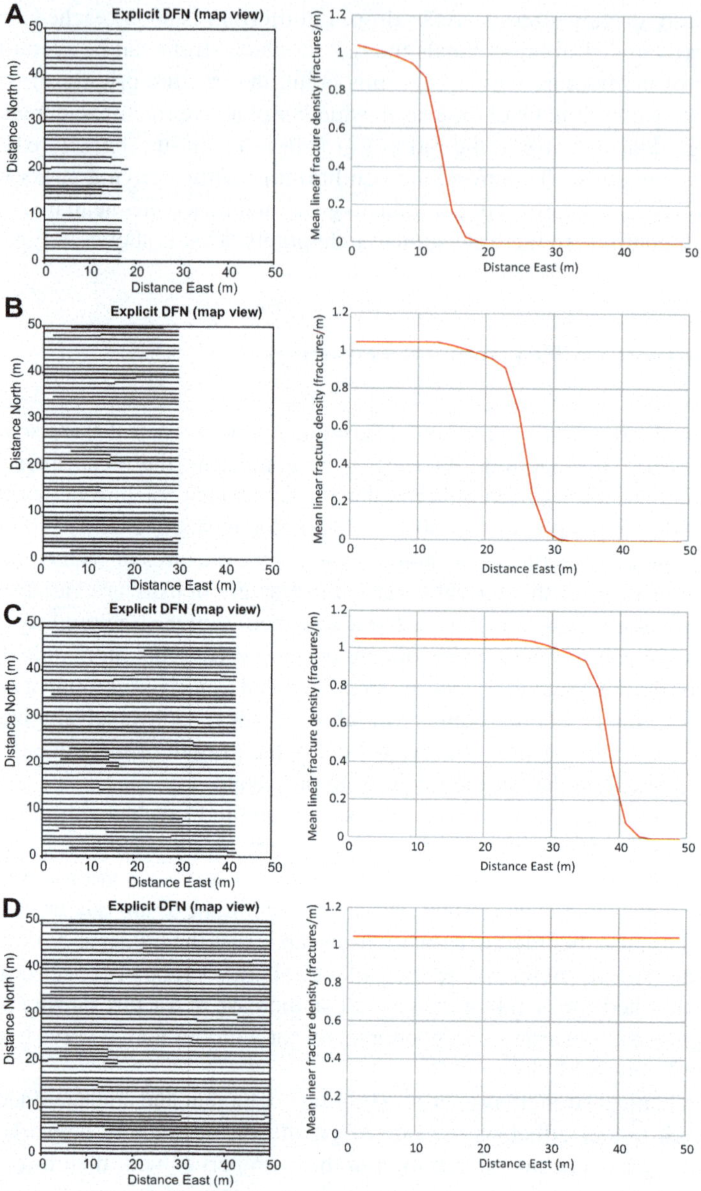

Fig. 8.8 Explicit DFN (left) and calculated total ${}_{MF}P_{32}$ profile (right) for a propagating fracture network in a layer with a linear gradient in Young's Modulus, from 10 GPa on the left to 10.5 GPa on the right, after 0.1398 ma (**a**), after 0.1415 ma (**b**), after 0.1432 ma (**c**) and after 0.4 ma (full saturation; **d**). The applied horizontal strain is uniform across the model, and the fractures are Mode 2 shear fractures; the DFNs are shown in plan view, with the black lines representing the centrelines of the fractures. Other input parameters are as listed in Table 8.2

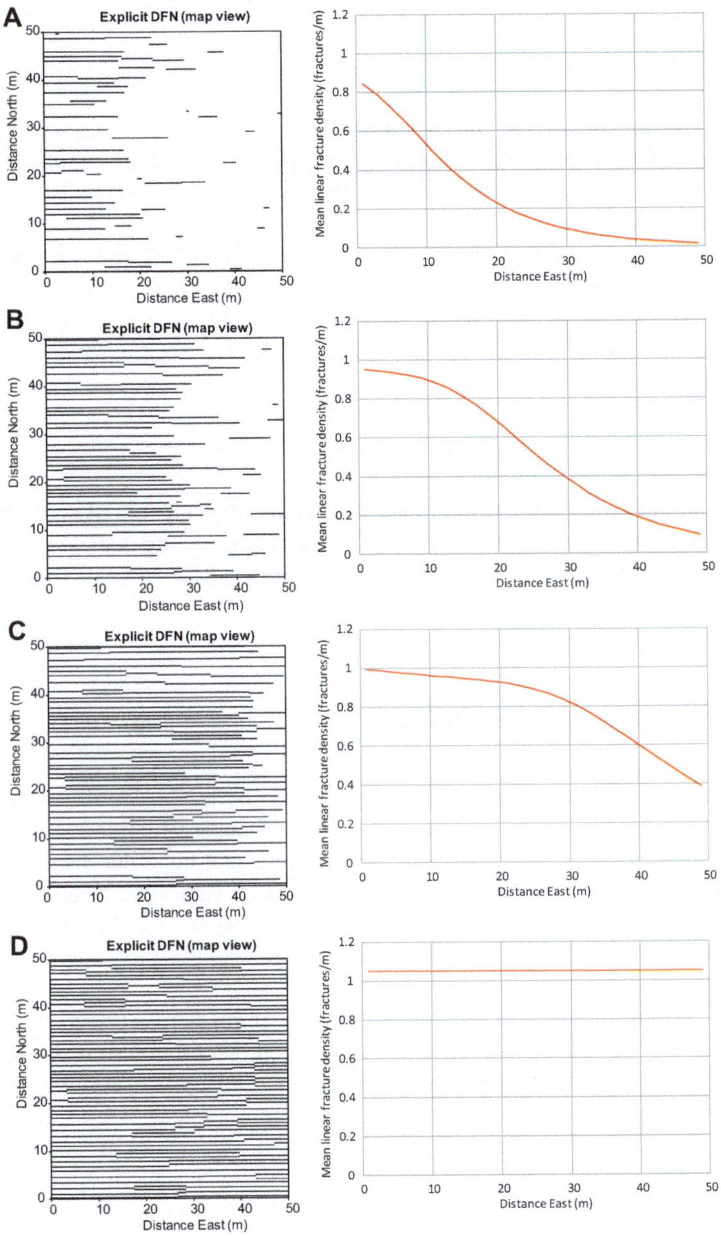

Fig. 8.9 Explicit DFN (left) and calculated total $_{MF}P_{32}$ profile (right) for a propagating fracture network in a layer with a linear gradient in crack surface energy, from 1000 J/m^2 on the left to 2000 J/m^2 on the right, after 0.1453 ma (**a**), after 0.1456 ma (**b**), after 0.1460 ma (**c**) and after 0.4 ma (full saturation; **d**). The applied horizontal strain is uniform across the model, and the fractures are Mode 2 shear fractures; the DFNs are shown in plan view, with the black lines representing the centrelines of the fractures. Other input parameters are as listed in Table 8.2

8.3 Controls on Volumetric Fracture Density and Fracture Length

As we noted in Sect. 1.1.1, the volumetric fracture density (P_{30}) and the fracture size distribution are also important controls on the fracture permeability and elastic moduli, but unlike the mean linear fracture density, these properties cannot be determined from borehole data.

They must, therefore, be calibrated against outcrops, where the fractures can be mapped out laterally, and the lengths of at least the layer-bound macrofractures can be measured directly. In fact, if we know the mean linear density of layer-bound macrofractures, $_{MF}P_{32}$, then the volumetric density, $_{MF}P_{30}$, and the mean length $\overline{L}$ are directly related since

$$_{MF}P_{32} = {}_{MF}P_{30}\bar{L}h \tag{8.3}$$

Thus, we can measure the mean linear density and mean length of macrofractures in outcrop and calibrate it against the volumetric macrofracture density predicted by the implicit models (or more precisely, the volumetric half-macrofracture density, given by $_{MF}P_{30} = 2_{MF}P_{30}$).

8.3.1 *Duration of Deformation*

The duration of the deformation episode can also be a key control on the mean macrofracture length since macrofractures that propagate for longer will have a greater length. This is most apparent when there are no other factors causing macrofractures to stop propagating, for example, when the fracture network comprises a single set of fractures with no stress shadows. As Fig. 8.6 shows, in this scenario, both the mean and maximum fracture length increase steadily with time.

The shape of the cumulative half-macrofracture length distribution in this scenario is controlled by the rate of macrofracture nucleation, which in turn is dependent on the subcritical fracture propagation index. The relationship between the subcritical fracture propagation index and macrofracture nucleation rate is investigated in detail in Welch et al. (2019). To summarise, a low subcritical fracture propagation index ($b \leq$ c.5) gives a rate of macrofracture nucleation that is both high and increases with time, while a moderate to high subcritical fracture propagation index ($b \geq$ c.10) gives a rate of macrofracture nucleation that starts low and decreases further with time. As a result, a low subcritical fracture propagation index produces a half-macrofracture population with a broad range of lengths, reflecting different nucleation times (Fig. 8.10a). By contrast, a high subcritical fracture propagation index produces a much narrower range of half-macrofracture lengths, since most fractures in the population nucleated at the same time and grew at the same rate (Fig. 8.10b).

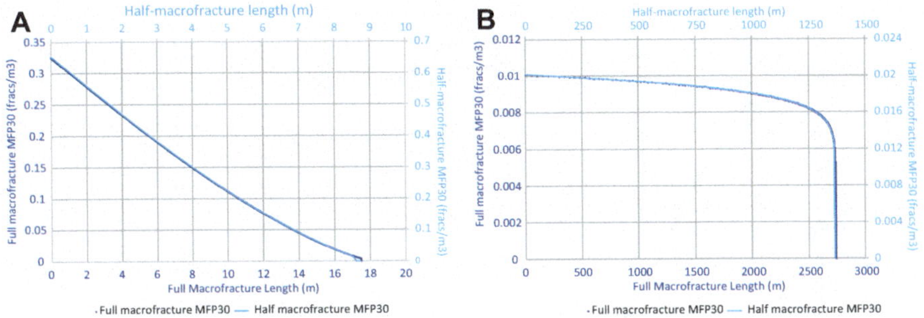

Fig. 8.10 Comparison of the cumulative macrofracture and half-macrofracture length distributions calculated for a saturated fracture network in a homogeneous layer with no stress shadows, with subcritical fracture propagation index $b = 5$ (**a**) and $b = 50$ (**b**). The continuous light blue lines represent the cumulative half-macrofracture volumetric density distributions calculated for the implicit fracture models, while the dark blue dots represent the cumulative volumetric density distribution of full macrofractures in the explicit DFNs. Other input parameters are as listed in Table 8.2

However, as we saw in Sect. 8.1.4, when the macrofractures have stress shadows or can intersect with another fracture set, the fracture network will typically reach saturation rapidly. The mechanisms of macrofracture deactivation will play a significant role in determining the macrofracture length distribution and volumetric density. As we saw in Sect. 5.1, the rate of fracture deactivation, both by stress shadow interaction and by intersection, increases as the fracture network grows. This will create a complex feedback system where the size of the existing fracture network is the primary control on the length of the fractures in that network.

Overall, higher macrofracture deactivation rates will generally result in a shorter mean half-macrofracture length, but the length distribution will have a much broader spread than in the scenario with no stress shadows, even if the subcritical fracture propagation index is high. We can see this if we compare Fig. 8.10 with 8.11, which shows the length distribution for a uniaxial fracture network with stress shadows and a subcritical propagation index $b = 20$.

It is important to note that when macrofracture deactivation rates are high, the length distribution of full macrofractures will not be the same as the length distribution of half-macrofractures. We can see this by overlaying the cumulative volumetric density distribution for full macrofractures at twice the length scale and half the density scale as the cumulative volumetric density distribution for half-macrofractures. The cumulative volumetric density distribution for full macrofractures is represented by the dark blue dots on Figs. 8.10 and 8.11. The models with no macrofracture deactivation in Fig. 8.10 show a perfect match between the full and half-macrofracture cumulative volumetric density distribution functions. By contrast, the model with stress shadow deactivation shown in Fig. 8.11 shows a clear and systematic discrepancy between the two cumulative volumetric density distribution functions: the cumulative half-macrofracture volumetric density distribution follows

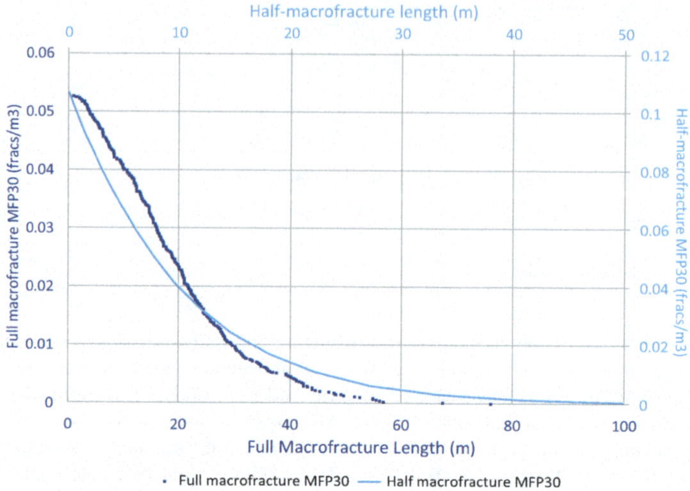

Fig. 8.11 Cumulative macrofracture and half-macrofracture length distribution calculated for a saturated network of Mode 2 macrofractures with stress shadows, with subcritical fracture propagation index $b = 20$. The continuous light blue line represents the cumulative half-macrofracture volumetric density distribution calculated for the implicit fracture model, while the dark blue dots represent the cumulative volumetric density distribution of full macrofractures in the explicit DFN. Other input parameters are as listed in Table 8.2

an almost exponential profile, while the cumulative full macrofracture volumetric density distribution is much more linear.

The reason for this is that, although the two tips of a given macrofracture propagate independently, their deactivation probabilities are correlated because they both nucleated at the same time. As a result, the lengths of the two co-joined half-macrofractures are not independent, and this is sufficient to distort the length distribution for full macrofractures. We cannot, therefore, derive the full macrofracture length distribution by simply doubling the lengths and halving the densities of the half-macrofracture length distribution. In fact, there is no method to calculate the implicit cumulative volumetric density distribution function for full macrofractures directly; instead, it is generally easier to calculate the length distribution of macrofractures in the explicit DFN (as we have done on Figs. 8.10 and 8.11). This effect becomes even more noticeable when we consider macrofracture deactivation due to the intersection in Sect. 8.3.4.

8.3.2 Subcritical Fracture Propagation Index and Initial Microfracture Population

When the majority of half-macrofractures become deactivated due to stress shadow interaction, the rate of fracture deactivation is controlled by the volumetric density of other macrofractures in the same set (see Sect. 5.1.2). The key control on mean macrofracture length will, therefore, be the rate of macrofracture nucleation: if macrofractures nucleate more frequently, there will be more macrofractures propagating at any given time, the rate of stress shadow interaction will increase and, hence, the mean distance that a macrofracture can propagate before being deactivated due to stress shadow interaction will be shorter. A similar effect was found in the models run by Olson (2004).

The key controls on macrofracture nucleation rate are the subcritical fracture propagation index, *b*, and the initial microfracture population:

- As noted in Sect. 8.3.1, a low subcritical fracture propagation index ($b \leq$ c.5) gives a higher rate of macrofracture nucleation than a high subcritical fracture propagation index ($b \geq$ c.10). Therefore, a higher subcritical fracture propagation index will result in longer macrofractures (Fig. 8.12). In practice, the subcritical fracture propagation index is likely to be 10 or greater for most consolidated rock types (see Atkinson 1984; Holder et al. 2001), but even in this range, we see a significant variation in the mean macrofracture length between models with $b =$ 10, 20 and 50.
- As would be expected, a higher initial microfracture population leads to a higher macrofracture nucleation rate, and hence shorter macrofractures. However, two parameters are controlling the initial microfracture population: a density coefficient, *B*, and a distribution coefficient, *c* (which controls the ratio of smaller to larger microfractures). While the macrofracture nucleation rate is linearly

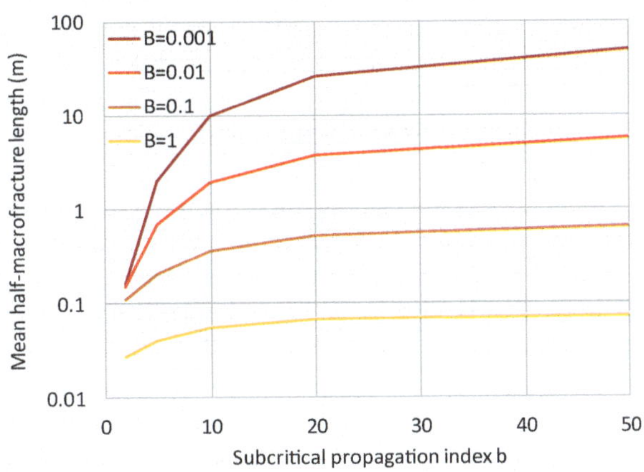

Fig. 8.12 Mean half-macrofracture lengths in a 1 m thick layer, calculated for different values of subcritical fracture propagation index *b* and initial microfracture density coefficient *B*; the initial microfracture distribution coefficient $c = 2$ in all cases

proportional to the density coefficient (Fig. 8.12), the relationship between the distribution coefficient and the macrofracture nucleation rate is more complex.

The subcritical fracture propagation index is constrained by a number of experimental studies (e.g. Atkinson 1984; Swanson 1984; Holder et al. 2001). However, the initial microfracture population is poorly constrained. A number of studies (e.g. Marrett and Allmendinger 1991, 1992; Westaway 1994; Ouillon et al. 1996; Marrett 1996; Bonnet et al. 2001) suggest that the microfracture distribution coefficient, c, is typically around 2. However, many of these studies are based on measurements of fracture displacement rather than fracture radius; furthermore, many are based on scanline data, and hence relate to mean linear fracture density rather than volumetric fracture density. This suggests that typical values of c may lie somewhere between 2 and 4. There is, however, a very little constraint on the microfracture density coefficient, B.

The best approach is, therefore, to constrain the values of B and c empirically, by calculating the predicted mean layer-bound macrofracture length at saturation for a range of B and c values and comparing these with the fracture lengths observed in outcrop examples. Figure 8.13a shows the mean half-macrofracture lengths predicted for a fracture set in a 1 m thick layer, for values of B ranging from 10^{-6} to 0.02, for values of c between 2 and 4 and for a subcritical fracture propagation index between 10 and 50. As we will see in the outcrop examples of Chap. 9, a typical mean length for Mode 1 layer-bound fractures in a 1 m thick layer is c.50 m (corresponding to a mean half-macrofracture length of c.25 m). If we assume a subcritical fracture propagation index $b \geq 20$ (as suggested by the Mode 1 fractures; see Sect. 8.4), the relationships shown on Fig. 8.13a imply a range of possible B values of 0.001–0.002 if $c = 2$, 0.0003–0.0008 if $c = 3$ and 0.00003–0.0003 if $c = 4$.

Additional constraints can be provided by extrapolating fracture growth back in time using (3.11), to determine the initial radius of the seed microfracture for the nucleating macrofractures (Fig. 8.13b), and by calculating the total mean linear density of the initial microfracture population using (3.15) (Fig. 8.13c). Although the initial mean linear microfracture density and microfracture radius will not generally be known, it is at least possible to check that these values lie within geologically realistic ranges. The B values quoted above correspond to an initial mean linear microfracture density range of 0.02–0.05 fractures/m if $c = 2$, 0.28–0.75 fractures/m if $c = 3$ and 1.88–18.84 fractures/m if $c = 4$, assuming a lower cutoff microfracture radius $r_{min} = 1$ cm (Fig. 8.13c).

When we extrapolate the microfracture growth back in time, we find for all plausible cases, the initial microfracture radii lie in the range of 10–40 cm (Fig. 8.13b). This implies that the macrofractures nucleate from small compaction- or diagenesis-related fractures, rather than intergranular cracks. In Sect. 10.1, we will see examples of such fractures in cores from a chalk reservoir, alongside examples of much larger fractures interpreted as layer-bound macrofractures.

If the layer-bound fractures are developing within a restricted area, for example, a fault-bounded block, a low macrofracture nucleation rate can also limit the total mean linear macrofracture density at saturation. This is because there may be insufficient seed microfractures within the block to nucleate enough macrofractures to fill the

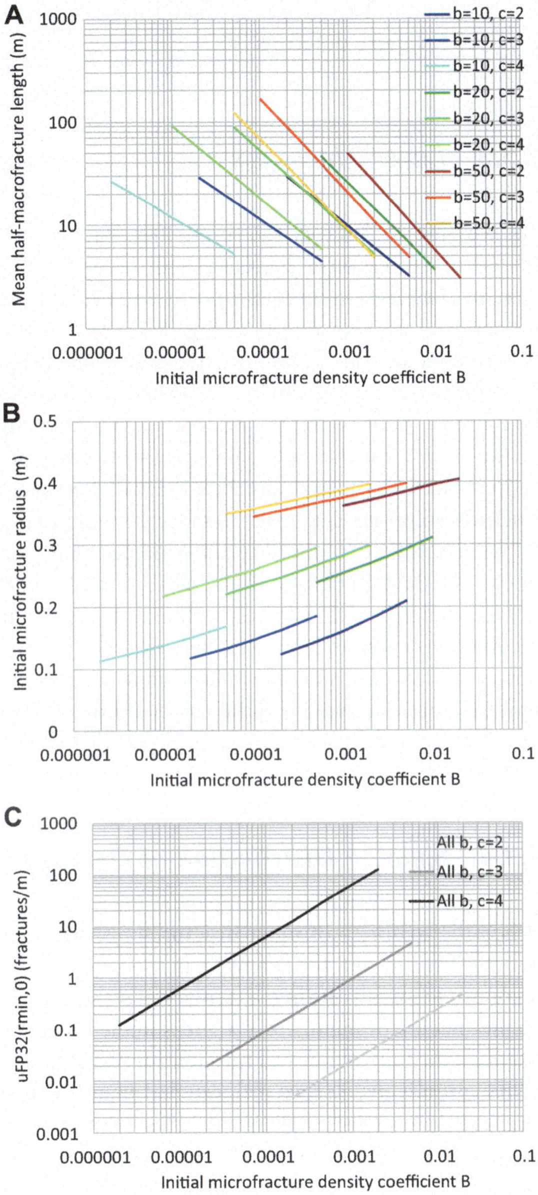

Fig. 8.13 Charts showing the calculated mean half-macrofracture length (**a**), the initial seed microfracture radius for macrofractures nucleating at 50% fracture saturation (**b**) and the total mean linear density of the initial microfracture population ${}_{\mu F}P_{32}(r_{min}, 0)$, assuming a lower cutoff microfracture radius $r_{min} = 1$ cm (**c**), for different values of subcritical fracture propagation index b, initial microfracture density coefficient B and initial microfracture distribution coefficient c

block to saturation, while macrofractures nucleating outside the block are prevented from propagating into fill the gaps by the surrounding faults. This does not appear to be the case for the outcrops studied in Chap. 9: even though some of them are heavily faulted, with fault-bounded compartments typically c.50 m in size, there is no observed reduction in mean linear fracture density even in the smaller fault-bounded compartments. This provides an additional constraint on the range of possible B and c values for these outcrops.

8.3.3 Layer Thickness

As we saw in Sect. 8.1.2, both the spacing and the mean length of the layer-bound macrofractures increase with layer thickness. However, while the fracture spacing increases in direct proportion to the layer thickness, the mean fracture length increases at a lower rate (see Fig. 8.1). This is a result of the conflicting effects of different factors on the fracture nucleation and propagation rate: in a thicker layer, the microfractures must grow larger to become layer-bound macrofractures, but there will be more of them (since the number of microfractures is proportional to the volume of the layer). Furthermore, the stress intensity at the macrofracture tips (which is proportional to the fracture height) will be greater in a thicker layer, causing the fractures to propagate faster than they would in a thinner layer.

As a result, the fracture network geometry does not scale directly with layer thickness; the aspect ratio of the layer-bound fractures (the ratio of fracture length to fracture height) will be lower in thicker layers. In the example shown in Fig. 8.1, the mean fracture length approximately doubles for a ten-fold increase in layer thickness; however, this may vary in other scenarios, especially if the subcritical propagation index is different. Note that the total volumetric fracture density decreases by a factor of c.200 for a ten-fold increase in layer thickness in Fig. 8.1.

8.3.4 Spacing Distribution of Orthogonal Fracture Sets

When the majority of macrofractures become deactivated due to intersection with another fracture set, for example, in the case of a secondary fracture set developing in response to an anisotropic extensional strain, the key control on fracture length will be the spacing of the macrofractures in the other set. In this scenario, the discrepancy between the half-macrofracture length distribution and the full macrofracture length distribution is even more pronounced than it is for a uniaxial fracture network with stress shadows (Sect. 8.3.1).

As discussed in Sect. 5.1.3, the spacing distribution of the macrofractures may be completely random (e.g. if there are no stress shadows or the mean linear macrofracture density is low), completely regular (e.g. a fully saturated macrofracture set with stress shadows) or semi-regular (partly random and partly regular, e.g. an unsaturated macrofracture set with stress shadows). When the primary fracture set has a regular spacing distribution, the secondary macrofracture set will have a very narrow full macrofracture length distribution since most secondary macrofractures will have length exactly equal to the spacing of the primary macrofractures. However, the half-macrofracture length distribution for the secondary macrofracture set will be broader

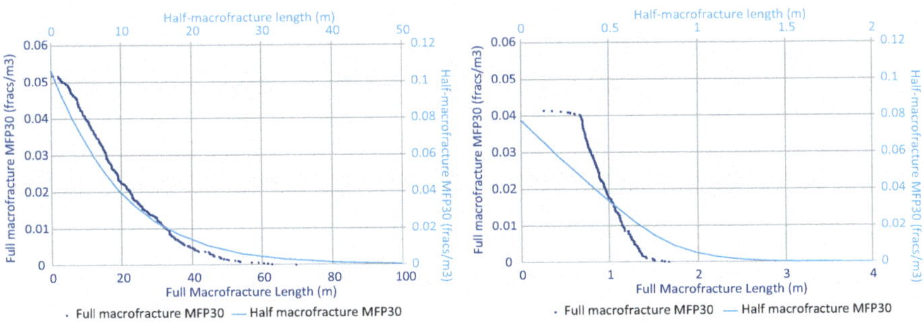

Fig. 8.14 Cumulative macrofracture and half-macrofracture length distribution calculated for a saturated primary set of Mode 2 macrofractures (**a**) and for a secondary fracture set terminating against the primary set (**b**). The continuous light blue lines represent the cumulative half-macrofracture volumetric density distributions calculated for the implicit fracture models, while the dark blue dots represent the cumulative volumetric density distribution of full macrofractures in the explicit DFNs. The macrofractures have friction coefficient 0.5, corresponding to a stress shadow width 0.67 m; note that most full secondary macrofracture lengths lie in the range of 0.67–1.34 m. Other input parameters are as listed in Table 8.2

since the secondary macrofractures can nucleate at any point between the primary macrofractures, and so the half-macrofracture lengths can have any value between zero and the maximum primary macrofracture spacing (see Fig. 8.14). In this instance, we can see very clearly that the lengths of the two co-joined half-macrofractures are not independent.

8.3.5 *Linking of Adjacent Layer-Bound Fracture Segments Across Relay Zones*

Some studies (e.g. Huggins et al. 1995; Nicol et al. 1995; Ferrill and Morris 2001; Walsh et al. 2003) have suggested that most large fractures and faults observed in outcrop and subsurface data formed by the linking of initially independent but parallel segments that propagated laterally towards each other, eventually connecting via the development of relay segments. We can replicate this phenomenon in the explicit DFN by combining individual macrofractures whenever they terminate against each other due to stress shadow interaction and adding an additional relay segment connecting the two macrofracture tips.

This has no impact on the mean linear macrofracture density ${}_{MF}P_{32}$ since linking the individual macrofractures does not change the total fracture area. However, it reduces the volumetric macrofracture density ${}_{MF}P_{30}$ while increasing the mean macrofracture length $\overline{L}$ and significantly broadening the range of the macrofracture length distribution (see Fig. 8.15). Indeed, it is possible to generate fractures of almost unlimited length by linking individual macrofractures; the main control on

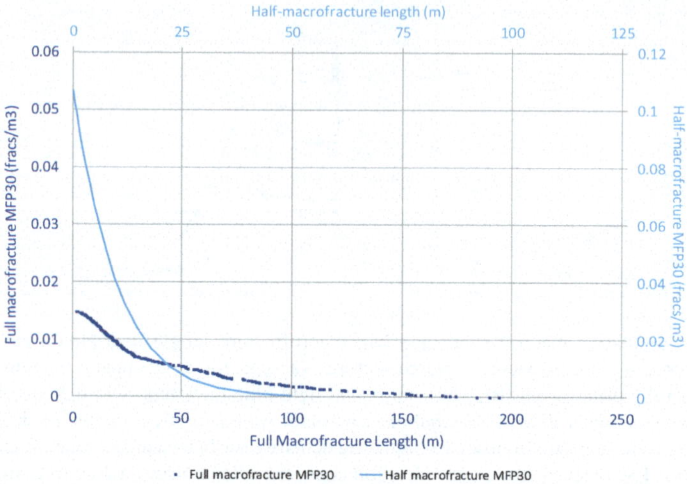

Fig. 8.15 Cumulative macrofracture and half-macrofracture length distribution calculated for a single saturated fracture set in a homogeneous layer, where individual macrofractures that terminate due to stress shadow interactions are linked by relay segments. The continuous light blue line represents the cumulative volumetric density distribution calculated for the initial half-macrofractures (i.e. before linking), while the dark blue dots represent the cumulative volumetric density distribution of full (linked) macrofractures in the explicit DFN. The model is 250 m wide and contains no internal barriers to fracture propagation. Other input parameters are as listed in Table 8.2

macrofracture length in this scenario is the location of barriers to fracture propagation (e.g. faults), or ultimately, the size of the model. Linking macrofractures in this way may also allow more accurate calculation of fluid flow through the fractured layer since it reproduces the long flow pathways that such connected fractures and faults can provide.

8.4 Controls on Fracture Mode

Because this method can simulate the growth of both Mode 1 and Mode 2 fractures, we can use it to predict which mode of fractures will develop in any given situation; this will depend on which fracture mode nucleates first and propagates fastest. Indeed it is possible to adapt the method to simulate the growth of both fracture modes simultaneously (the main modification required is to the calculation of stress shadow interaction probability since the Mode 1 and Mode 2 fractures have different stress shadow widths).

The main control on fracture mode is the effective vertical stress, σ_v'. The results from numerous models suggest that in general, Mode 1 fractures will usually develop when $\sigma_v' \leq$ c.0.2 MPa and Mode 2 fractures will usually develop when $\sigma_v' \geq$ c.1 MPa. This means that we can expect Mode 1 fractures to develop only at very shallow depths

(σ_v' of 0.2 MPa corresponds to a depth of c.16 m at hydrostatic fluid pressure) or at high fluid overpressure (an overpressure of c.24 MPa is required to reduce σ_v' to 0.2 MPa at 2000 m depth).

The situation at intermediate effective vertical stress (e.g. $\sigma_v' = 0.5$ MPa) is more interesting. Figure 8.16a shows the driving stress for Mode 1 fractures and for Mode 2 fractures with different friction coefficients, as the horizontal extensional strain increases. Fractures of a given mode will start to nucleate and propagate when the driving stress for that mode becomes positive. A simple analysis would suggest that Mode 2 fractures will develop under these conditions: the driving stress for Mode 2 fractures with $\mu_{fr} = 0.5$ becomes positive when $\varepsilon_{h\min} = \text{c.}-3 \times 10^{-5}$, while the driving stress for Mode 1 fractures does not become positive until $\varepsilon_{h\min} = \text{c.}-4.7 \times 10^{-5}$, so we might expect the layer to become fully saturated with Mode 2 fractures before any Mode 1 fractures can nucleate.

As Fig. 8.17a shows, this is indeed what will happen, if the subcritical fracture propagation index is low to moderate (e.g. $b \leq 10$). However, if the subcritical fracture propagation index is higher, then the rate of half-macrofracture propagation at this stage will be very low ($<1 \times 10^{-10}$m/year for $b = 50$), and hence almost no Mode 2 fractures will actually form (compare Fig. 8.16b, c). As the horizontal extensional strain increases, the driving stress and propagation rate of both Mode 1 and Mode 2 half-macrofractures will increase, but the propagation rate of Mode 1 half-macrofractures will increase faster, overtaking that of $\mu_{fr} = 0.5$ Mode 2 half-macrofractures at c.1×10^{-9}m/year ($\varepsilon_{h\min} = \text{c.}-8.5 \times 10^{-5}$). In this scenario, a Mode 1 fracture network will develop first, as shown in Fig. 8.17b.

It should, however, be noted that, even if the Mode 1 fracture network reaches saturation first, it may still be possible for some Mode 2 fractures to develop at a later stage. This is because, as we saw in Sect. 8.1.3, the stress shadow around Mode 1 fractures is wider than that around Mode 2 fractures. There may, therefore, be some gaps preserved between the Mode 1 fractures that are wide enough for Mode 2 fractures to nucleate in later.

8.5 Controls on Anisotropy and Fracture Connectivity

Anisotropy and fracture connectivity are two of the key parameters that characterise the fracture network as a whole, rather than the individual fractures. They are, therefore, especially suited for calibrating methods, such as this, which simulate the evolution of the fracture network as a whole. They are also particularly useful in calculating fracture permeability, as we will discuss in Sect. 8.7.

The anisotropy of a fracture network describes the relative predominance of the different fracture sets that comprise the network. This can be qualitative or quantitative.

Qualitatively, we can define three end-member types of fracture network, illustrated in Fig. 8.18:

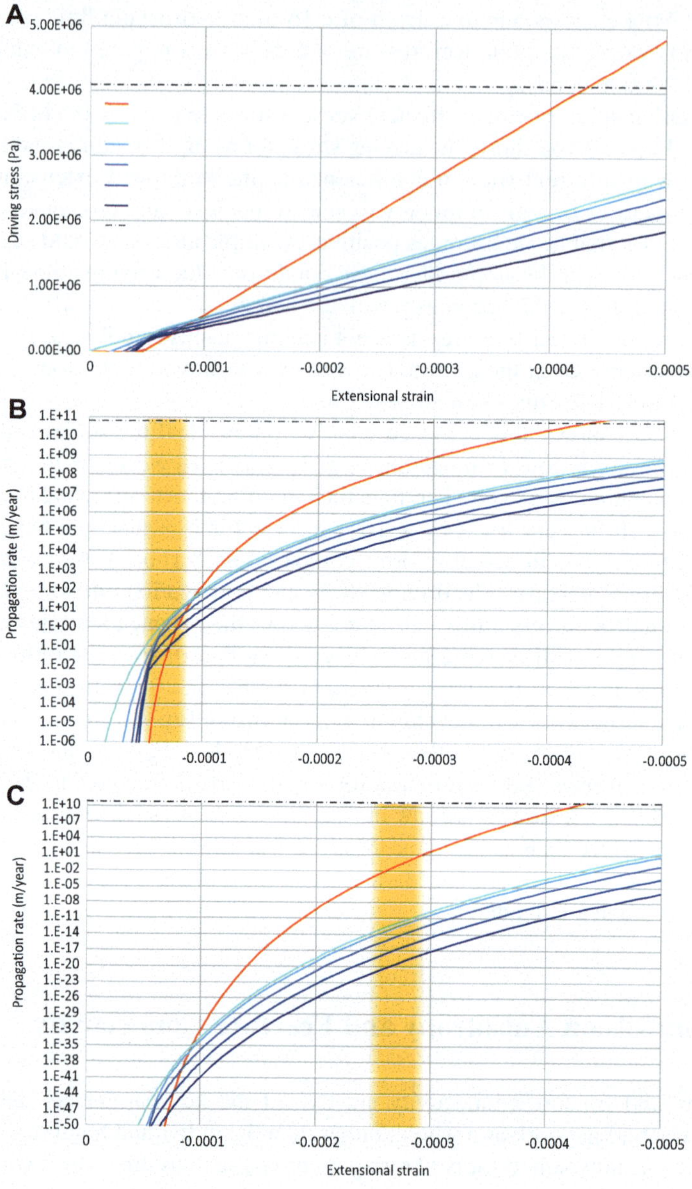

Fig. 8.16 Charts showing the growth in driving stress σ_d (**a**) and resulting half-macrofracture propagation rate at subcritical fracture propagation index $b = 10$ (**b**) and $b = 50$ (**c**) as horizontal extensional strain increases, for Mode 1 macrofractures (red line) and Mode 2 macrofractures with different values of friction coefficient μ_{fr}, (blue lines), developing in a 1 m thick layer at vertical effective stress $\sigma_v{'} = 0.5$ MPa. The orange shading indicates the period during which the fracture networks shown in Fig. 8.17 developed; note that this only happens when the fracture propagation rate exceeds c.10^{-2} m/year. Other input parameters are as listed in Table 8.2

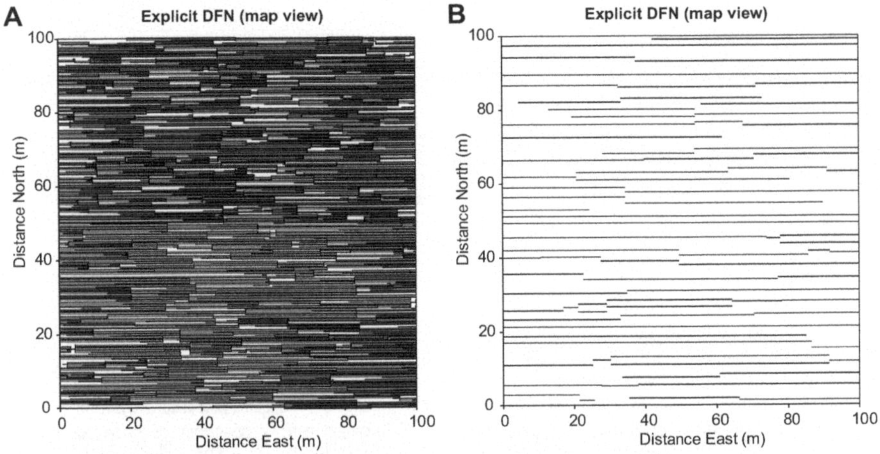

Fig. 8.17 Explicit DFNs generated for a saturated fracture network forming at vertical effective stress $\sigma_v' = 0.5$ MPa, with subcritical fracture propagation index $b = 10$ (**a**) and $b = 50$ (**b**) (corresponding to the scenarios shown in Fig. 8.16b, c respectively). The DFNs are shown in plan view, so the vertical Mode 1 fractures appear as black lines, while the inclined Mode 2 fractures appear as rectangles shaded in red. Note that even though both fracture networks are saturated, the fracture density is higher for the Mode 2 fractures in (**a**) because they have narrower stress shadows than the Mode 1 fractures in (**b**); $_{MF}P_{32} = 1.06$ fractures/m for **a** compared with 0.35 fractures/m for **b**. Other input parameters are as listed in Table 8.2

- A uniaxial fracture network comprises only one fracture set (e.g. Figure 8.18a, b). Therefore, at any one point, all fractures will be parallel, even though the orientation of the fracture set may vary laterally, causing the fractures to curve (see Sect. 8.6).
- A fully anisotropy fracture network comprises a primary set of long fractures, and a secondary set of short connecting fractures (Fig. 8.18c). In a fully anisotropy fracture network, most or all of the secondary fractures terminate against primary fractures, but none of the primary fractures terminate against the secondary fractures.
- In an isotropic fracture network, all the fracture sets have identical densities and length distributions (Fig. 8.18e, f). Fractures from any set may terminate against fractures from any other set.

We can also observe intermediate fracture networks, such as the partially anisotropic network shown in Fig. 8.18d. In this example, we can still identify a predominant orientation and primary and secondary fracture sets, but the secondary fractures are often longer than in the fully anisotropic network, and some of the primary fractures terminate against secondary fractures.

Quantitatively, we can describe fracture anisotropy by comparing parameters relating to the different fracture sets. In the fracture models generated by our method, which contain at most two orthogonal fracture sets denoted Hmin and Hmax (striking

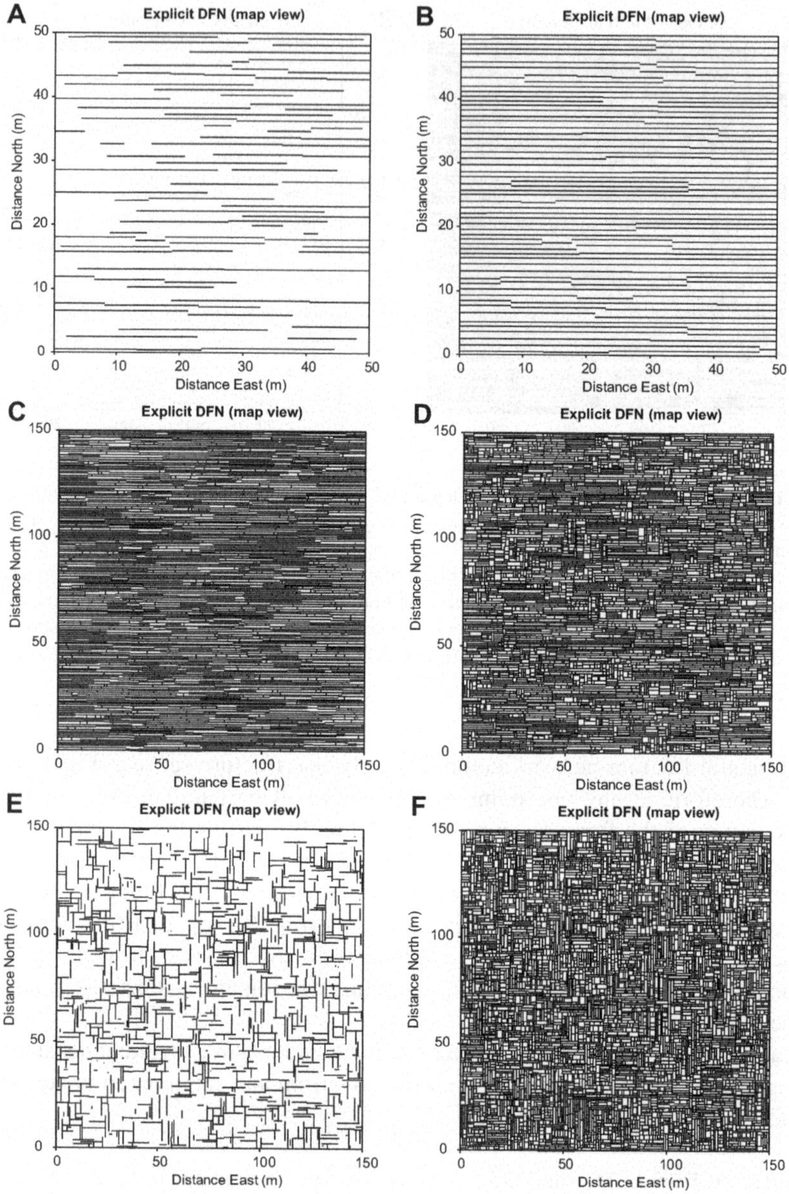

Fig. 8.18 Comparison of explicit DFNs representing an unsaturated uniaxial fracture network (**a**; from Fig. 8.5**b**), a saturated uniaxial network (**b**; from Fig. 8.5**d**), a fully anisotropic network (**c**; from Fig. 8.14), a partially anisotropic fracture network (**d**), an unsaturated isotropic fracture network (**e**) and a saturated isotropic fracture network (**f**). The DFNs are shown in plan view, with the black lines representing the centrelines of the fractures. Other input parameters are as listed in Table 8.2

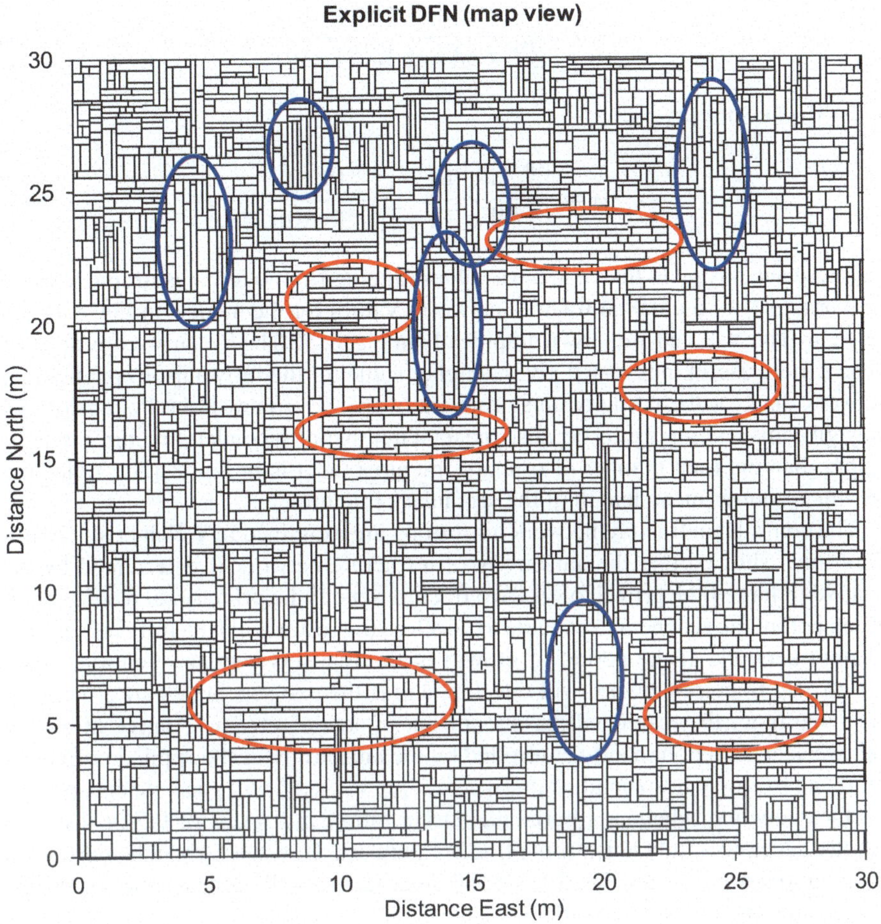

Fig. 8.19 Explicit DFN representing a saturated isotropic fracture network, generated in a homogeneous layer by an isotropic horizontal applied strain (shown in map view). Note that while the overall fracture network is isotropic, with similar cumulative density distribution functions for E-W and N-S striking fractures, it is locally anisotropic, with one or the other fracture set predominant. This is purely the result of initial random variability in the local fracture populations that become amplified by the blocking effect. Red ellipses mark areas where the E-W fractures predominate and blue ellipses mark areas where the N-S fracture predominates. Similar effects are seen on Fig. 8.18f

perpendicular to the minimum and maximum horizontal stress directions, respectively), this can be reduced to a simple ratio. There are, however, several different properties that we can use to calculate this ratio. Two of the most useful are the total volumetric half-macrofracture density, ${}_{MF}P_{30}(0)$, and the total mean linear half-macrofracture density, ${}_{MF}P_{32}(0)$; thus, we can define a P_{30} anisotropy, $\mathbb{I}_{30}$:

$$\mathbb{I}_{30} = \frac{\left| {}^{\text{HMin}}_{\text{MF}}\text{P}_{30}(0) - {}^{\text{HMax}}_{\text{MF}}\text{P}_{30}(0) \right|}{{}^{\text{HMin}}_{\text{MF}}P_{30}(0) + {}^{\text{HMax}}_{\text{MF}}\text{P}_{30}(0)} \tag{8.4}$$

and a P_{32} anisotropy, $\mathbb{I}_{32}$:

$$\mathbb{I}_{32} = \frac{\left| {}^{\text{HMin}}_{\text{MF}}\text{P}_{32}(0) - {}^{\text{HMax}}_{\text{MF}}\text{P}_{32}(0) \right|}{{}^{\text{HMin}}_{\text{MF}}\text{P}_{32}(0) + {}^{\text{HMax}}_{\text{MF}}\text{P}_{32}(0)} \tag{8.5}$$

These anisotropy indices can have values ranging from 0 for an isotropic fracture network, to 1 for a uniaxial fracture network. Values for the models in Fig. 8.18 are listed in Table 8.1, and are shown graphically in Fig. 8.20. The P_{32} anisotropy appears to be a more useful index, as it gives a higher value for the fully anisotropic fracture network than for the partially anisotropic network. By contrast, the P_{30} anisotropy may give misleading results because the total number of secondary fractures may be higher than the total number of primary fractures (as it is in Fig. 8.18c), reflecting the much shorter mean length of the secondary fractures.

The controls on fracture network anisotropy are examined in Welch et al. (2019). This study showed that the anisotropy of a fracture network is determined by the relative timing of the growth of the two fracture sets. If the primary set reaches saturation before the secondary set starts to develop, the result will be a fully anisotropic fracture network. If the period of growth of the two sets overlap, then a partially anisotropic fracture network will develop, and if the two sets both start growing at the same time, the result will be an isotropic fracture set. The relative timing of the two sets is predominantly controlled by the anisotropy of the applied horizontal strain: an isotropic horizontal strain (where $\dot{\varepsilon}_{h\text{min}} = \dot{\varepsilon}_{h\text{max}}$) will generate an isotropic fracture network, whereas a horizontal strain ratio $\dot{\varepsilon}_{h\text{max}}/\dot{\varepsilon}_{h\text{min}}$ of 0.5 or less will usually generate a fully anisotropic fracture network. However, when the horizontal strain ratio lies between 0.5 and 1, several other factors will influence the anisotropy of the resulting fracture network:

Table 8.1 Anisotropy and connectivity indices for five models, representing an unsaturated uniaxial fracture network (from Fig. 8.5b), a saturated uniaxial network (from Fig. 8.5d), a fully anisotropic network (from Fig. 8.14), a partially anisotropic model (from Fig. 8.18d) and an isotropic model (from Fig. 8.18f)

Model	Uniaxial, unsaturated (Fig. 8.5b)	Uniaxial, saturated (Fig. 8.5d)	Fully anisotropic (Fig. 8.14)	Partially anisotropic (Fig. 8.18d)	Isotropic, saturated (Fig. 8.18f)
$\mathbb{I}_{30}$	1	1	0.16	0.23	0.00
$\mathbb{I}_{32}$	1	1	0.93	0.58	0.00
I nodes	42%	0%	0%	0%	0%
R nodes	58%	100%	58%	16%	7%
Y nodes	0%	0%	42%	84%	93%

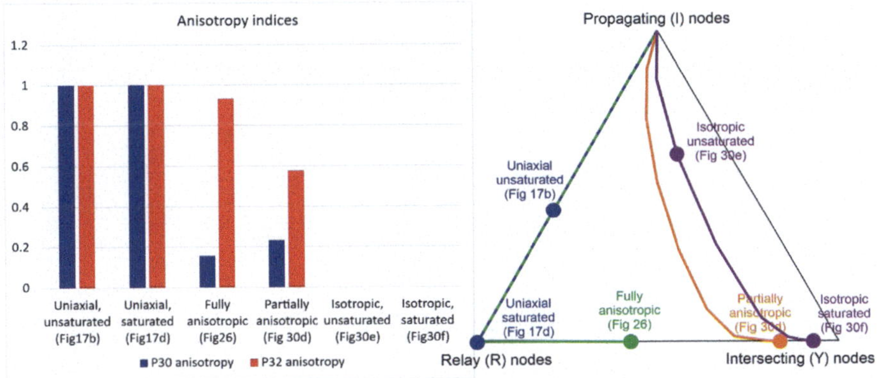

Fig. 8.20 Plots showing anisotropy indices (left) and connectivity indices (right) for six models, representing an unsaturated uniaxial fracture network (from Fig. 8.5b), a saturated uniaxial network (from Fig. 8.5d), a fully anisotropic network (from Fig. 8.14), a partially anisotropic model (Fig. 8.18d), an unsaturated isotropic model (Fig. 8.18e) and a saturated isotropic model (Fig. 8.18f). The lines on the ternary plot show the evolution of fracture connectivity as the fracture networks develop

- **Vertical effective stress**: If the vertical effective stress is low, macrofractures can start to nucleate and propagate at a lower horizontal strain. The secondary fracture set can thus start to develop before the primary fracture set has reached saturation. By contrast, if the vertical effective stress is high, the secondary fracture set will not start to develop until much later, after the primary set has reached saturation. We would, therefore, expect fracture networks developed at shallow burial depth or with high fluid overpressure to be more isotropic than those developed at depth at hydrostatic fluid pressure would.
- **Initial horizontal stress**: If the initial horizontal stress is low, then less applied horizontal strain is required for macrofractures the start nucleating and propagating, and the secondary fracture set can start to develop before the primary fracture set has reached saturation. If the initial horizontal stress is high, then more applied horizontal strain is required for macrofractures the start nucleating and propagating, and the secondary fracture set will not start to develop until much later after the primary set has reached saturation. We would, therefore, expect that a critically stressed layer, i.e. a layer in which optimally oriented Mode 2 shear fractures lie on the failure envelope, will develop a more isotropic fracture network, while a layer in viscoelastic equilibrium, where the initial horizontal effective stress is equal to the vertical effective stress, will develop a more anisotropic fracture network. For this reason, we used a critically stressed layer, rather than a layer in viscoelastic equilibrium, when generating the partially anisotropic fracture network shown in Fig. 8.18d. The initial horizontal stress state is discussed in more detail in Sect. 6.1.
- **Strain relaxation**: The models described so far in this book have all assumed that horizontal strain is entirely accommodated by either elastic strain in the rockmass

Table 8.2 Input parameters for the models illustrated in Figs. 8.1, 8.4, 8.5, 8.6, 8.8, 8.9, 8.10, 8.11, 8.14, 8.15, 8.17, 8.18 and 8.19

Property	Figure 8.1a	Figure 8.1b	Figure 8.1c	Figure 8.1d	Figure 8.4a	Figure 8.4b	Figure 8.4c	Figure 8.4d	Figure 8.5	Figure 8.6	Figure 8.8
Thickness h (m)	0.1	1	10	100	1	1	1	1	1	1	1
Model size (m)	5	50	500	5000	50	50	50	50	50	50	50
Effective vertical stress $\sigma_v{}'$ (MPa)	24.5	24.5	24.5	24.5	0.12	24.5	24.5	24.5	24.5	24.5	24.5
Min horizontal strain rate $\dot{\varepsilon}_{\mathrm{hmin}}$ (/ma)	−0.01	−0.01	−0.01	−0.01	−0.01	−0.01	−0.01	−0.01	−0.01	−0.01	−0.01
Max horizontal strain rate $\dot{\varepsilon}_{\mathrm{hmax}}$ (/ma)	0	0	0	0	0	0	0	0	0	0	0
Strain relaxation time constant (ma)	None	None	None	None	None	None	None	None	None	None	None
Stress shadows?	Yes	Yes	Yes	Yes	Yes	Yes	Yes	Yes	Yes	No	Yes
Subcritical index b	10	10	10	10	20	20	20	20	10	10	10

(continued)

Table 8.2 (continued)

Property	Figure 8.1a	Figure 8.1b	Figure 8.1c	Figure 8.1d	Figure 8.4a	Figure 8.4b	Figure 8.4c	Figure 8.4d	Figure 8.5	Figure 8.6	Figure 8.8
Initial μF density coefficient *B*	0.001	0.001	0.001	0.001	0.001	0.001	0.001	0.001	0.001	0.001	0.001
Initial μF distribution coefficient c	2	2	2	2	2	2	2	2	2	2	2
Young's Mod E (GPa)	10	10	10	10	10	10	10	10	10	10	10 to 10.5
Poisson's ratio ν	0.25	0.25	0.25	0.25	0.25	0.25	0.25	0.25	0.25	0.25	0.25
Friction coefficient μ_{fr}	0.5	0.5	0.5	0.5	0.5	0	0.5	1	0.5	0.5	0.5
Crack surface energy (J/m^2)	1000	1000	1000	1000	1000	1000	1000	1000	1000	1000	1000
Initial compaction factor υ	1	1	1	1	1	1	1	1	1	0.073 (critical)	1
Fracture mode	2	2	2	2	1	2	2	2	2	2	2

(continued)

Table 8.2 (continued)

Property	Figure 8.9	Figure 8.10a	Figure 8.10b	Figure 8.11	Figure 8.14	Figure 8.15	Figure 8.17a	Figure 8.17b	Figure 8.18d	Figure 8.18e	Figure 8.19
Thickness *h* (m)	1	1	1	1	1	1	1	1	1	1	0.1
Model size (m)	50	50	250	150	150	250	100	100	150	150	30
Effective vertical stress σ_v' (MPa)	24.5	24.5	24.5	24.5	24.5	24.5	0.5	0.5	24.5	24.5	1.47
Min horizontal strain rate $\dot{\varepsilon}_{hmin}$ (/ma)	−0.01	−0.01	−0.01	−0.01	−0.01	−0.01	−0.01	−0.01	−0.01	−0.01	−0.01
Max horizontal strain rate $\dot{\varepsilon}_{hmax}$ (/ma)	0	0	0	0	−0.005	0	0	0	−0.008	−0.01	−0.01
Strain relaxation time constant (ma)	None	None	None	None	None	None	None	None	0.002	None	None
Stress shadows?	Yes	No	No	Yes	Yes	Yes	Yes	Yes	Yes	Yes	Yes
Subcritical index *b*	10	5	50	10	10	10	10	50	10	10	10

(continued)

Table 8.2 (continued)

Property	Figure 8.9	Figure 8.10a	Figure 8.10b	Figure 8.11	Figure 8.14	Figure 8.15	Figure 8.17a	Figure 8.17b	Figure 8.18d	Figure 8.18e	Figure 8.19
Initial μF density coefficient *B*	0.001	0.001	0.001	0.001	0.001	0.001	0.001	0.001	0.001	0.001	0.001
Initial μF distribution coefficient *c*	2	2	2	2	2	2	2	2	2	2	2
Young's Mod E (GPa)	10	10	10	10	10	10	10	10	10	10	10
Poisson's ratio ν	0.25	0.25	0.25	0.25	0.25	0.25	0.25	0.25	0.25	0.25	0.25
Friction coefficient μ_{fr}	0.5	0.5	0.5	0.5	0.5	0.5	0.5	0.5	0.5	0.5	0.5
Crack surface energy (J/m^2)	1000 to 2000	1000	1000	1000	1000	1000	1000	1000	1000	1000	1000
Initial compaction factor υ	1	0.073 (critical)	0.073 (critical)	1	1	1	1	1	0.073 (critical)	1	1
Fracture mode	2	2	2	2	2	2	2	1	2	2	1

or displacement on the fractures. In practice, however, most rocks also undergo some degree of time-dependent inelastic deformation over geological timescales, due to processes such as grain sliding, creep and pressure solution. This "strain relaxation" limits the differential stress that can build up in the layer, causing the fracture sets to develop more slowly. As a result, it is more likely that the primary fracture set will not have reached saturation before the secondary set starts to develop. We, therefore, expect the fracture networks to be more isotropic when strain relaxation rates are high. For this reason, we applied strain relaxation with a time constant 0.002 ma (i.e. the horizontal elastic strain drops to 1/e of its initial value within 0.002 ma) when generating the partially anisotropic fracture network shown in Fig. 8.18d. The impact of strain relaxation on the evolution of fracture networks is explored in more detail in Welch et al. (2019).

However, even a fully isotropic fracture network, generated by an isotropic horizontal strain ($\dot{\varepsilon}_{h\min} = \dot{\varepsilon}_{h\max}$), will have patches of local anisotropy. These develop as a result of "blocking", where a fracture that nucleated early forms a shield, blocking the propagation of perpendicular fractures into a certain area and allowing other parallel fractures to grow without interruption. This positive feedback process causes small, random anisotropies to become amplified as the fracture network grows. As a result, a fracture network that is fully isotropic on a large scale may appear to be anisotropic on a small scale. We can see this in Fig. 8.19: although the overall populations of N-S striking and E-W striking fractures in this model are identical, in many parts of the model one of the two sets has become predominant, while the other set comprises only short connecting fractures.

The fracture connectivity describes the proportion of fractures in a network that are connected to other fractures and the type of connection. In the case of layer-bound fractures, it can be reduced to a two-dimensional topological problem describing the fracture "nodes", which comprise the tips and intersection points of the fractures. A common approach is to classify the fracture nodes into three types: I nodes, representing isolated fracture tips; Y nodes, representing points where one fracture intersects and terminates against another fracture; and X nodes, where two fractures cross but neither terminate (Manzocchi 2002; Mäkel 2007). The fracture connectivity is represented by the ratio of the three types of nodes, which can be plotted on a ternary diagram.

This scheme can be adapted slightly for use with the fracture models generated by our method. There will be no X nodes in our models, since we assume that fractures can never cross another fracture, but will always terminate against any fractures they intersect, forming a Y node. We can, however, distinguish between two other types of fracture tip: truly isolated tips that are not associated with any other fractures, and tips resulting from stress shadow interactions. The latter is often associated with deformed relay zones, and may even be connected to the other tip by short relay segments (Lamarche et al. 2018). There may, therefore, be fluid transmissibility directly between the two fractures, in which case it will be more appropriate to consider these tips as permeability baffles in the fracture network, rather than as dead-ends. It is, therefore, useful to classify the truly isolated tips separately from the stress shadow related tips, so we will label the former as I nodes and latter as R nodes.

We can determine the relative proportion of the different node types directly from the implicit model: the number of I nodes will be given by the total volumetric density of active half-macrofractures ${}^{a}_{MF}P_{30}(0)$, the number of R nodes by the total volumetric density of static half-macrofractures deactivated due to stress shadow interaction ${}^{sII}_{MF}P_{30}(0)$ and the number of Y nodes by the total volumetric density of static half-macrofractures deactivated due to intersection ${}^{sIJ}_{MF}P_{30}(0)$. The connectivity indices for several of the models presented previously are listed in Table 8.1 and are shown graphically on a ternary plot in Fig. 8.20.

The principal controls on fracture connectivity are the duration of deformation and the anisotropy of the fracture network. Since macrofractures nucleate with two active tips, fracture networks start with a predominance of I nodes, i.e. near the uppermost vertex of the ternary plot in Fig. 8.20. As the half-macrofractures become deactivated, the proportion of R and Y nodes increases and the fracture network moves down the triangle. Fully saturated fracture networks contain no active tips and hence lie on the lower edge of the triangle. The exact proportion of R and Y nodes will depend on the anisotropy of the fracture network: a fully saturated uniaxial fracture network will contain only R nodes, while an isotropic network will comprise mostly Y nodes (although there may still be some R nodes). An anisotropic network will contain a mixture of Y and R nodes, depending on the degree of anisotropy. There is also a tendency for anisotropic fracture networks, in particular, to evolve towards the bottom right vertex of the triangle, reflecting an increasing proportion of Y nodes, as increasing numbers of secondary fractures terminate against the primary fractures.

8.6 Controls on Fracture Orientation

The strike of the fractures in the explicit DFN generated by this method is controlled entirely by the minimum and maximum horizontal strain orientation specified in the input data, while the implicit fracture data is completely independent of fracture orientation. Strictly speaking, therefore, the fracture orientation is not a property that we can use to calibrate the method.

Nevertheless, fracture orientation is a key property of any fracture network since it determines important attributes such as preferred flow direction and preferred failure orientation. It is, therefore, important to replicate the fracture orientation accurately in any fracture models, especially if they will be used for flow modelling or for engineering evaluation. In this section, therefore, we discuss various methods we can use to determine the orientation (and magnitude) of the horizontal strain.

The best method to determine the strain orientation and magnitude depends on the cause of the fracturing. We will examine three scenarios: regional fracture sets, fractures related to larger geological structures such as folds or diapirs and fault-related fractures. These scenarios will be revisited in the real-world examples of Chaps. 9 and 10.

8.6.1 Regional Fractures

When modelling regional fracture sets formed in response to a regional tectonic strain, for example, in a rift basin, it may be sufficient to apply a uniform strain across the entire model. In this scenario, the fracture network will mostly be controlled by the geometry of the layer (e.g. variations in depth or layer thickness) or by lateral variations in the mechanical properties.

The strain orientation in such cases may be obtained from regional tectonic studies or from the orientation of larger faults. Typical mean strain rates across active rift basins are in the range of 0.001–0.01/ma, for example, the North Viking Graben experienced a total strain of c.15% (Roberts et al. 1997) applied over c.20 ma (Fjeldskaar et al. 2004), giving a strain rate of 0.0075/ma; the Bristol Channel Basin experienced a total strain 6–11% applied over c.20 ma, giving a strain rate of 0.0003–0.0055/ma (Kamerling 1979); the Wessex Basin experienced a mean strain rate of 0.003–0.005/ma (Chadwick 1986); and the St George's Channel Basin experienced a mean strain rate over several episodes of Triassic and Jurassic rifting of c.0.006/ma (Welch and Turner 2000).

8.6.2 Fractures Related to Large Geological Structures

Fracture sets often form in response to more local strains, generated by processes such as large-scale folding or diapirism. In this scenario, we would expect to see a lateral variation in the orientation and magnitude of the strain across the structure, which will control the orientation, distribution and geometry of the fracture network. This may also result in variation in the fracture anisotropy across the structure, reflecting variation in the relative magnitudes of the minimum and maximum horizontal strains.

We can often obtain a first pass approximation for the strain across the structure from the curvature of the layer we are modelling. This is a technique that is often used for conditioning stochastic fracture models (Mäkel 2007). It is based on the assumption that the extensional strain experienced by the layer is proportional to the positive (convex-upward) curvature of the top of the layer. The minimum and maximum strain rates can thus be calculated from the magnitude of the maximum and minimum curvature, using a suitable negative multiplier, while the orientation of minimum horizontal strain will be the orientation of maximum curvature.

While curvature has the advantage that it is quick and easy to calculate, it only gives an empirical approximation for the strain. More rigorous results can be obtained by more detailed kinematic or dynamic modelling of the relevant large-scale structure. For example, in Sect. 10.1, we will show how Petersen et al. (2018) used back-stripping and kinematic analysis to calculate the strain experienced by a brittle chalk layer overlying a salt pillow in the North Sea. Structural restoration is another technique that can be used to calculate the strain experienced by a layer in response to folding (e.g. Sanders et al 2004).

8.6.3 Fault-Related Fractures

Fractures that develop around larger faults often reflect the orientation of the local stress field that develops around those faults (e.g. Rawnsley et al. 1992; Bourne and Willemse 2001; Maerten et al. 2016). The orientation and magnitude of both the stress and the elastic strain generated by slip on larger faults can be calculated using the boundary element method (Maerten 1999). This technique is often used for conditioning stochastic fracture models (e.g. Dee et al. 2007; Freeman et al. 2015).

The magnitude and orientation of the local stress field around a large fault may be highly variable over short distances—often less than the length of the fractures. This can generate some very complex fracture geometries, with curved or even concentric circular fractures, especially around tips, bends and splays in the larger faults (see Welch et al. 2015). In this case, it is important that the gridblocks in the model are sufficiently small to capture the variations in stress and strain and in particular to keep the change in strain orientation across the gridblock boundaries to a minimum (<c.45°) while still respecting the constraints on gridblock size outlined in Sect. 7.1. If this is not possible, then there will be a trade-off between accurately replicating the geometry of the fracture network and calculating the implicit fracture densities. Smaller gridblocks will favour the former, while larger gridblocks will favour the latter. We will see that this becomes an issue in the Pegwell Bay model described in Sect. 9.3.

8.7 Controls on Fracture Porosity and Permeability

Fracture porosity and permeability are often of particular interest to modellers since they are a key control on fluid flow in the subsurface and in some cases on the total volume of fluids present. They are hence essential parameters to populate dynamic flow models. They are also useful parameters for calibrating subsurface fracture models against hydrocarbon production data, well test data or tracer test data.

They are, however, difficult properties to calculate accurately since they require knowledge of the fracture aperture, which is not generally well constrained. Although Eq. (4.19) does give a value for the aperture of Mode 1 macrofractures (see Sect. 4.4), this represents the dynamic aperture, under the conditions of fracture propagation. The fracture aperture under present-day conditions may be very different due to strain relaxation and fluid leakoff. There is no expression for the dynamic aperture of Mode 2 macrofractures since they can propagate even when they are closed.

One approach is to assign a uniform aperture for all fractures based on subsurface measurements or estimates from well test data. Fracture aperture is difficult to measure in the subsurface but can sometimes be estimated from borehole image data (see Luthi and Souhaité 1990; Hornby and Luthi 1992). Note that we cannot use measurements from fractures in core or outcrop analogues to estimate fracture aperture in the subsurface since they are not measured under subsurface stress conditions. Typical values for subsurface fracture aperture are in the range 0.1–2 mm. Since fracture aperture is often dependent on fracture orientation (e.g. Barton et al. 1995), we can refine this technique by assigning different apertures to the different fracture sets; indeed, some sets may be closed and assigned an aperture of zero.

A more sophisticated approach is to use the Barton-Bandis method (Bandis et al. 1983), which gives an empirical estimate of the aperture of fractures in the subsurface based on the fracture properties (e.g. the surface roughness) and the normal stress acting on the fracture. The advantages of this technique are that it can model the effect of fracture "propping" (where a fracture is held open by asperities on its surface) and that it accounts for both the current in situ stress and the fracture orientation. It can also be applied to both Mode 1 and Mode 2 fractures. The disadvantage is that it requires a number of input parameters (e.g. the Joint Roughness Coefficient, the stiffness normal to the fracture and the maximum fracture closure) that are poorly constrained.

Once we have an estimate for the fracture aperture, we can multiply it by the mean linear fracture density P_{32}, which represents the total fracture area per unit volume, to calculate fracture porosity. Since the fracture porosity is directly proportional to P_{32}, we can use the P_{32} data contained within the implicit fracture model to calculate a detailed breakdown of the fracture porosity, for example, between microfractures and macrofractures, between different fracture sets, by fracture size or by fracture connectivity. Even if we do not have confidence in our estimate of fracture aperture, the P_{32} data in the implicit fracture model can be useful in revealing the likely distribution of fracture porosity.

Fracture permeability is more difficult to determine, even if the fracture aperture is known with some confidence. The permeability of a single fracture can be calculated using Pouseille's Law, although this only gives an approximation since it assumes flow between parallel smooth plates, so it does not account for roughness and friction on the fracture surface, or for the tortuosity of actual fractures (Neuzil and Tracy 1981). However, the permeability of an entire fracture network is much harder to calculate, as it depends not just on the number of fractures but also on factors such as the fracture length distribution, the fracture connectivity and the anisotropy of the network. Fracture permeability is, therefore, highly non-linear with respect to fracture density. It often rises suddenly when a certain fracture density is reached, representing the point at which the fracture network becomes fully connected (the "percolation threshold"; Odling et al. 1999; Manzocchi 2002). Furthermore, fracture permeability generally varies with orientation, so must be expressed as a permeability tensor.

Although some methods exist for estimating the permeability of fracture networks (e.g. Oda's method: Oda 1984; Dershowitz et al. 2000), it is often not possible to apply them in the subsurface due to a lack of data to characterise the fracture network. An alternative approach is to run gridblock-scale numerical models of flow through the fracture network to calculate the permeability tensor empirically (e.g. Sarda et al. 2002; Jonoud et al. 2011), although this also requires a detailed model of the subsurface fracture network. Our intention is that the implicit models generated using our method will provide sufficient information to calculate the fracture permeability using methods such as Oda's, while the explicit DFNs that can be generated using our method can be used as input for numerical flow models to calculate the fracture permeability. This is discussed further in Sect. 11.2.2.

Another advantage of our method for generating fracture models is that it provides an opportunity to examine how fracture permeability evolves as the fracture network develops. This is beyond the scope of this study but is examined in more detail in Kadeethum et al. (2019), Torres et al. (2020) and Salimzadeh et al. (2019).

References

Atkinson, B.K. 1984. Subcritical crack growth in geological materials. *Journal of Geophysical Research* 89 (B6): 4077–4114.

Bai, T., and D.D. Pollard. 2000. Fracture spacing in layered rocks: a new explanation based on the stress transition. *Journal of Structural Geology* 22: 43–57.

Bai, T., D.D. Pollard, and H. Gao. 2000a. Explanation for fracture spacing in layered materials. *Nature* 403: 753–756.

Bai, T., D.D. Pollard, and M.R. Gross. 2000b. Mechanical prediction of fracture aperture in layered rocks. *Journal of Geophysical Research* 105 (B1): 707–721.

Bandis, S.C., A.C. Lumsden, and N.R. Barton. 1983. Fundamentals of rock joint deformation. *International Journal of Rock Mechanics, Mining Science & Geomechanics Abstracts* 20: 249–268.

Barton, C.A., M.D. Zoback, and D. Moos. 1995. Fluid flow along potentially active faults in crystalline rock. *Geology* 23: 683–686.

Bonnet, E., O. Bour, N.E. Odling, P. Davy, I. Main, P. Cowie, and B. Berkowitz. 2001. Scaling of fracture systems in geological media. *Reviews of Geophysics* 39: 347–383.

Bourne, S.J., and E.J.M. Willemse. 2001. Elastic stress control on the pattern of tensile fracturing around a small fault network at Nash Point, UK. *Journal of Structural Geology* 23: 1753–1770.

Chadwick, R.A. 1986. Extension tectonics in the Wessex Basin, southern England. *Journal of the Geological Society, London* 143: 465–488.

Chang, C., M.D. Zoback, and Abbas Khaksar. 2006. Empirical relations between rock strength and physical properties in sedimentary rocks. *Journal of Petroleum Science and Engineering* 51: 223–237.

Dee, S.J., G. Yielding, B. Freeman, D. Healy, N.J. Kusznir, N. Grant, and P. Ellis. 2007. Elastic dislocation modelling for prediction of small-scale fault and fracture network characteristics. In: *Fractured reservoirs*, eds. L. Lonergan, R.J.H. Jolly, K. Rawnsley, and D.J. Sanderson, vol. 270, 139–155. London: Geological Society, Special Publications.

Dershowitz, B., P. LaPointe, T. Eiben, and L. Wei. 2000. Integration of discrete feature network methods with conventional simulator approaches. In *SPE49069, presented at SPE annual technical conference and exhibition*, New Orleans, Louisiana.

Ferrill, D.A., and A.P. Morris. 2001. Displacement gradient and deformation in normal fault systems. *Journal of Structural Geology* 23: 619–638.

Fjeldskaar, W., M. Ter Voorde, H. Johansen, P. Christiansson, J.I. Faleide, and S.A.P.L. Cloetingh. 2004. Numerical simulation of rifting in the northern Viking Graben: the mutual effect of modelling parameters. *Tectonophysics* 382: 189–212.

Freeman, B., D.J. Quinn, C.G. Dillon, M. Arnhild, and B. Jaarsma. 2015. Predicting subseismic fracture density and orientation in the Gorm Field, Danish North Sea. In: *Industrial structural geology: principles, techniques and integration*, vol. 421, eds. F.L. Richards, N.J. Richardson, S.J. Rippington, R.W. Wilson, and C.E. Bond, 231–244. London: Geological Society, Special Publications.

Holder, J., J.E. Olson, and Z. Phillip. 2001. Experimental determination of subcritical crack growth parameters in sedimentary rock. *Geophysical Research Letters* 28: 599–602.

Hornby, B.E., and S.M. Luthi. 1992. An integrated interpretation of fracture apertures computed from electrical borehole scans and reflected Stoneley waves. In: *Geological applications of wireline logs II*, vol. 65, eds. A. Hurst, C.M. Griffiths, and P.F. Worthington, 185–198. London: Geological Society, Special Publication.

Huggins, P., J. Watterson, J.J. Walsh, and C. Childs. 1995. Relay zone geometry and displacement transfer between normal faults recorded in coal-mine plans. *Journal of Structural Geology* 17: 1741–1755.

Jonoud, S., O.P. Wennberg, A. Larsen, and G. Casini. 2011. Capturing the effect of fracture heterogeneity on multiphase flow. In *SPE147834, presented at SPE reservoir characterisation and simulation conference and exhibition*, Abu Dhabi, UAE.

Kadeethum, T., S. Salimzadeh, and H.M. Nick. 2019. An investigation of hydromechanical effect on well productivity in fractured porous media using full factorial experimental design. *Journal of Petroleum Science and Engineering* 181: 106233.

Kamerling, P. 1979. The geology and hydrocarbon habitat of the Bristol Channel Basin. *Journal of Petroleum Geology* 2: 75–93.

Lamarche, J., A. Chabani, and B.D.M. Gauthier. 2018. Dimensional threshold for fracture linkage and hooking. *Journal of Structural Geology* 108: 171–179.

Luthi, S.M., and P. Souhaité. 1990. Fracture apertures from electrical borehole scans. *Geophysics* 55: 821–833.

Maerten, L. 1999. Mechanical interaction of intersecting normal faults: theory, field examples and applications. Ph.D. thesis, Stanford University.

Maerten, L., Maerten, F., Lejri, M., and Gillespie, P. 2016 . Geomechanical paleostress inversion using fracture data. *Journal of Structural Geology* 89: 197–213. https://doi.org/10.1016/j.jsg.2016.06.007.

Mäkel, G.H. 2007. The modelling of fractured reservoirs: constraints and potential for fracture network geometry and hydraulics analysis. In *Structurally complex reservoirs*, ed. Jolley, S.J., D. Barr, J.J. Walsh, R.J. Knipe, vol. 292, 375–403, Geological Society, London, Special Publications.

Manzocchi, T. 2002. The connectivity of two-dimensional networks of spatially correlated fractures. *Water Resources Research* 38: 1162–1181.

Marrett, R. 1996. Aggregate properties of fracture populations. *Journal of Structural Geology* 18: 169–178.

Marrett, R., and R.W. Allmendinger. 1991. Estimates of strain due to brittle faulting: sampling of fault populations. *Journal of Structural Geology* 13: 735–738.

Marrett, R., and R.W. Allmendinger. 1992. Amount of extension on "small" faults: an example from the Viking graben. *Geology* 20: 47–50.

Neuzil, C.E., and J.V. Tracy. 1981. Flow through fractures. *Water Resources Research* 17: 191–199.

Nicol, A., J.J. Walsh, J. Watterson, and P.G. Bretan. 1995. Three-dimensional geometry and growth of normal faults. *Journal of Structural Geology* 17: 847–862.

Oda, M. 1984. Permeability tensor for discontinuous rock masses. *Geotechnique* 35: 483–495.

Odling, N.E., P. Gillespie, B. Bourgine, C. Castaing, J.-P. Chiles, N.P. Christensen, E. Fillion, A. Genter, C. Olsen, L. Thrane, R. Trice, E. Aarseth, J.J. Walsh, and J. Watterson. 1999. Variations in fracture system geometry and their implications for fluid flow in fractured hydrocarbon reservoirs. *Petroleum Geology* 5: 373–384.

Olson, J.E. 2004. Predicting fracture swarms—the influence of subcritical crack growth and the crack-tip process zone on joint spacing in rock. In *The initiation, propagation and arrest of joints and other fractures*, eds. Cosgrove, J.W., T. Engelder, 73–88, Geological Society, London, Special Publications 231.

Ouillon, G., C. Castaing, and D. Sornette. 1996. Hierarchical geometry of faulting. *Journal of Geophysical Research* 101 (B3): 5477–5487.

Petersen, K.D., and O.R. Clausen. 2018. Forward and reverse dynamic and kinematic modeling of the evolution of the Kraka Field. Unpublished DHRTC report.

Pollard, D.D., and P. Segall. 1987. Theoretical displacements and stresses near fractures in rocks: with applications to faults, joints, veins, dikes and solution surfaces. In *Fracture mechanics of rock*, ed. B.K. Atkinson, 277–349. London: Academic Press.

Rawnsley, K.D., Rives, T., Petti, J.P., Hencher, S.R., and Lumsden, A.C. 1992. Joint development in perturbed stress fields near faults. *Journal of Structural Geology* 14(8–9): 939–951. https://doi.org/10.1016/0191-8141(92)90025-R.

Roberts, A.M., E.R. Lundin, and N.J. Kusznir. 1997. Subsidence of the Vøring Basin and the influence of the Atlantic continental margin. *Journal of the Geological Society, London* 154: 551–557.

Salimzadeh, S., M. Grandahl, M. Medetbekova, and H.M. Nick. 2019. A novel radial jet drilling stimulation technique for enhancing heat recovery from fractured geothermal reservoirs. *Renewable Energy* 139: 395–409.

Sanders, C., M. Bonora, D. Richards, E. Kozlowski, C. Sylwan, and M. Cohen. 2004. Kinematic structural restorations and discrete fracture modeling of a thrust trap: a case study from the Tarija Basin, Argentina. *Marine and Petroleum Geology* 21: 845–855.

Sarda, S., L. Jeannin, R. Basquet, and B. Bourbiaux. 2002. Hydraulic characterisation of fractured reservoirs: simulation on discrete fracture models. In *SPE77300, presented at SPE reservoir simulation symposium*, Houston, Texas.

Swanson, P.L. 1984. Subcritical crack growth and other time- and environment-dependent behaviour in crustal rocks. *Journal of Geophysical Research* 89: 4137–4152.

Torres, L.F.L., S. Salimzadeh, and H.M. Nick. 2020. Interplay of heat transport and fracture aperture in fractured reservoirs. In *Proceedings of the 4th EAGE naturally fractured reservoir workshop*, Ras Al Khaimah, UAE.

Walsh, J.J., W.R. Bailey, C. Childs, A. Nicol, and C.G. Bonson. 2003. Formation of segmented normal faults: a 3-D perspective. *Journal of Structural Geology* 25: 1251–1262.

Welch, M.J., C. Souque, R.K. Davies, and R.J. Knipe. 2015. Using mechanical models to investigate the controls on fracture geometry and distribution in chalk. In: *Fundamental controls on fluid flow in carbonates*, vol. 406, eds. S.M. Agar, and S. Geiger, 281–309. London: Geological Society, Special Publications.

Welch, M.J., and J.P. Turner. 2000. Triassic-Jurassic development of the St. George's Channel basin, offshore Wales. *UK. Marine and Petroleum Geology* 17: 723–750.

Welch, M.J., M. Lüthje, and A.C. Glad. 2019. Influence of fracture nucleation and propagation rates on fracture geometry: insights from geomechanical modelling. *Petroleum Geoscience* 25: 470–489.

Westaway, R. 1994. Quanititative analysis of population of small faults. *Journal of Structural Geology* 16: 1259–1273.

Chapter 9
Calibration Against Outcrop

Abstract In this chapter, we use observations from three fractured outcrops in the UK (Nash Point, Robin Hood's Bay and Pegwell Bay) to generate discrete fracture network models (DFNs). We show that these DFNs can replicate the complexity observed in the real fracture networks. We use only basic observations of the geometry of large-scale faults (>100 m in size) controlling their formation and simple assumptions regarding the mechanical properties and stress state at the time of deformation. This is sufficient for us to generate DFNs that replicate the fracture spacing (i.e. mean linear microfracture density, P_{32}), the fracture length distributions, the spatial distribution of the fractures and the fracture anisotropy observed in the different outcrops. Furthermore, our application to these field analogues begins to reveal further questions as to the role of fluid pressure and stress shadows in generating fracture corridors.

Keywords Fracture modelling · Discrete fracture networks · Naturally fractured reservoir · Outcrop analogue · Mechanical stratigraphy · Geomechanics · Fracture corridor

In Chap. 1, we discussed why we expect that fracture networks generated by geomechanical simulation will much better emulate the geometry of real natural fracture networks than fracture networks generated by stochastic or kinematic methods. In Chap. 8, we examined the various geometric and geomechanical parameters that control the evolution of fracture networks and that we can adjust in order to calibrate the outputs of the simulations. In this chapter, we will demonstrate this using fractured outcrops.

Fractured outcrops present an ideal opportunity to test and calibrate the new method, since it is often possible not only to map out the fracture network in considerable detail but also to characterise it quantitatively, measuring parameters such as volumetric fracture density, mean linear fracture density, fracture length and spacing distributions and fracture connectivity. By comparing the fracture patterns to the larger scale structure, it is also usually possible to determine the cause of fracturing

© The Editor(s) (if applicable) and The Author(s), under exclusive license to Springer Nature Switzerland AG 2020

M. J. Welch et al., *Modelling the Evolution of Natural Fracture Networks*,
https://doi.org/10.1007/978-3-030-52414-2_9

(e.g. regional tectonic strain, fold- or fault-related strain, etc.), and hence to calculate the magnitude and orientation of the horizontal strain across the outcrop.

We can, therefore, build and run detailed simulations of the growth of the fracture networks, including effects such as lateral variations in strain magnitude and orientation. For input parameters that are poorly constrained (e.g. the initial microfracture density, B), we can run multiple simulations to determine the optimal values to replicate the observed fracture network quantitatively. This can help us to determine realistic ranges of values for these parameters.

We examine three coastal outcrops from the UK where it is possible to map layer-bound fractures laterally across a reasonably large area and relate these to the larger-scale structure. These outcrops are located at Nash Point in South Wales, Robin Hood's Bay in northeast England and Pegwell Bay in southeast England (Fig. 9.1). For each outcrop, we draw on previous studies detailing the controls on fracture evolution and attempt to replicate the observed fracture patterns with the new method. In evaluating the results, we consider two questions:

1. Are the modelled fracture predictions representative of natural fracture patterns?
2. Can they be generated with data typically available in the subsurface?

We show that in addition to matching the observed fracture orientations, our models provide qualitatively similar fracture density, length and spacing distributions and similar patterns of fracture anisotropy. They, therefore, provide a fracture network that is comparable topologically to a natural fracture set and better able to capture the effect of fracture network geometry on flow (Rives et al. 1992).

9.1 Nash Point

Using the Nash Point outcrop, we will demonstrate a workflow to generate fracture models that capture the key characteristics of the fracture network geometry. Using published field data from the area, we compare the explicit DFNs that we generate with the natural fracture pattern observed on aerial photography of the area.

9.1.1 Description of the Outcrop

Nash Point lies on the southern coast of Wales, in the East Bristol Channel, a Mesozoic extensional basin comprising an E-W trending, S-dipping half-graben c. 150 km long and 5 km wide (Tappin et al. 1994). The East Bristol Channel was explored for hydrocarbons in the 1970s (Lloyd et al. 1973; Kamerling 1979). Although most of the basin lies offshore, the northern and southern flanks are exposed onshore in South Wales and North Somerset. Both areas contain excellent coastal exposures of metre to 100 m-scale faults and fractures and have been extensively mapped and studied

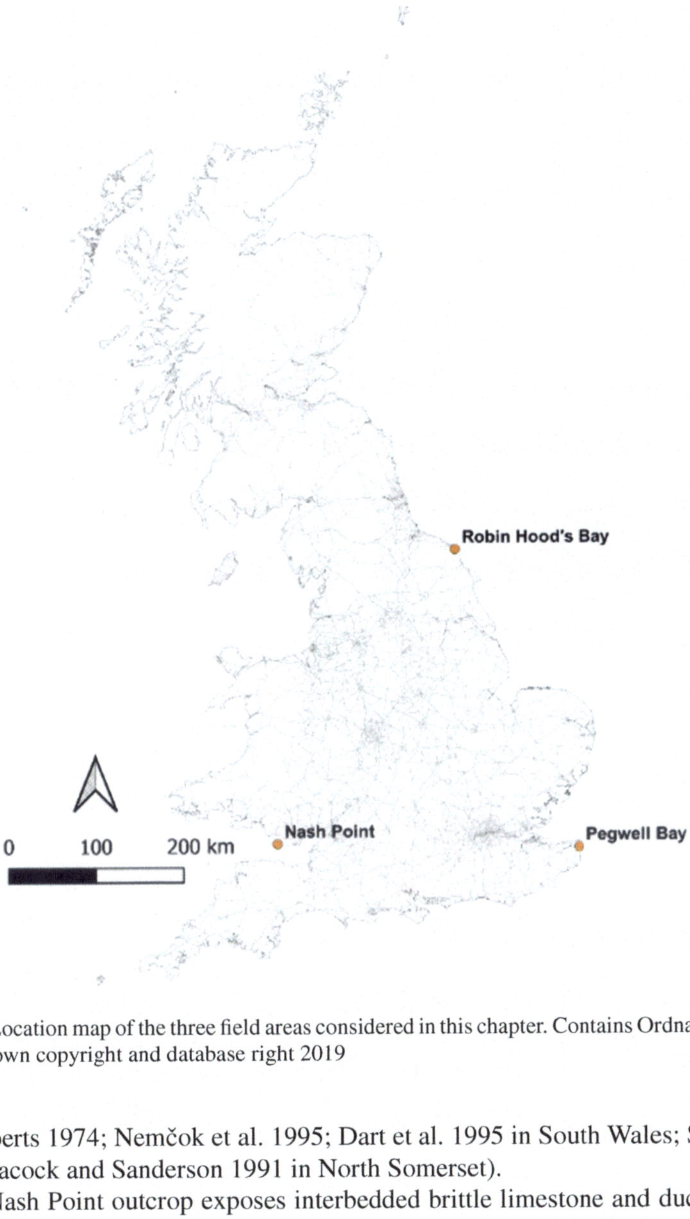

Fig. 9.1 Location map of the three field areas considered in this chapter. Contains Ordnance Survey data © Crown copyright and database right 2019

(e.g. Roberts 1974; Nemčok et al. 1995; Dart et al. 1995 in South Wales; Skar et al. 2017; Peacock and Sanderson 1991 in North Somerset).

The Nash Point outcrop exposes interbedded brittle limestone and ductile shale layers. The limestone layers are up to 1 m thick and are heavily fractured with layer-bound Mode 1 dilatant fractures. Rawnsley et al. (1992) identified the outcrop as an example of fracturing controlled by the perturbed stress field around larger-scale faults. This hypothesis was complemented by numerical modelling by Bourne and Willemse (2001) and more recently extended to consider the effect of spatially varying fluid pressure and poroelasticity by Maerten et al. (2018). Both of these

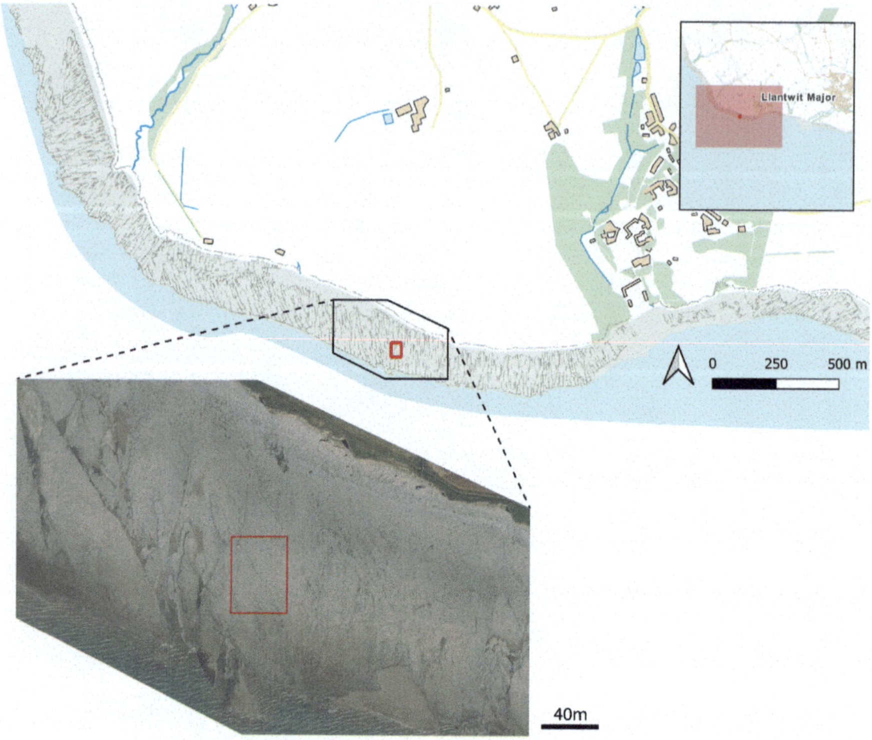

Fig. 9.2 Location map of the study area with vertical orthorectified aerial photography of the fractured area of the foreshore. Contains Ordnance Survey data © Crown copyright and database right 2019. (Aerial photography: Getmapping Plc. Vintage 2018.)

modelling studies confirmed the importance of the fault-related stress perturbations in controlling the fracture geometry.

The excellent preservation of fractures in the foreshore provides us with the opportunity to visually compare our DFNs with the real fracture network, imaged using orthorectified aerial photography (Figs. 9.2 and 9.3). This allows us to calibrate the input parameters including the layer thickness, subcritical fracture propagation index and initial microfracture density and to evaluate whether we can generate representative fracture models.

9.1.2 Building the Model

The geometry of the model is derived from the orthorectified images presented in Maerten et al. (2018). The controlling faults were digitized and used to define the faults of a geocellular grid.

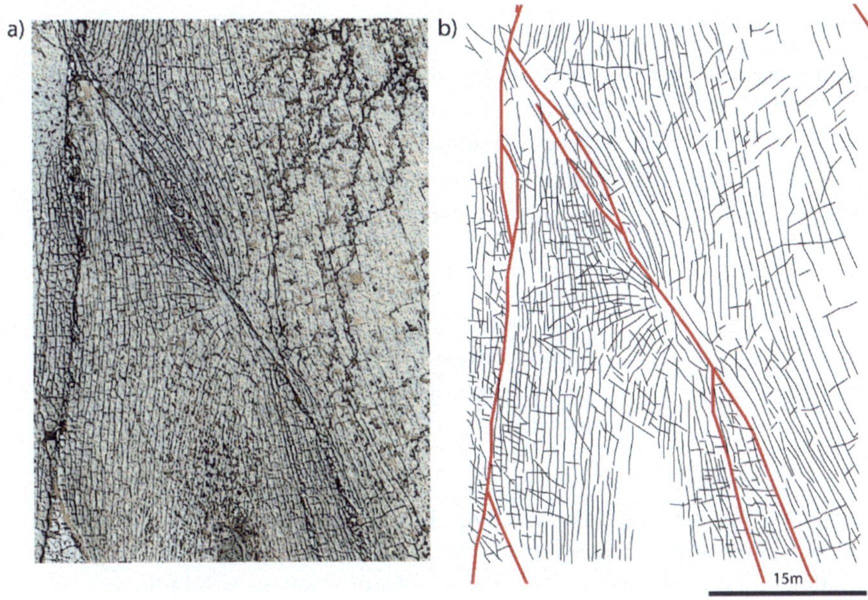

Fig. 9.3 **a** Reproduction of the orthorectified aerial photo used for analysis by Maerten et al. (2016) (their Fig. 6), alongside **b** our own interpretation of the image. Note the omission of some areas obscured by weathering.

Following the methods of Bourne and Willemse (2001) and Maerten et al. (2018), we assume the fracture distribution is controlled by the geometry of the main faults in the area (shown in red on Fig. 9.3), and we use a boundary element method to predict the stress distribution around those faults. For this calculation, we assume a burial depth of 1 km at the time of fracturing, in line with previous studies. An additional fluid overpressure of 7.7 MPa is applied, to replicate the Mode 1 fractures observed. We assume that the orientation of maximum horizontal compressive stress is represented by the intersection of the two dominant faulting directions, which is at 175°. To generate strike-slip displacement on the faults, we assume a maximum compressional horizontal strain of +0.0002/m in this direction and a maximum extensional horizontal strain of −0.0002/m perpendicular to this, at 085°. Finally, we assume uniform mechanical properties with values for typical limestones. From this data, we calculate the regional (or far-field) stress after 1 ma by calculating the overburden stress from an average overburden density (2250 kg m^{-3}) and horizontal stresses using (6.2) from Sect. 6.1. The results are summarised in Table 9.1. This far-field stress was used as the input to calculate the local elastic strains around the faults using the boundary element method. The local strain tensors generated by this model were then used to determine the local horizontal strain rates and orientation for each gridblock in the fracture simulation.

Using these assumptions, we can commence parameter testing considering the results of Chap. 8. Since the fractures mapped in the outcrop show a semi-regular

Table 9.1 Tectonic stresses and physical properties assumed for modelling the Nash Point area

Depth	1000 m
Maximum horizontal strain rate	+0.0002/ma
Minimum horizontal strain rate	−0.0002/ma
Fluid overpressure	7.7 MPa
Maximum horizontal stress orientation, σ_{azi}	175°
Young's modulus	10 GPa
Poisson's ratio	0.25
Biot's coefficient	1
Crack surface energy	1000 J/m^2
Friction coefficient	0.5
Critical fracture propagation rate	2000 m/s
Effective vertical stress	4.525 MPa

spacing, we assume that these fractures have stress shadows. From our earlier work, we know that the spacing of fractures with stress shadows is primarily controlled by layer thickness, while the lengths of fractures within the network is primarily controlled by the subcritical fracture propagation rate index and the initial microfracture density.

9.1.3 Model Results

To calibrate the bedding thickness we run the model using a series of different thicknesses and compare their predicted fracture spacing to that observed in outcrop (Fig. 9.4). We observe both increased fracture spacing and increased mean fracture length in the thicker layers, consistent with the predictions from Sect. 8.1.

The prevalence of smaller fractures also varies across the model, with spatial variance in the relative number of shorter and longer fractures. Visually, one can appreciate that in areas with a greater proportion of shorter fractures, the volumetric fracture density is also increased. Conversely, in areas dominated by longer fractures, the volumetric fracture density is decreased. On balance, we have selected a thickness of 0.5 m to offer a representative fracture spacing in areas dominated by longer fractures, similar to observations from the outcrop.

Parameter testing for the initial microfracture density coefficient, B, and subcritical fracture propagation index, b, is somewhat more difficult due to the greater sensitivity of the output fracture population to the variation in these parameters. Figures 9.5 and 9.6 illustrate some of the model runs used for this calibration.

Figure 9.5 shows the effect that large (order of magnitude) variations in these two parameters will have on the resulting fracture network. At high initial microfracture densities, fractures are unable to propagate far before truncating against each other. At lower values, fracture spacing becomes much greater and when combined with a

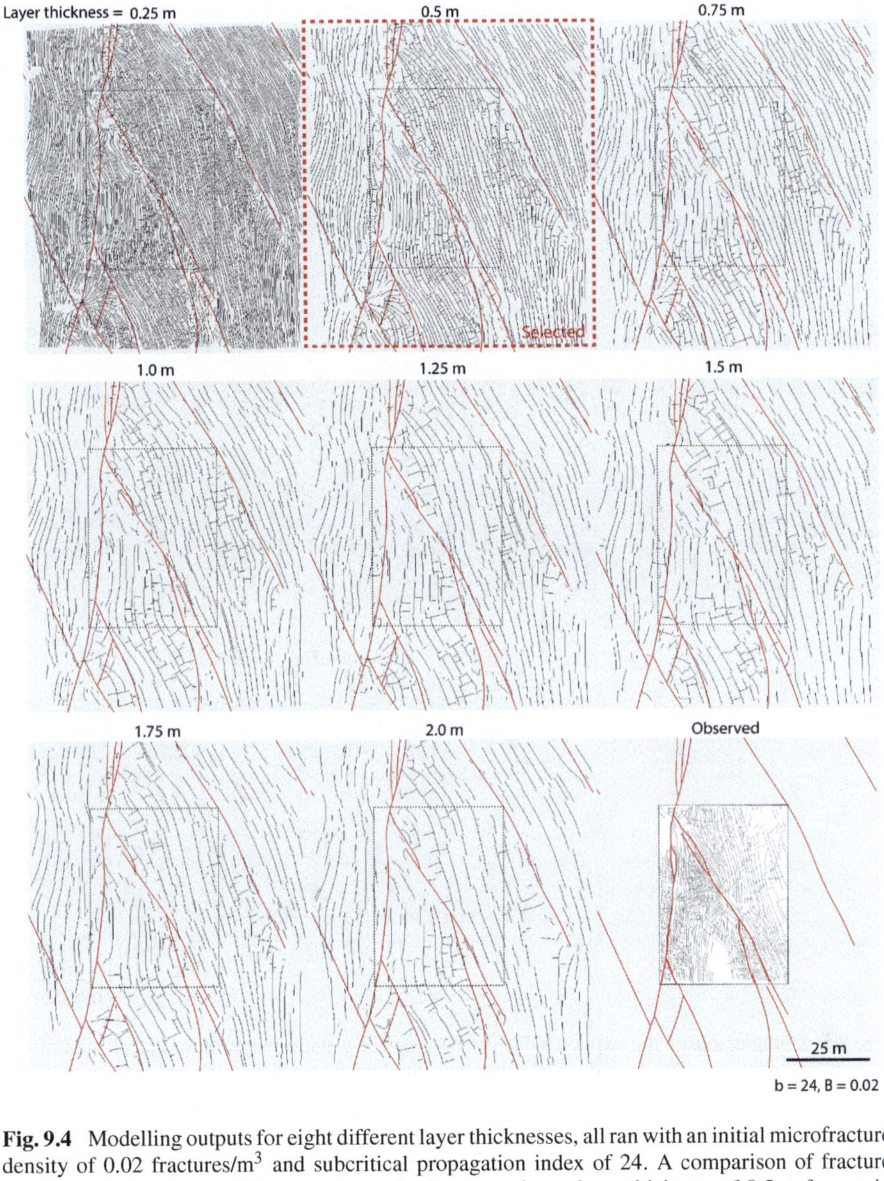

Fig. 9.4 Modelling outputs for eight different layer thicknesses, all ran with an initial microfracture density of 0.02 fractures/m^3 and subcritical propagation index of 24. A comparison of fracture spacing with the observed fracture pattern leads us to select a layer thickness of 0.5 m for use in subsequent modelling of this area

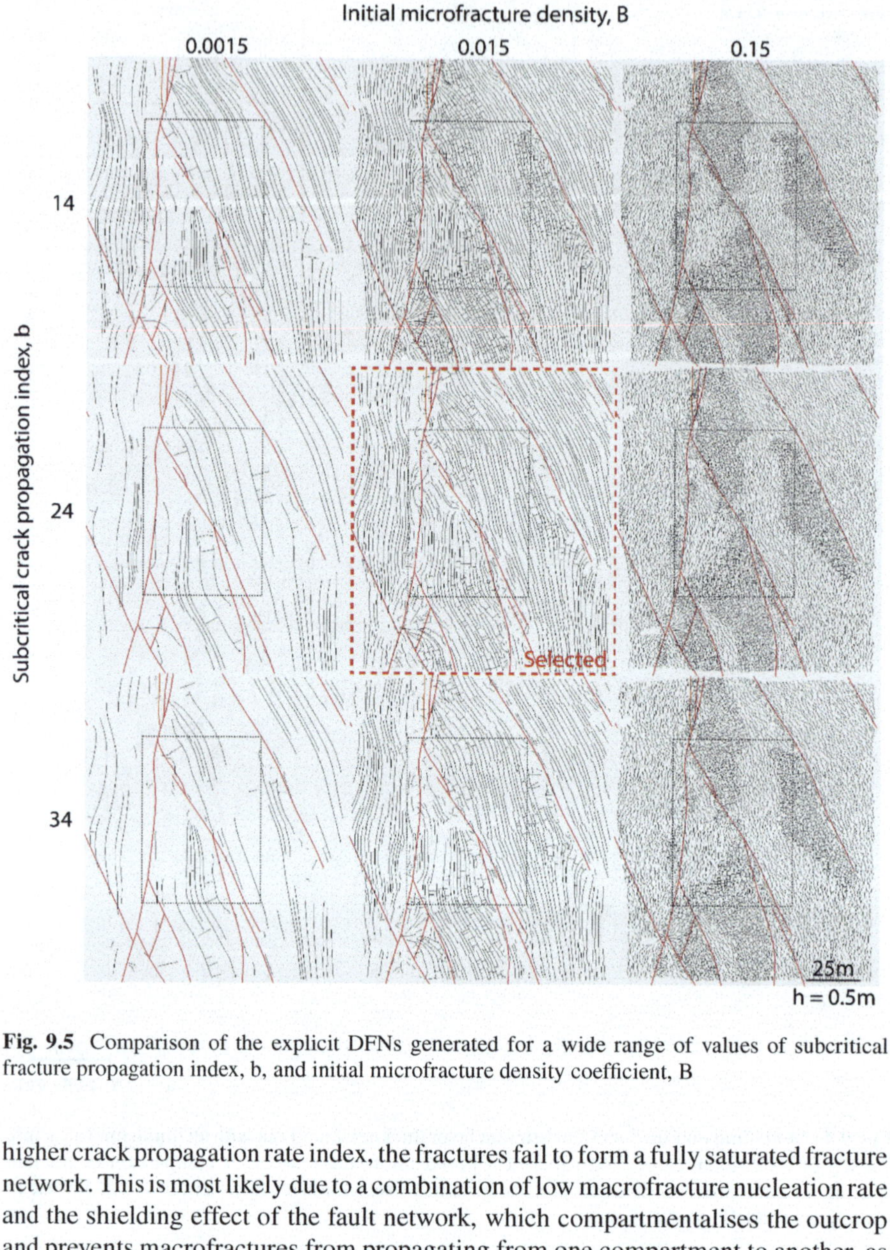

Fig. 9.5 Comparison of the explicit DFNs generated for a wide range of values of subcritical fracture propagation index, b, and initial microfracture density coefficient, B

higher crack propagation rate index, the fractures fail to form a fully saturated fracture network. This is most likely due to a combination of low macrofracture nucleation rate and the shielding effect of the fault network, which compartmentalises the outcrop and prevents macrofractures from propagating from one compartment to another, as described in Sect. 8.3.2. This illustrates the importance of selecting broadly correct values for these parameters.

Figure 9.6 shows the sensitivity of the resulting fracture network to smaller variations in these parameters. The impact of these smaller variations in the input parameters on the output DFNs is much smaller and all the models shown here broadly match

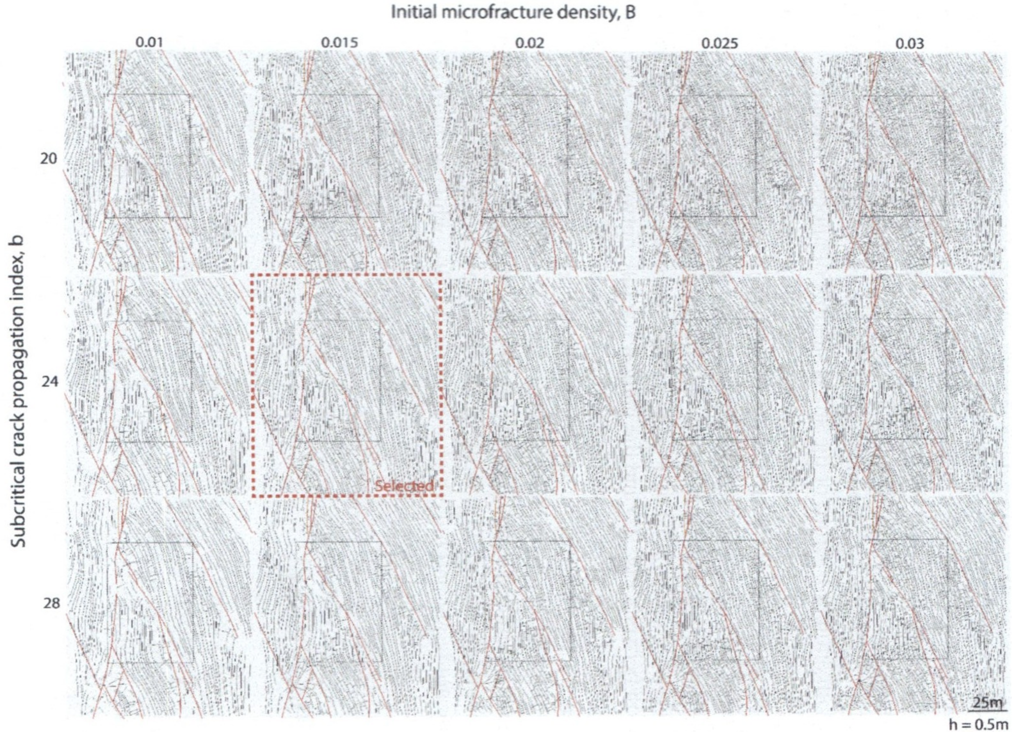

Fig. 9.6 Explicit DFNs generated to calibrate the values for the initial microfracture density coefficient, B, and subcritical fracture propagation index, b, that best match the observed fracture geometry. These models illustrate the spatial variation in the network anisotropy and connectivity as a result of the influence of these parameters on the fracture length distribution

the observed fracture geometry, although we can still identify the best-fit model with values of B = 0.015 and b = 24.

Large initial microfracture densities result in a greater number of fractures overall and particularly a greater number of small fractures between larger fractures in areas of preferable strain conditions. At lower densities, the fracture-length distribution is more heavily dominated by longer fractures.

Considering the subcritical crack propagation index, the difference between models is most clear when comparing the value of low initial microfracture density; those with a low propagation index feature less continuity in the longer fractures, while those with the highest indices have longer fractures that nearly span the length of the model. This is consistent with the principle that these fractures will have grown faster and with less microfracture initiation they will be less likely to terminate due to intersection with another fracture or the stress shadow of another fracture.

This type of spatial variation in the fracture intensity and relative prevalence of primary and secondary fractures is directly observed in outcrop, Fig. 9.3. Such variation in anisotropy and connectivity is expected to significantly influence the permeability and hydraulic conductivity of any fracture network. However, such variation is not considered by most approaches currently used in practice. Drawing visual comparison between Fig. 9.6 and the natural fracture network (Fig. 9.3b), we see a similar degree of variation in the fracture density across the image. Areas of greater and lesser fracture anisotropy and connectivity appear to correspond.

In areas of higher initial microfracture density, the subcritical propagation index acts to determine the duration over which secondary fractures may propagate before terminating through interaction with larger fractures. Thus, at low indices, areas of lower anisotropy are larger and have higher fracture density, while, at high indices, these areas are smaller with lower fracture density.

9.2 Robin Hood's Bay

Robin Hood's Bay is located on the coast of northeast England, between Scarborough and Whitby. The locality features a broad wave cut platform on the foreshore, extending up to 300 m out to sea at low tide. The location has continued to be of interest since early geological investigations (Kendall 1816; Bird and George 1822) and has proven to be a globally significant site for defining the stratigraphy of the Lower Jurassic (Hesselbo et al. 2000).

9.2.1 Description of the Outcrop

The beds of the study area comprise of interbedded layers of brittle limestone and ductile shales, with the limestone layers typically 0.5–1 m thick. The limestone beds

Table 9.2 Modelling parameters used for Robin Hood's Bay

Depth	1000 m
Maximum horizontal compressive stress, σ_H	15 MPa
Minimum horizontal compressive stress, σ_h	0.5 MPa
Vertical compressive stress, σ_v	30 MPa
Fluid overpressure	7.7 MPa
Young's modulus	10 GPa
Poisson's ratio	0.25
Biot's coefficient	1
Crack surface energy	1000 J/m^2
Friction coefficient	0.5
Critical fracture propagation rate	2000 m/s
Effective vertical stress	4.525 MPa
Initial micro-fracture density	0.001/m^2
Subcritical crack propagation index	10

calculate the overburden pressure and subsequently the Poisson's ratio to derive strain rates in the same manner as for Nash Point.

In this case, the local ratio of major to minor strain magnitude and the strain orientation is determined from the boundary element model, making this test purely a test of the approaches' ability to predict strain anisotropy.

In addition to mode 1 fracturing, other evidence for overpressure includes mineralised horizontal fractures along the coast (e.g. Emery 2016) and a mature hydrocarbon system evidenced by matured bituminous material in a local borehole drilled in 1949 (Fisons 1 Borehole, BGS ref: NZ90SW2, British National Grid: 494783, 504040).

Note that we have also applied strain relaxation in this model, with a relaxation time constant of 0.01 ma. The reason for this is that, as shown by Welch et al. (2019) and discussed in Sect. 8.5, strain relaxation enhances the contrast in fracture anisotropy that will result from variation in the applied horizontal strain since it reduces the horizontal stress and slows the rate of fracture propagation.

9.2.3 Model Results

We have run models for localities 3, 5 and 8 of Rawnsley et al. (1992), as illustrated in Fig. 9.9 and compared the results with the observed spatial variation in the anisotropy of the fracture network. At localities 3 and 8, we observed greater anisotropy in both the outcrop and the modelled fracture network with primary fracture sets oriented NNW-SSW and secondary fracture sets oriented ENE-WSW to E-W, while in locality 5, we see a much more isotropic fracture pattern with both fracture sets containing both long and short fractures. It is interesting to note that, in both the model results and

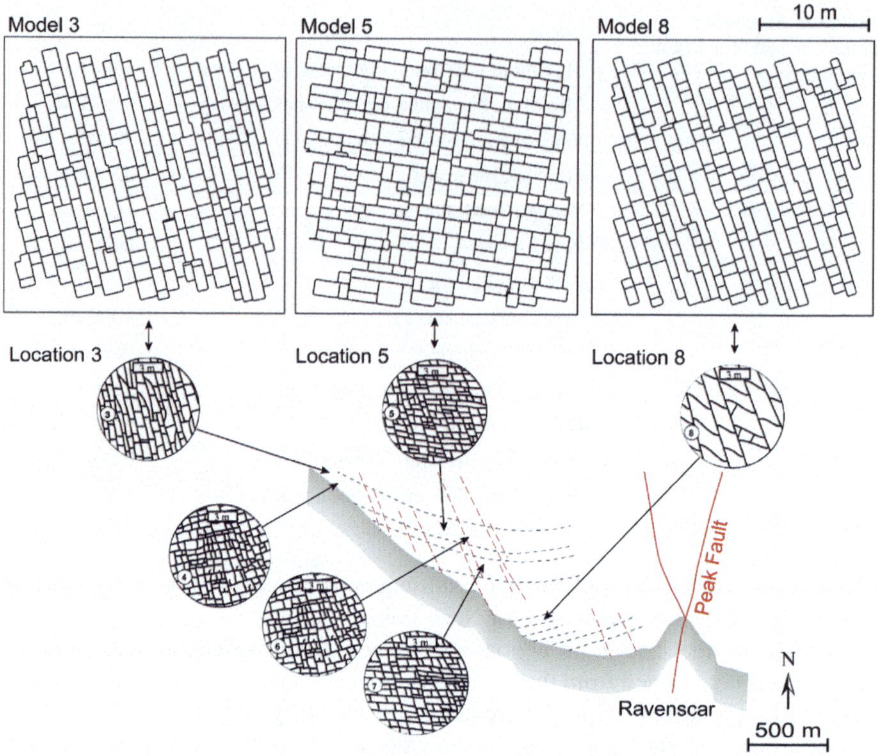

Fig. 9.9 Comparison of three modelled fracture outputs compared to the fracture patterns observed by Rawnsley et al. (1992). Figure modified from Rawnsley et al. (1992), their Fig. 3

the outcrop, we do see local anisotropy, with locally dominant fracture orientations. This is likely to be a result of the "shielding" effect described in Sect. 8.5.

We should also note that in several locations (especially location 8), the two fracture sets are slightly oblique suggesting a degree of strike-slip deformation. This is not replicated in the models since strike-slip displacement is not yet implemented in the algorithm; however, the modelling output continues to capture connectivity between major and minor fracture trends and the fact that they do truncate one another.

Overall, the Robin Hood's Bay model demonstrates our ability to build fracture models that represent small-scale variability from minimal large-scale observations. Considering only the large-scale structure of one fault that is observable in seismic offshore, we have been able to generate a large-scale model that predicts fracture orientation and smaller-scale models that capture natural variability and the observed fracture network anisotropy.

The scale of this example is similar to many subsurface problems, where a small number of large controlling features are likely to be well constrained; however, the variability of interest is occurring at a much smaller scale. In these examples, the inset models are equivalent in size to the immediate area surrounding a borehole. We suggest that this implies that the level of detail of the models should be sufficient to enhance studies investigating flow between boreholes as the expected deviations are likely to average out at a larger scale. Work to investigate how these models may be used to consider representative elementary volumes of fractured systems is underway.

9.3 Pegwell Bay

Pegwell Bay in southeast England comprises a cliff section and wavecut platform exposing Coniacian to Santonian age chalk from the Upper Chalk Formation (Shepherd-Thorn 1988). The cliff section trends E-W and is approximately 20–30 m high and 2 km long, although the best exposure of the wavecut platform is found at the easternmost part, where it is up to 80 m wide. The outcrop lies on the southern limb of the Thanet anticline, a drape fold interpreted as the result of differential compaction over a large pre-existing south-dipping normal fault. This fault was subsequently reactivated in strike-slip by a period of Neogene to Quaternary NE-SW-oriented extension (Vandycke 2002).

9.3.1 Description of the Outcrop

The chalk in this area has a typical porosities of 35% or greater, and an inferred maximum burial depth of c.500 m (Carter and Mallard 1974; Jones 1999). It contains many different types of deformation features, indicating multiple deformation stages. These range from deformation bands and flint-filled fractures, interpreted as early compaction-related deformation features (Souque et al. 2019), to brittle open fractures and faults with plume and hackle structures or slickensides, respectively, interpreted as post-lithification features formed during the late deformation episodes (Ameen 1995).

The easternmost end of the outcrop was mapped in detail by Souque et al. (2019), as shown in Fig. 9.10. A number of NNW-SSE striking faults, which are exposed in the cliff section but can also be traced along with the wavecut platform, cut the area. These are generally steep (dips range from 60° to vertical) and extend throughout the height of the cliff. These faults show a normal offset of 20 cm–1 m, but slickensides in the cliff section suggest that the predominant displacement on these faults, at least in the latest stage, was left-lateral strike-slip. Souque et al. (2019) infer that these faults were reactivated in strike-slip during the latest Neogene to Quaternary age NE-SW-oriented extension.

A number of brittle open fractures in this outcrop cluster together to form fracture corridors. These corridors are typically 1–5 m wide and up to 100 m long and contain parallel unfilled open fractures, interpreted as Mode 1 fractures, which are spaced

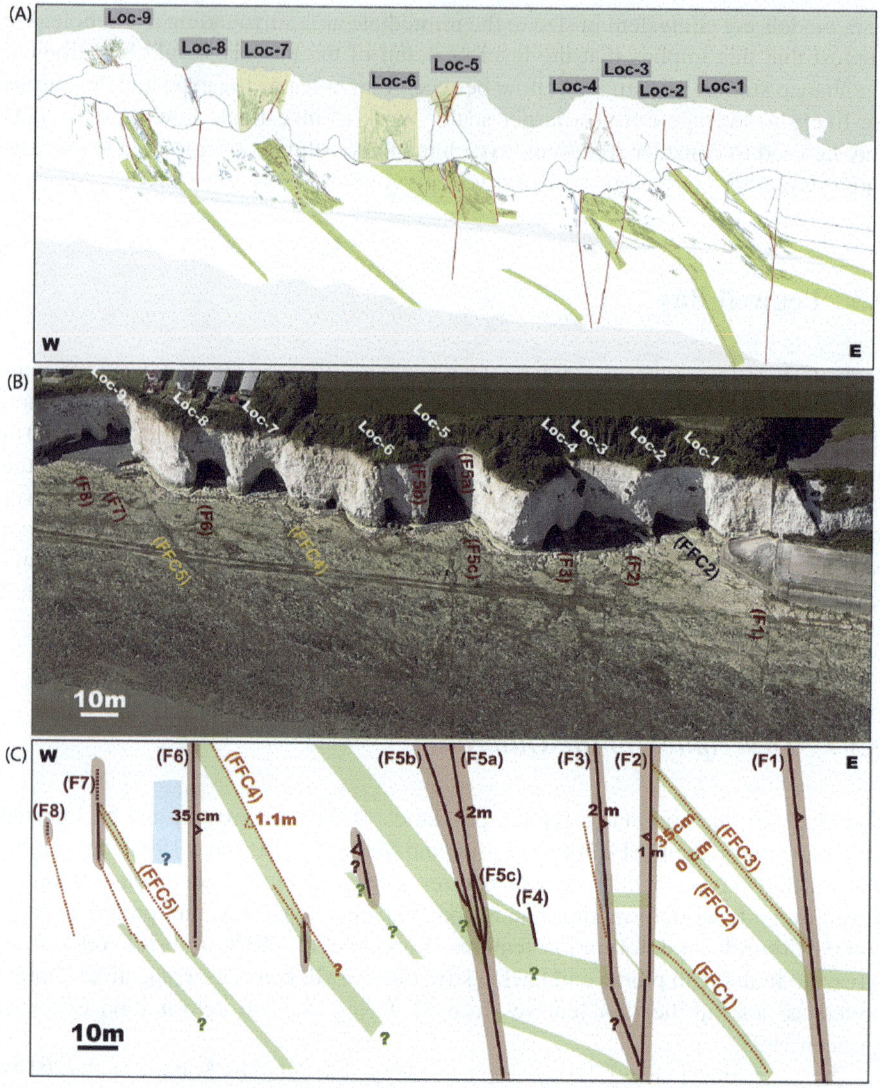

Fig. 9.10 Field observations (**a**), aerial photo (**b**) and simplified map (**c**) showing the larger strike-slip faults (in red) and the fracture corridors (in green) at the eastern end of the Pegwell Bay outcrop, from Souque et al. (2019), their Fig. 7. Fault displacements are shown where these could be estimated. Areas marked? were difficult to access or obscured by drift, and are therefore mapped with lower confidence

Fig. 9.11 A fracture corridor in the wavecut platform at Pegwell Bay. The green lines indicate the edge of the fracture corridor, which is approximately 2 m wide

from 2 to 20 cm apart (Fig. 9.11). The corridors often terminate at tips, bend and splay in the larger strike-slip faults but are oblique to them; they generally strike NW-SE, although their orientation may vary in the immediate vicinity of the larger faults (Souque et al. 2019). They thus provide potential flow pathways between the faults.

Other brittle fractures form localised damage zones around the faults, often with complex geometries. Finally, there is a population of background fractures outside the fracture corridors, although the density of these fractures is much lower than in the fracture corridors, so it is easy to delineate the corridors.

The stratigraphy is fairly uniform throughout the cliff section. However, bed-parallel joints are common and form mechanical boundaries typically 1–2 m apart. While the larger faults cut across them, most of the open fractures that comprise the fracture corridors terminate at these bed-parallel joints and thus can be considered as layer-bound fractures (Ameen 1995). Note, however, that the fracture corridors themselves may extend much further through the stratigraphy.

Mechanical modelling using boundary element, finite element and discrete element models suggests that the fracture corridors nucleated at local stress concentrations around the tips, bends and splays in the larger strike-slip faults, and then propagated outwards following the regional stress orientation. This modelling also suggests that they formed in a highly overpressured environment (Welch et al. 2015; Eshiet et al. 2018).

9.3.2 Building the Model

We built a static geomodel comprising a single planar mechanical layer, 1 m thick, and extending across the area mapped in Fig. 9.10. The model included all the mapped strike-slip faults, which were extended 30 m above and below the top of the modelled layer (Fig. 9.12). The fault dips were based on the measurements from the cliff section, extrapolated across the wavecut platform. The layer was then discretised into 1 m × 1 m gridblocks, giving a total of 150 × 80 gridblocks.

Since no direct measurements of mechanical properties from Pegwell Bay were available, the model was populated with uniform mechanical property values typical of chalk from southeast England, from Bell et al. (1999). Young's Modulus was set to 5 GPa, Poisson's ratio to 0.25, friction coefficient to 0.5 and crack surface energy to 1000 J/m^2.

The local stress and elastic strain around the faults were calculated using the boundary element method (Maerten 1999). Based on Vandycke (2002), we assume a uniaxial regional strain, with $\dot{\varepsilon}_{hmin}$ oriented at 045° and a strain rate of −0.002/ma. This generates left-lateral strike-slip movement on the faults and large stress and strain anomalies around the fault tips, bends and splays, as shown in Fig. 9.12. We assume a burial depth of 500 m but a fluid overpressure of 6 MPa, resulting in a very low effective vertical stress of 0.13 MPa, and we assume a critical initial horizontal stress (see Sect. 6.1).

The main objective in this example is to replicate the fracture corridors observed in outcrop. To do this, we need to ensure that sufficient layer-bound macrofractures are able to nucleate in the high strain zones around the larger faults to act as seeds for the fracture corridors and that these are then able to propagate rapidly across the regions of lower strain between the larger faults. At the same time, we must minimise the rate

Fig. 9.12 Calculated horizontal strain rates and orientation in Pegwell Bay, resulting from regional NE-SW extension with local variation around the larger mapped faults. The pink lines show the direction of maximum extensional strain (i.e. $\dot{\varepsilon}_{hmin}$, since the extensional strain is negative); the length of the lines is proportional to the strain magnitude. The inclined planes show the main faults, as mapped on Fig. 9.10a

of macrofracture nucleation outside of the high strain zones around the larger faults, to reduce the population of background fractures. We, therefore, ran several models varying the subcritical fracture propagation index b and the initial microfracture density coefficient B to determine which values could best reproduce the situation described above.

A further constraint on the input parameters is provided by the observation that the fractures comprising the fracture corridors are Mode 1 fractures, thus implying that a subcritical fracture propagation index b $\geq$ c.20 (based on the results in Sect. 8.4).

There is, however, a key difference between this example and the Nash Point and Robin Hood's Bay examples: the spacing of fractures in the fracture corridors (2–20 cm) is much smaller than the width of the stress shadows that would be expected in a 1 m thick layer (c.1.5 m for Mode 1 fractures). We, therefore, used the evenly distributed stress scenario described in Sect. 6.2 in these models. In this scenario, the strain accommodated by displacement on the fractures results in a reduction in stress throughout the layer, rather than just in a narrow stress shadow zone around each fracture. There is therefore no limit on the spacing between adjacent fractures.

9.3.3 Model Results

In Fig. 9.13, we examine in detail the development of a fracture corridor nucleating at the tip of one of the strike-slip faults, for different values of subcritical fracture propagation index b and initial microfracture density coefficient B. To some extent, all these models are able to replicate the development of the fracture corridors, as they all show clustering of the layer-bound macrofractures into elongated zones which mostly extend outwards from the tips of the strike-slip faults. The subcritical propagation index b does not have a great effect on the model results, as long as it is sufficiently high to allow the development of Mode 1 fractures; similar fracture network geometries are observed for b = 20 (Fig. 9.13a, c, e) and b = 50 (Fig. 9.13b, d, f). Matching the observed fracture density is mostly achieved by varying the initial microfracture density coefficient B. However, it is difficult to match the observed fracture density both inside and outside the corridors. The best matches for the fracture density within the corridors is obtained when B = 0.01 (Fig. 9.13a, b); however, this gives fracture corridors that are wider than those observed in outcrop. A high value of B also causes a large number of layer-bound macrofractures to nucleate outside of the high-stress zones around the faults, and hence the background fracture density is higher in these models. We can reduce the fracture corridor width and prevent the nucleation of fractures outside the high-stress zones by reducing B to 0.001 (Fig. 9.13e, f); however, this will also reduce the number of fractures nucleating within the high-stress zones, and hence the density of fractures within the fracture corridors will be too low. The best trade-off is achieved with B = 0.003 (Fig. 9.13c, d), which gives a reasonable match for the observed fracture corridor width.

However, a slightly better match can be obtained if we allow the initial microfracture density to vary laterally, so it is higher in the high-stress zones around the

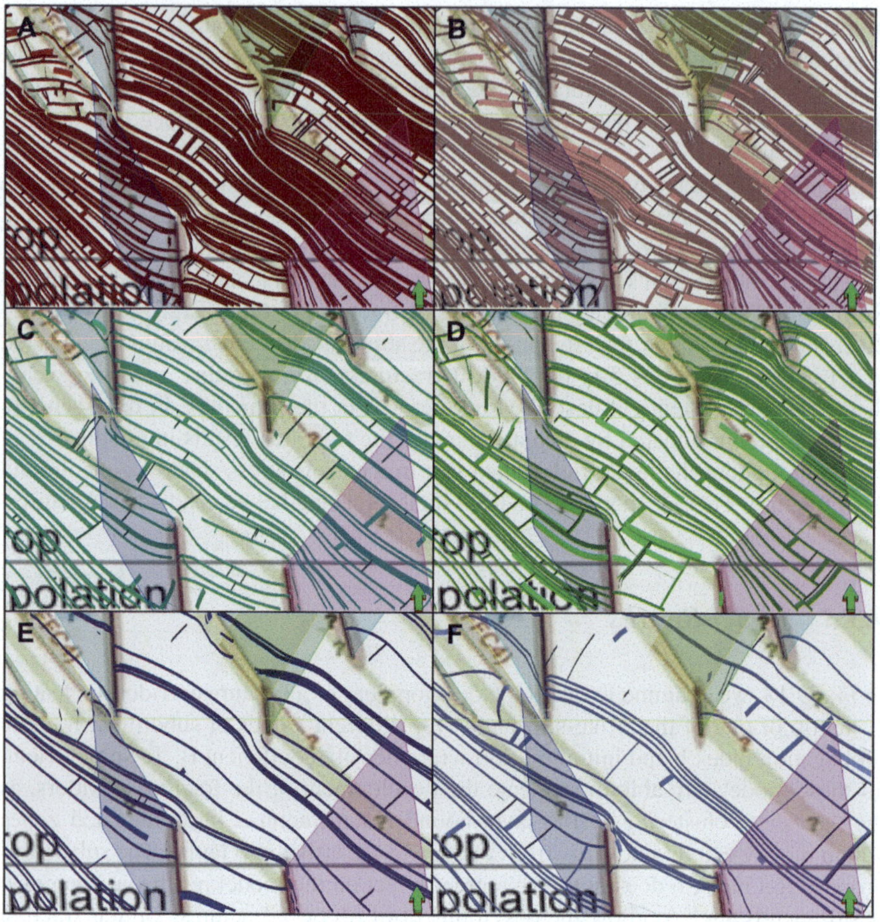

Fig. 9.13 Close-ups showing the development of a fracture corridor around the tip of one of the strike-slip faults after 1 ma, with b = 50, B = 0.01 (**a**), b = 20, B = 0.01 (**b**), b = 50, B = 0.003 (**c**), b = 20, B = 0.003 (**d**), b = 50, B = 0.001 (**e**), and b = 20, B = 0.001 (**f**). The green shading indicates the fracture corridors mapped by Souque et al. (2019)

strike-slip faults and lower in the areas between the faults. A number of factors could cause this, such as damage caused by movement on the faults or brecciation and fracturing due to high-pressure fluids injected along the faults. We, therefore, ran an additional model in which the initial microfracture density is based on the minimum horizontal strain rate $\dot{\varepsilon}_{\mathrm{hmin}}$:

$$B = 10^5 \times (-\dot{\varepsilon}_{\mathrm{hmin}})^3 \tag{9.1}$$

This gives B values ranging from c.0.0008 away from the strike-slip faults to c.0.1 in the high-stress zones around the faults (Fig. 9.14).

Nemčok, M., R. Gayer, and M. Miliorizos. 1995. Structural analysis of the inverted Bristol Channel Basin: Implications for the geometry and timing of fracture porosity. *Geological Society Special Publication* 88: 355–392. https://doi.org/10.1144/GSL.SP.1995.088.01.20.

Peacock, D.C., and D.J. Sanderson. 1991. Displacements, segment linkage and relay ramps in normal fault zones. *Journal of Structural Geology* 13 (6): 721–733. https://doi.org/10.1016/0191-8141(91)90033-F.

Rawnsley, K.D., T. Rives, J.-P. Petit, S.R. Hencher, and A.C. Lumsden. 1992. Joint development in perturbed stress fields near faults. *Journal of Structural Geology* 14: 939–951.

Rives, T., M. Razack, J.-P. Petit, and K.D. Rawnsley. 1992. Joint spacing: Analogue and numerical simulations. *Journal of Structural Geology* 14: 925–937.

Roberts, J.C. 1974. Jointing and minor tectonics of the Vale of Glamorgon between Ogmore-By-Sea and Lavernock Point, South Wales. *Geological Journal* 9 (2): 97–114.

Shephard-Thorn, E.R. 1988. *Geology of the country around Ramsgate and Dover*. Memoir for 1:50000 Geological Sheets 274 and 290 (England and Wales). British Geological Survey.

Skar, T., S.S. Berg, R.H. Gabrielsen, and A. Braathen. 2017. *Fracture networks of normal faults in fine-grained sedimentary rocks: Examples from Kilve Beach, SW England*, vol. 439, no. 1, 289–306. Geological Society. London: Special Publication. https://doi.org/10.1144/SP439.10.

Souque, C., R.J. Knipe, R.K. Davies, P. Jones, M.J. Welch, and J. Lorenz. 2019. Fracture corridors and fault reactivation: Example from the Chalk, Isle of Thanet, Kent, England. *Journal of Structural Geology* 122: 11–26.

Tappin, D.R., R.A. Chadwick, A.A. Jackson, R.T.R. Wingfield, and N.J.P. Smith. 1994. United Kingdom offshore regional report: The Geology of Cardigan Bay and the Bristol Channel. HMSO for the British Geological Survey.

Vandycke, S. 2002. Palaeostress records in Cretaceous formations in NW Europe: Extensional and strike-slip events in relationships with Cretaceous-Tertiary inversion tectonics. *Tectonophysics* 357: 119–136.

Welch, M.J., C. Souque, R.K. Davies, and R.J. Knipe. 2015. Using mechanical models to investigate the controls on fracture geometry and distribution in chalk. In *Fundamental controls on fluid flow in carbonates*, ed. S.M. Agar, and S. Geiger, vol. 406, 281–309. Geological Society. London: Special Publications.

Welch, M.J., M. Lüthje, and A.C. Glad. 2019. Influence of fracture nucleation and propagation rates on fracture geometry: Insights from geomechanical modelling. *Petroleum Geoscience* 25: 470–489.

Zwart, H.J. 1951. Breuken en diaklazen in Robin Hood's Bay (England). *Geologie en Mijnbouw* 13 (1): 1–4.

Chapter 10
Application to the Subsurface

Abstract In this chapter, we apply our method to two subsurface examples: the Kraka oilfield in the Danish North Sea and an onshore geothermal prospect in the Drenthe province in the Netherlands. Production from the Kraka field is highly dependent on the natural fracture network of the chalk reservoir. We test the method using two sets of strain input data, one based on a regional extension and one based on detailed backstripping and show how we can use the geometry of the salt pillow underlying the Kraka field, and the large-scale faults observed in seismic data, to create large-scale, detailed fracture models covering the entire field. We demonstrate a good match to the fractures observed on calibration data such as core, borehole images and high-quality three-dimensional seismic volumes. The Drenthe prospect has limited calibration data, so we generate a set of multiple geologically realistic fracture network models encompassing the uncertainty that can be used in scenario-based risk analysis. We then examine lateral variability in the resulting fracture networks and to identify areas that are more likely to develop more or less anisotropic fracture networks, and areas that are likely to have longer or shorter fractures. Finally, we consider the implications for fluid flow within the geothermal prospect.

Keywords Fracture modelling · Discrete fracture networks · Naturally fractured reservoir · Reservoir modelling · Geothermal energy · Geomechanics

Calibrating and validating subsurface fracture models are usually much more difficult since the calibration data is often very limited compared to outcrops. Typically, the only direct information about fractures in the subsurface comes from wells (either from core or borehole images). On the scale of our models, well data is essentially one-dimensional and gives information only on mean linear fracture density, fracture orientation and fracture spacing distribution. There will be no direct information on fracture length, height, geometry or connectivity: even the best quality seismic data is unable to resolve individual fractures, although advanced processing and interpretation techniques, such as ant-tracking (Pederson et al. 2002) and 4D seismic, may reveal large fracture corridors or small (metre-scale) throw faults.

© The Editor(s) (if applicable) and The Author(s), under exclusive license to Springer Nature Switzerland AG 2020

M. J. Welch et al., *Modelling the Evolution of Natural Fracture Networks*,
https://doi.org/10.1007/978-3-030-52414-2_10

However, the subsurface is also where accurate fracture models are most useful, to understand and simulate fluid flow through reservoir or aquifer units, or to assess the stability of mines, tunnels and other engineering structures. A good approach is, therefore, to use appropriate outcrop analogues to calibrate the key lithological parameters (especially subcritical propagation index, initial microfracture density and Young's Modulus) and then use fracture density and orientation data from wells to calibrate the total strain applied in the subsurface models. It may also be possible to use well test or tracer data for further indirect calibration of the fracture network.

Here, we will present two examples of subsurface models: an oilfield producing from a fractured chalk reservoir, offshore Denmark, and a geothermal prospect in fractured limestone onshore in the Netherlands.

10.1 The Kraka Oilfield, Offshore Denmark

The Kraka oilfield is located in the Danish sector of the North Sea, within the Central Graben, and is operated by Total E&P Denmark (TEPDK). The structure comprises an elongated salt pillow, oriented NE-SW, approximately 8 km by 5 km in size (Rasmussen et al. 2005). Production is mainly from the Danian age Ekofisk Formation (Jorgensen and Anderson 1991), a fractured chalk with porosity ranging from 20 to 35% and matrix permeability typically <3mD (Klinkby et al. 2005). The Ekofisk Formation is overlain by Palaeogene shales, which provide a seal to the Kraka oilfield. It is underlain by more chalk from the Maastrichtian age Tor Formation; however, these units are separated by a hard ground that appears to act as a flow barrier. We can, therefore, infer that most of the fractures in the Ekofisk Formation do not extend either up into the Palaeogene shales or down into the Tor Formation and can be considered as layer-bound fractures.

10.1.1 Calibration Data

Borehole images (specifically resistivity images) are available from seven wells in the Ekofisk Formation, of which five are horizontal. Core is also available from four of these wells, as well as a further four wells without borehole images. Good quality seismic data (including 4D) is also available across the entire structure and was used by TEPDK to construct a static geomodel, including a number of seismic-scale faults, which strike predominantly NNE-SSW. Finally, mechanical property tests (Brazilian tensile strength tests, uniaxial compression tests and multistage triaxial compression tests) were carried out on ten core samples with varying porosity.

Fracture logging of both core and borehole images was carried out by Aabø et al. (2019). The majority of fractures identified on borehole images are conductive and are interpreted as open fractures. However, a number of resistive fractures are also identified; these are interpreted as cemented fractures. The majority of fractures observed in the core are small (with either one or both tips lying within the core, and lengths in the range 10–50 cm). These fractures are often associated with other features such as stylolites or chert nodules (see Fig. 10.1a, b), and may have formed

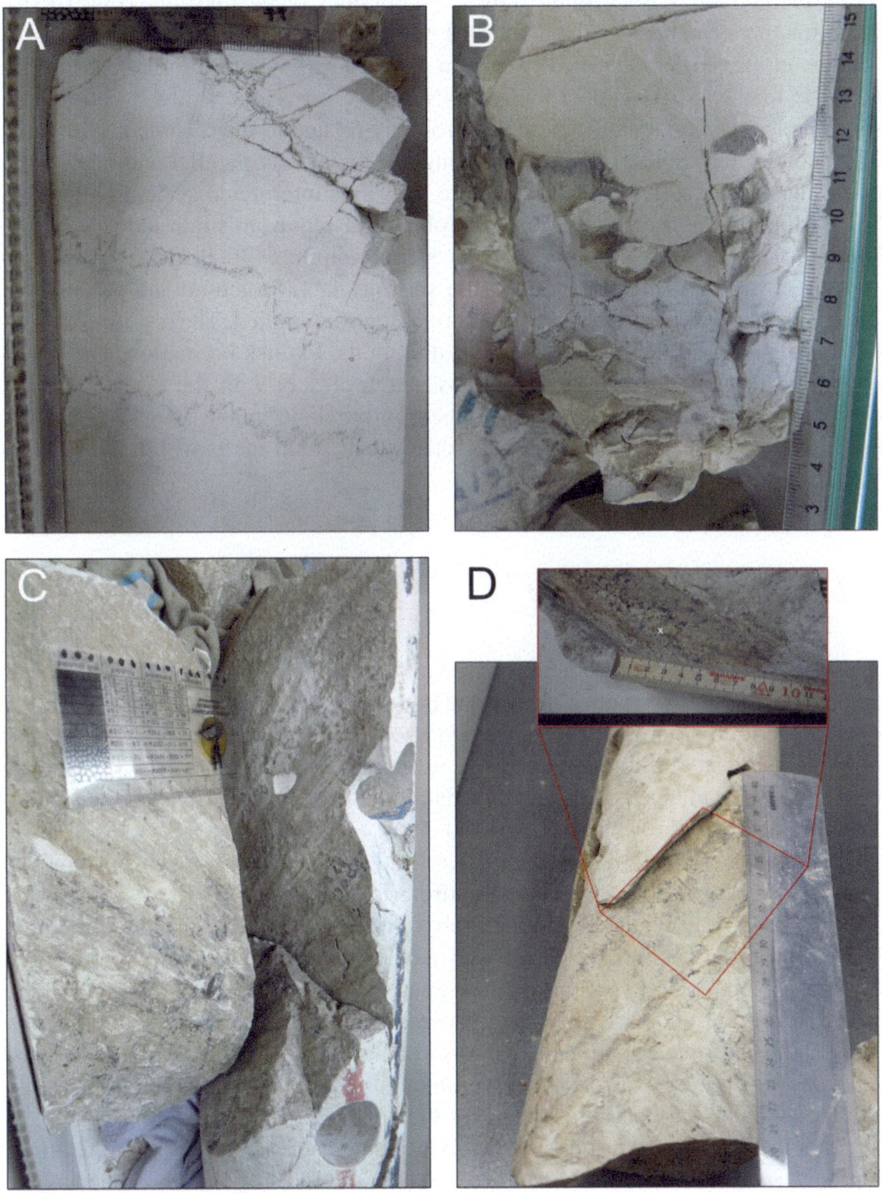

Fig. 10.1 Examples of small fractures associated with a stylolite (**a**) and a chert nodule (**b**), and large fractures showing slickensides (**c**) and a brecciated zone (**d**) in core from the Kraka field

as a result of local stress concentrations that developed around these features during compaction or diagenesis. However, there are also a number of larger fractures that cut completely across the core, forming core breaks. Crystalline calcite cement and/or slickensides are often observed on the walls of these fractures (Fig. 10.1c) and, in one case, the fracture is surrounded by a 2–3 cm thick brecciated zone (Fig. 10.1d). These fractures can be correlated with the resistive fractures on borehole images.

Aabø et al. (2019) identified two main fracture sets within the Ekofisk Formation. The first set strikes consistently c.NNE-SSW across the entire structure, and is interpreted as a regional fracture set, probably related to the seismic-scale faults. The second set comprises fractures striking approximately parallel and perpendicular to the contours of the Kraka structure. It is thus interpreted as being related to the development of the salt pillow and probably results from strain induced by the doming. Most fractures from both sets are steeply-dipping (70–90°).

Aabø et al. (2019) also examined the seismic data. Structural sharpening was carried out using eXchroma, and then ant-tracking was applied. The resulting dataset reveals a detailed structure of lineations within in the Ekofisk Formation. The orientation and distribution of these lineations match closely with the orientation and distribution of the resistive fractures logged on borehole images. These lineations may, therefore, represent large fracture corridors or small faults within the Ekofisk Formation.

10.1.2 Building the Model

The fracture model presented here is based on the static geomodel of the Kraka field built by TEPDK. In this model, the top and bottom surfaces of the Ekofisk Formation were mapped across the entire Kraka structure on three-dimensional seismic data and calibrated against well tops. In addition, several internal horizons and 32 seismic-scale faults were mapped on the seismic data. The faults are mostly steeply-dipping normal faults, with throws in the range 10–50 m.

The Ekofisk Formation was then discretised into a pillar grid with gridblocks approximately 50 × 50 m, so that the entire grid contains c.150 × 200 gridblocks. This grid was then populated with petrophysical data, including porosity, based on seismic inversion calibrated against well data. The porosity property shows both stratigraphic and lateral variation, with highest porosities (up to 40%) found in the upper Ekofisk Formation. It also reveals several very low-porosity patches at the top of the Ekofisk Formation, typically c.0.5–1 km in size, which may be tight cemented zones. However, these patches are very thin (c.1 m thick) and are underlain by chalk with the much higher porosities typical of the upper Ekofisk Formation.

Using the mechanical test data, it was possible to establish an empirical relationship between porosity and Young's Modulus for the Ekofisk chalk, and hence, we were able to use the porosity property data to calculate Young's Modulus across the Kraka structure (Frederic Amour pers. comm.). The upper Ekofisk chalk typically has moderate stiffness values of c.2–2.5 GPa, increasing to c.7–8 GPa in the lower

Ekofisk Formation, reflecting the lower porosity of the chalk here (Fig. 10.2). The low-porosity patches at the top of the Ekofisk may have stiffness values up to 16 GPa, but as discussed above these are very thin so will probably have only a minor effect on the overall stiffness of the formation in these areas. To represent the overall elastic response of the Ekofisk Formation, we, therefore, averaged the stiffness vertically within each gridblock. Note also that the Ekofisk Formation is underlain by a thin (c.1 m thick) layer of very high stiffness (up to 28 GPa) representing the top Maastrichtian hardground (shown in red on the cross-section on Fig. 10.2).

Other mechanical properties (Poisson's ratio, friction coefficient and crack surface energy) either could not be determined or showed a less clear relationship with porosity, so we populated the entire grid with uniform values for these properties, consistent with the mechanical test results: 0.25 for Poisson's ratio, 0.5 for friction coefficient and 1000 J/m^2 for crack surface energy. Based on the results from the outcrop models, the subcritical fracture propagation index b was set to 20, the initial microfracture density B to 0.001 and the initial microfracture distribution coefficient c to 2.

As noted above, the fractures observed in well data are thought to form either in response to regional ESE-WNW-oriented extension, at the same time as the seismic-scale faults developed, or in response to local strain related to the growth of the Kraka salt pillow. We, therefore, ran models simulating both types of deformation to compare with the observed fracture patterns. In both cases, we assumed fractures

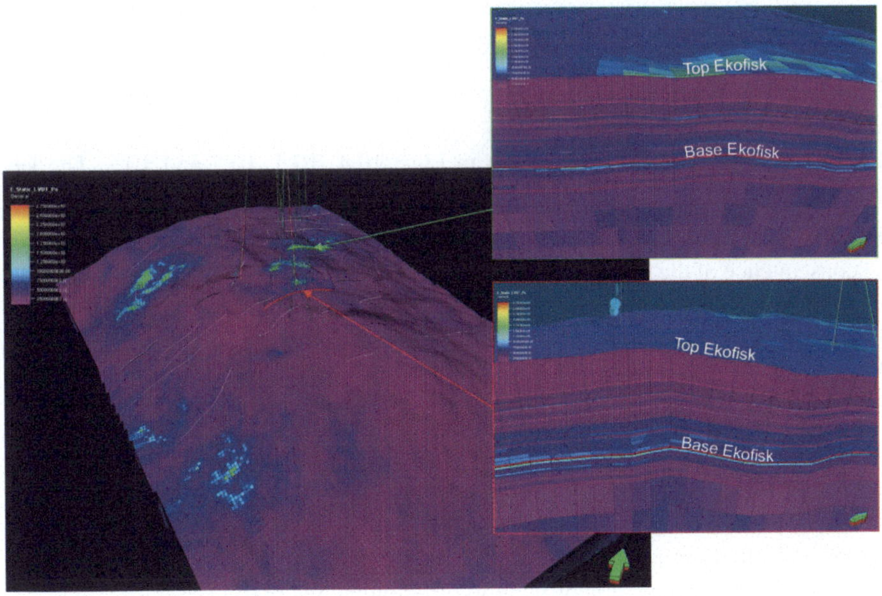

Fig. 10.2 Young's Modulus calculated from porosity, using an empirical relationship derived from mechanical property test data. The Kraka structure is shown with 5× vertical exaggeration

developed at approximately the current depth of burial (c.1750–1950 m) and with hydrostatic fluid pressure.

To model fracture development in response to regional ESE-WNW-oriented extension, we applied a uniaxial horizontal strain at a rate of 0.005/ma oriented at 120°; this represents a typical strain rate for an extensional basin (Welch et al. 2019). However, since the seismic-scale faults were also active at this time, we used the boundary element method to calculate the local variation in elastic strain around the faults, in response to this regional strain (see Maerten 1999). Thus, the horizontal strain magnitude and orientation vary locally around the seismic-scale faults in these models (Fig. 10.3a).

A group at Aarhus University, led by Ole Rønø Clausen and Kenni Petersen, calculated the strain related to the growth of the salt pillow. They carried out detailed backstripping of the Tertiary strata overlying the Ekofisk Formation, using multiple intra-Tertiary horizons picked on the three-dimensional seismic data. They have developed an algorithm to calculate the evolution of the strain tensor as the overburden layers are progressively removed and the sedimentary column decompacted. This is done at regularly spaced points on the model grid to generate strain tensors across the entire structure, revealing the local variations in strain orientation and magnitude (Fig. 10.3b). Sensitivity testing has shown that the three-dimensional strain tensor calculated using this method is equivalent to strain tensors calculated using full three-dimensional restoration in equivalent areas (Petersen and Clausen 2018).

10.1.3 Model Results

The results of the models are shown in Figs. 10.4, 10.5, 10.7 and 10.8. The key observations are:

- Both models predict the development of Mode 2 inclined shear fractures. This is a consequence of the high vertical effective stress, resulting from the depth of burial and the lack of fluid overpressure. This is consistent with the inclined fractures with slickensides observed in core and borehole images.
- Away from the seismic-scale faults, the DFN generated by horizontal strain related to the growth of the salt pillow shows a good match to the resistive fractures observed on borehole images, in terms of both orientation and density (see Fig. 10.8). This suggests that the resistive fractures represent large, layer-bound shear fractures and that away from the faults, these are mostly generated by strain related to the growth of the salt pillow (although it should be noted that in some of the wells, this strain has the same orientation as the ESE-WNW-oriented regional extension, and hence, it is difficult to differentiate the two).
- However, close to the seismic-scale faults, the DFN generated by local fault-related strain related to the regional ESE-WNW extension gives the closest match to the resistive fractures observed on borehole images (see Fig. 10.7).

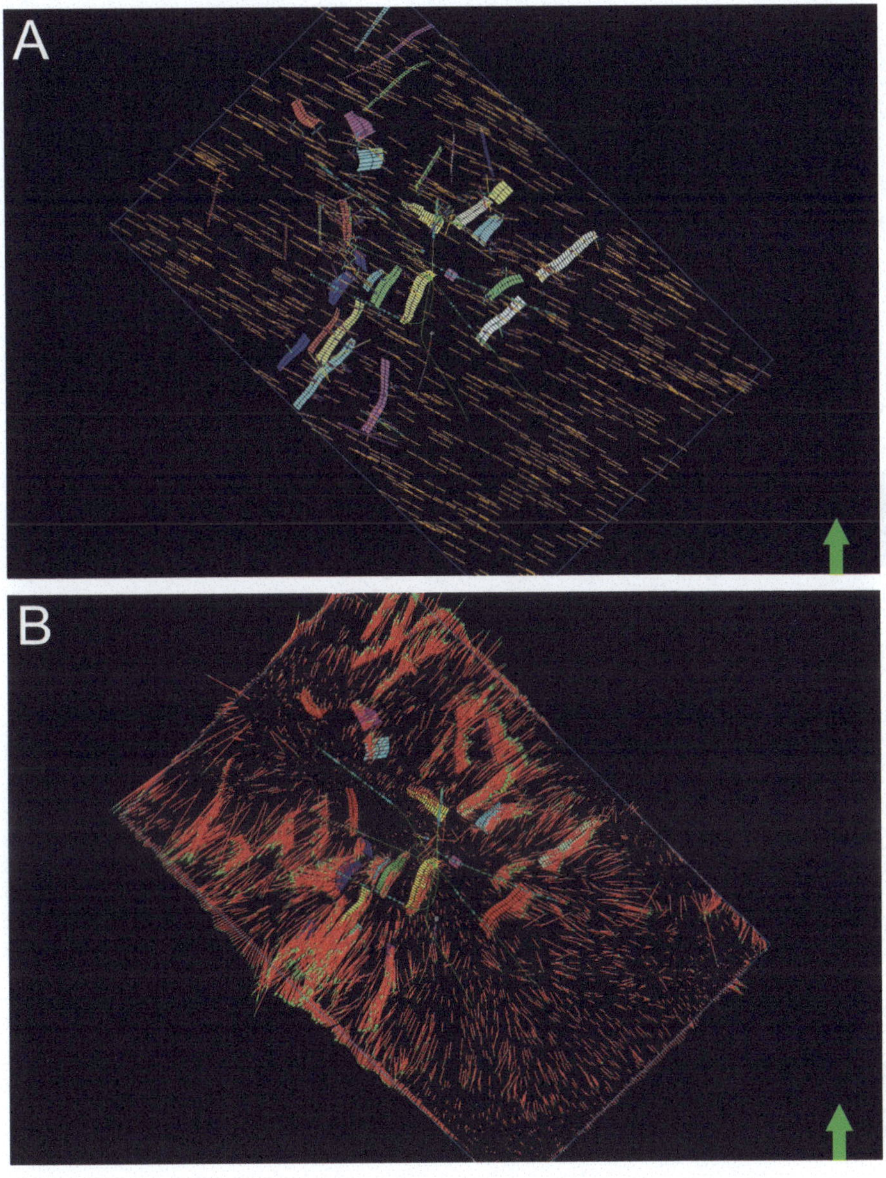

Fig. 10.3 Calculated horizontal strain rates and orientation related to regional ESE-WNW extension with local variation around the seismic-scale faults (**a**) and growth of the salt pillow (**b**). The arrows show the direction of maximum extensional strain (i.e. $\varepsilon_{h\min}$, since the extensional strain is negative); the length of the arrows is proportional to the strain magnitude

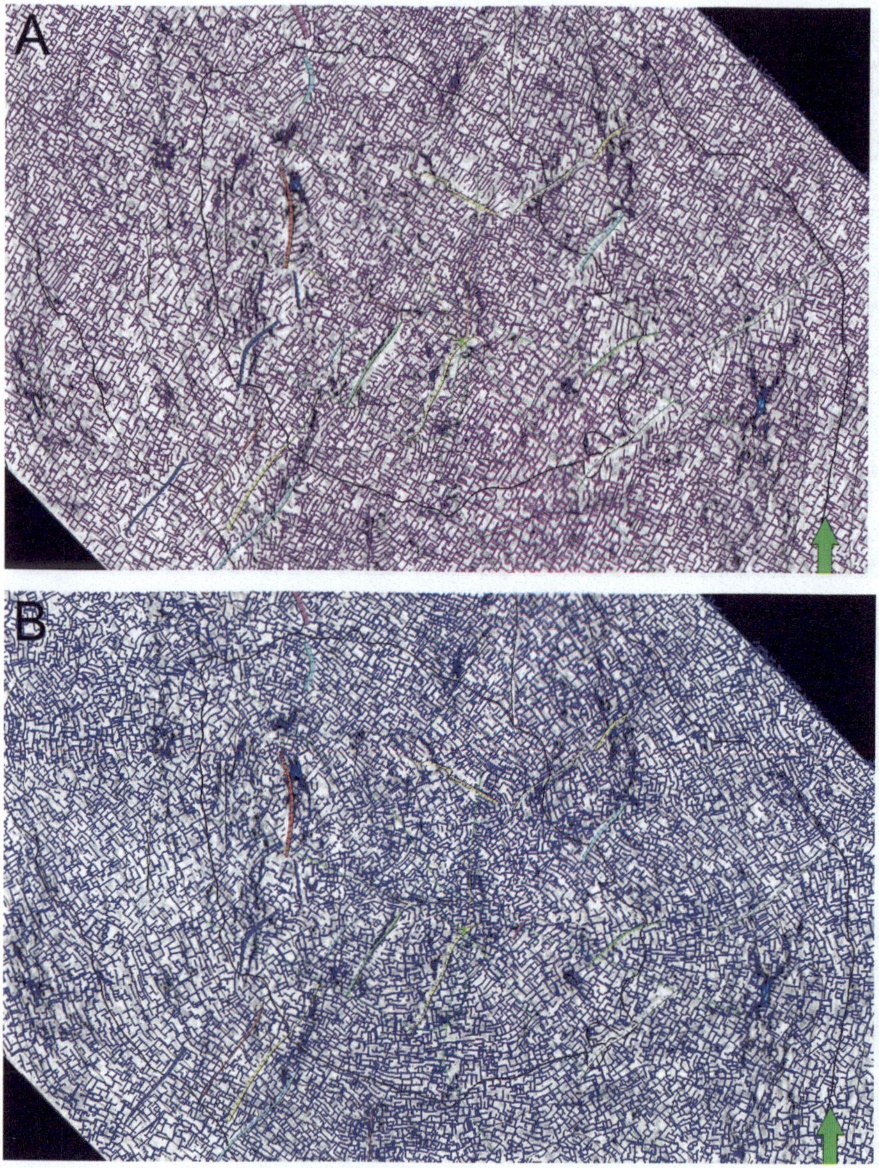

Fig. 10.4 DFN fracture geometry calculated for horizontal strain rates and orientation related to regional ESE-WNW extension with local variation around the seismic-scale faults (**a**) and growth of the salt pillow (**b**). The purple and blue lines show the traces of the fractures on a surface representing the centre of the Ekofisk Formation, superimposed onto the ant-tracked seismic data

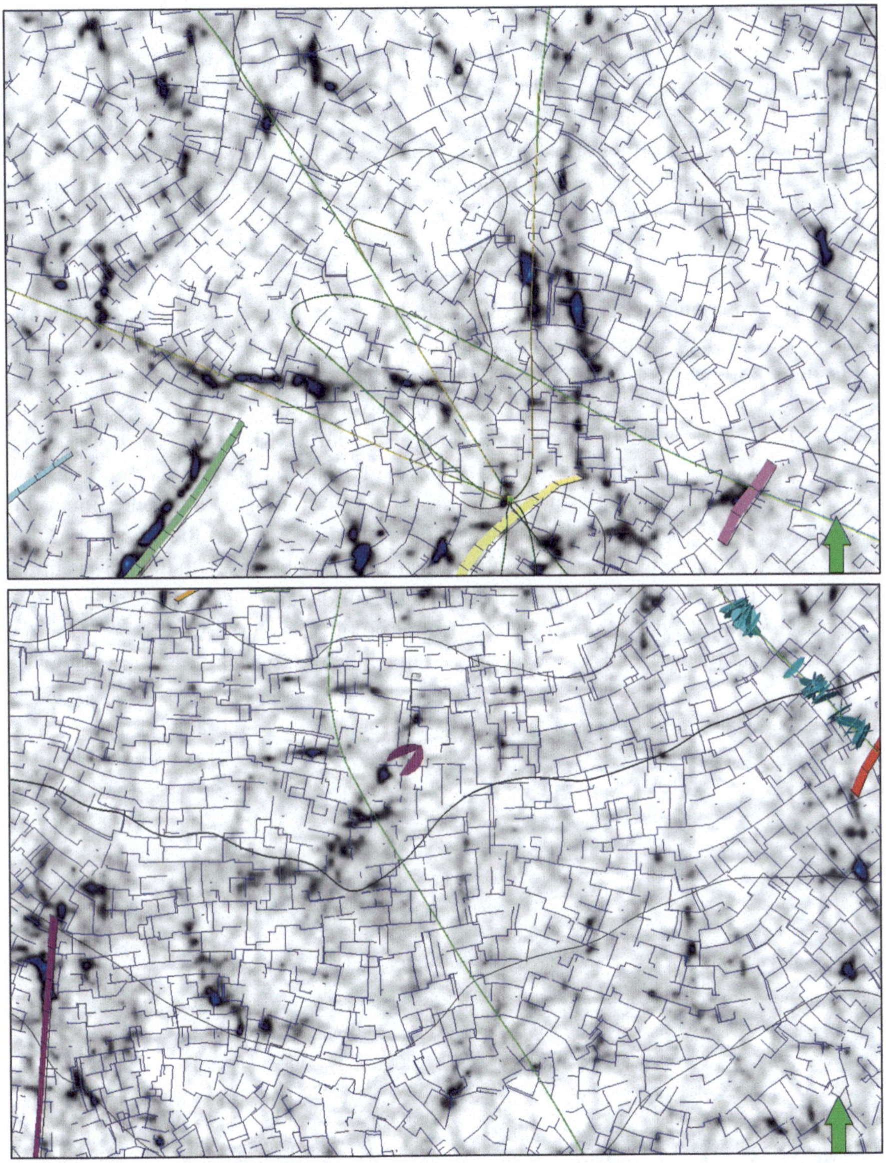

Fig. 10.5 Close-ups comparing the DFN fracture geometry calculated for horizontal strain rates and orientation related to the growth of the salt pillow (blue lines) and the ant-tracked seismic data on a surface representing the centre of the Ekofisk Formation. There is a close match in orientation in many areas but significant discrepancies in other areas. The light blue disks in the lower close-up show resistive fractures picked on borehole images and the pink disks represent fractures logged in the core

This suggests that the seismic-scale faults are surrounded by "damage zones" of fractures generated in response to the local fault-related elastic strain.

- Across many parts of the Kraka structure, the geometry of the fracture network generated by horizontal strain related to the growth of the salt pillow is also consistent with the lineations observed in the ant-tracked seismic data. In some areas, quite complex patterns develop in the DFN, resulting from strain patterns related to local periclinal or superimposed folding (Fig. 10.5). In other areas, however, neither of the fracture models matches the seismic lineations, and in most areas, the seismic lineations are more widely spaced than the modelled fractures. This is consistent with the observations comparing the ant-tracked seismic data to the resistive fractures on borehole images, which not only showed a correlation in many cases but also showed that not all resistive fractures corresponded to lineations on the ant-tracked seismic data (Aabø et al 2019; see also Figs. 10.6, 10.7 and 10.8). This suggests that, while some of the larger Mode 2 layer-bound fractures (which may have throws of up to a few metres) may be resolved on the ant-tracked seismic data, it does not resolve all such the fractures, and some lineations observed on the ant-tracked seismic data may be unrelated to the fracturing.
- The density of the small (10–50 cm length) fractures observed on the core is much higher than the predicted density of the layer-bound fractures. It is likely that these represent the initial "seed" microfracture population assumed by the conceptual model of fracture growth described here. This interpretation is consistent both with

Fig. 10.6 Close-up showing the DFN fracture geometry calculated near to two seismic-scale faults, calculated using ESE-WNW extension strain with local variation around the faults, superimposed onto the ant-tracked seismic and well data. Light blue disks show the resistive fractures picked on borehole images

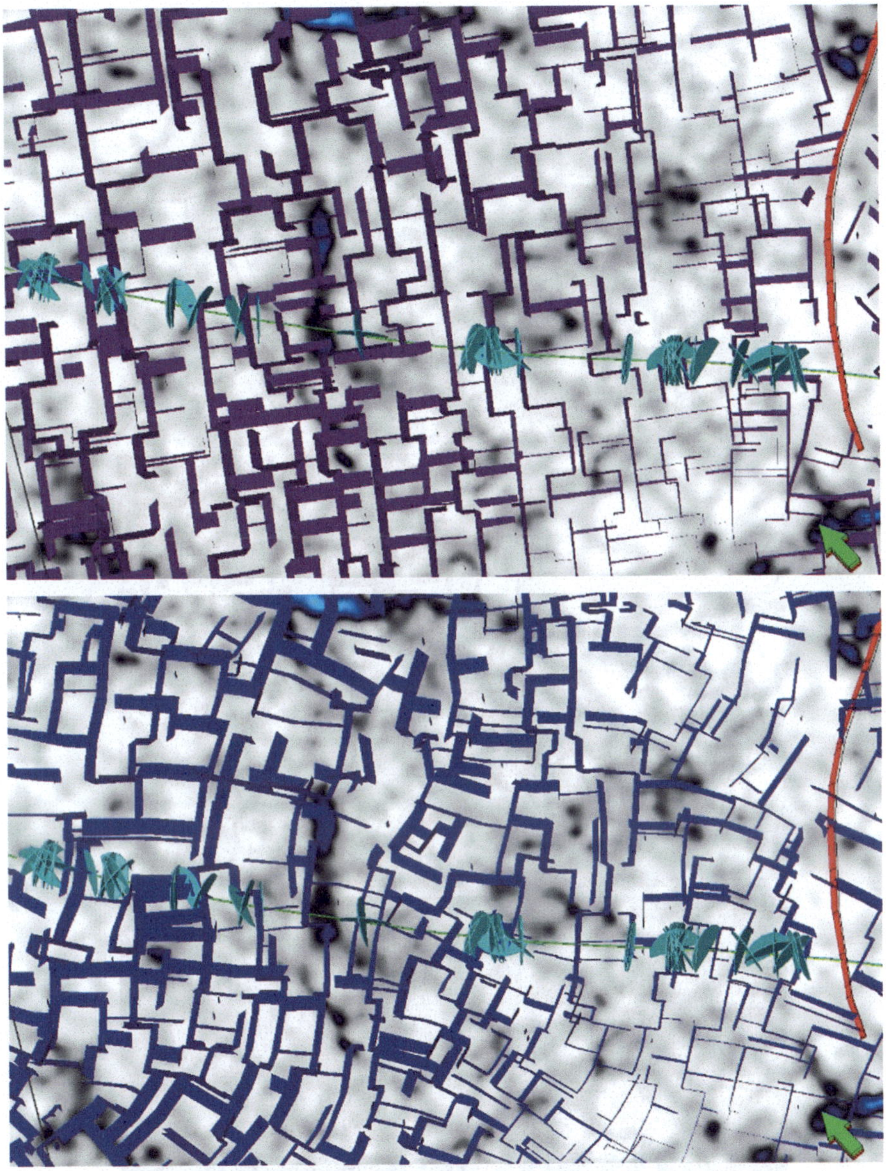

Fig. 10.7 Close-ups comparing the DFN fracture geometry calculated near to a seismic-scale fault using ESE-WNW extension strain with local variation around the faults (top) and strain related to growth of the salt pillow (bottom), superimposed onto the ant-tracked seismic and well data. Light blue disks show the resistive fractures picked on borehole images. Note that the DFN calculated for ESE-WNW extension strain gives a closer match to the observed fractures near the seismic-scale fault (in red at the right of the images), but the DFN calculated for strain related to growth of the salt pillow gives a closer match away from the seismic-scale fault

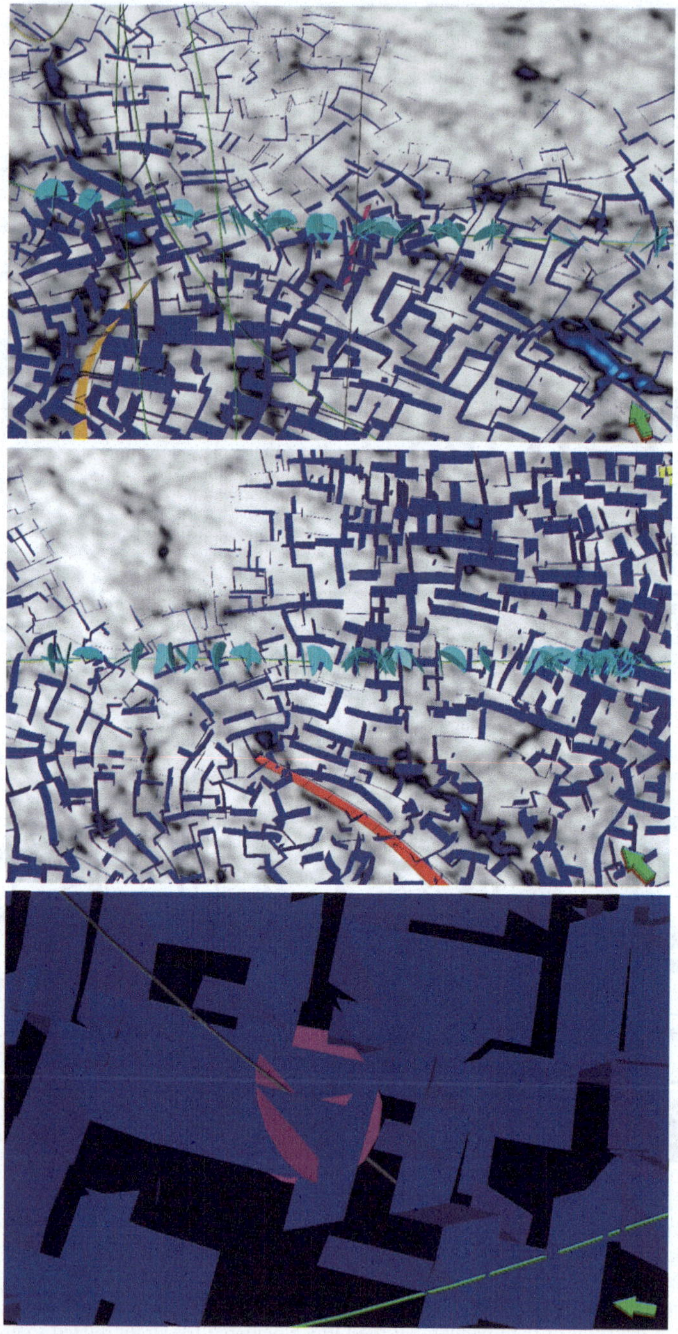

Fig. 10.8 Close-ups showing the DFN fracture geometry calculated for horizontal strain rates and orientation related to the growth of the salt pillow, superimposed onto the ant-tracked seismic and well data. Light blue disks show the resistive fractures picked on borehole images, while purple disks show the large slickensided fractures identified in the core

the size of these observed fractures (in Sect. 8.3.2 we note that the seed microfractures must have a size of around 10–50 cm prior to the onset of deformation), and with the inference that they are formed around stylolites, chert nodules or other heterogeneities during compaction or diagenesis, and hence precede the tectonic strain.

Overall, we can, therefore, be confident that the DFNs that we can generate using our method give a reasonably good approximation to the unmappable network layer-bound fractures in the Ekofisk Formation at Kraka, which are so important in controlling fluid flow through the oilfield.

10.2 Geothermal Prospect in the Drenthe Province, Netherlands

In this section, we consider the area of the Drenthe Aa Valley, Netherlands, shown in Fig. 10.9. In the Netherlands, there is an increasing focus on CO_2 sequestration and deep geothermal energy. While the focus for CO_2 sequestration is now mainly offshore, northern provinces continue to be prospective both for storing CO_2 (van Ansem et al. 2009) and harnessing deep geothermal energy. The influence of fracturing on flow and seal capacities is critically important for either application, making their understanding fundamental to successful exploration and development.

In the context of this book, we consider the area to be a useful example of how large-scale fracture networks may be considered using realistic subsurface datasets. In this area, a seismic volume as well as gamma ray and sonic well logs are available throughout the Upper Cretaceous due to wells targeting deeper hydrocarbon resources.

Anomalous heatflows are expected related to the Anloo salt diapir (Daniilidis and Herber 2017) providing a thermal anomaly of interest to geothermal exploration. Additionally, there is geochemical evidence for connectivity between deep and shallow groundwater systems (Elshehawi et al. 2019). This is of great significance to local water abstraction and could be evidence of fracture enhanced fluid flow.

We consider the likelihood of the presence of a fracture network and the effectiveness of such a network to promote hydraulic conductivity. We will analyse the likely orientation and degree of anisotropy in the network to illustrate its potential connectivity. In the absence of calibration data, we will then show how this workflow could be used to quickly generate multiple geologically realistic fracture network models for an area approaching 60 square kilometres, which could be used in probabilistic or scenario-based risk analysis of the geothermal reservoir.

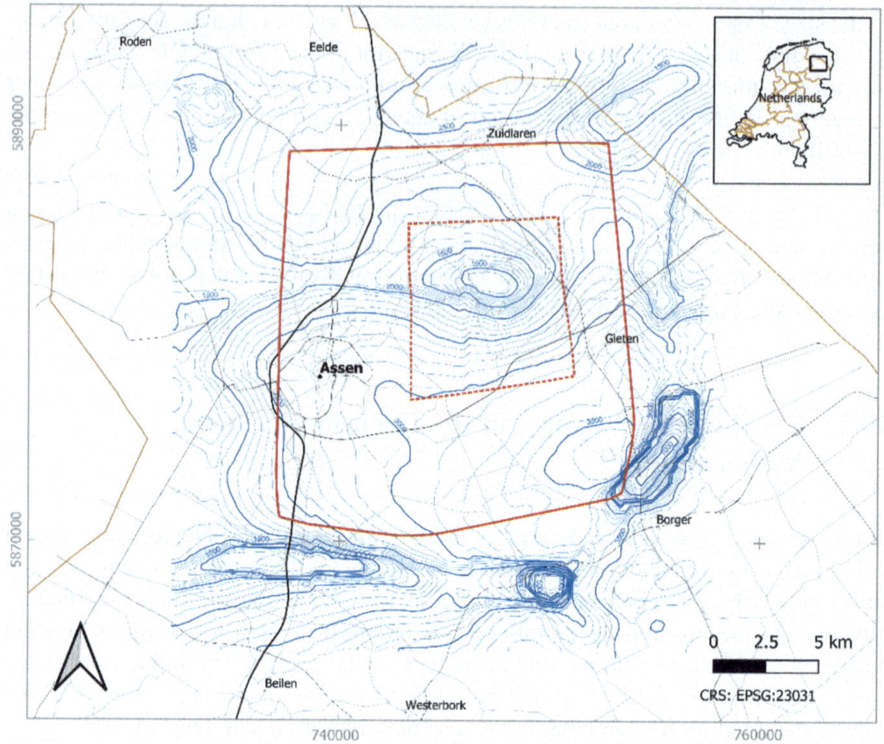

Fig. 10.9 Location map for the area of interest in the Netherlands, overlain with surface contours of the Top Zechestein Formation (blue), highlighting the presence of the Anloo Salt Diapir, around 10 km northeast of Assen. The dotted and the solid rectangles outline the area of interest used for fracture modelling in this chapter for detailed and larger-scale models, respectively

10.2.1 Geological Model and Strain Estimation

Located in the northeastern part of the Lower Saxony Basin, the area has undergone inversion during the Laramide and Subhercynian (Heijnen et al. 2010). The Upper Cretaceous Chalk Group that we are targeting is overlain by fluvio-deltaic clastics of the Cenozoic North Sea Group and underlain by marine clastics of the Lower Cretaceous Rijnland Group, clays and evaporites of the Jurassic Altena and Germanic Triassic Groups and evaporites of the Permian Zechstein Formation (Smit et al. 2018). Movement of the salt has developed significant ridges and domes, including the Anloo Salt Diapir, illustrated in Fig. 10.10. As was the case with the Kraka field, this movement is likely to have caused fracturing in the more brittle sections of the overlying units.

We have built a three-dimensional geological model for the area using onshore time-migrated three-dimensional seismic data (survey L3NAM1992B, acquired in 1992), seismic picks and a velocity model, all shared publicly by TNO (Gunninck

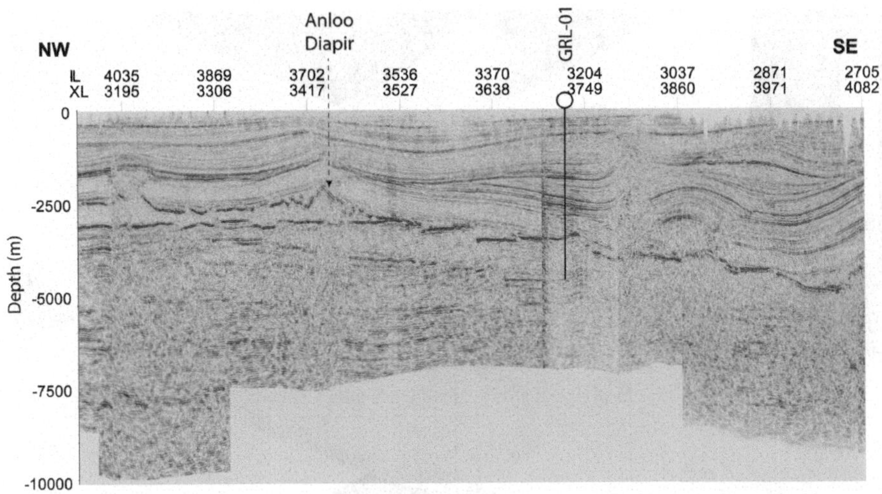

Fig. 10.10 Seismic intersection extracted from the depth-converted three-dimensional seismic volume, illustrating the location of well GRL-01 used for calibration of the mechanical and stratigraphic models, and the Anloo salt diaper investigated in our area of interest

et al. 2013; Hummelman et al. 2019). Using the supplied interval velocities, we depth-convert the interpreted seismic horizons to build a corner point geocellular grid. Comparing the depth-converted seismic horizons to well observations and regional stratigraphic markers (from TNO), we are satisfied that the data provides sufficient constraint on geometric variation across the structure to estimate strain development.

We applied the same backstripping algorithm used for the Kraka field (see Sect. 10.1.2) to estimate the strain caused by the growth of the salt diapirs. Results from this algorithm (Fig. 10.11) illustrate that maximum strain accumulation has occurred in the areas overlying upwards salt movement, while the salt minibasin centres are generally more structurally quiescent.

10.2.2 Log Analysis and Extraction of Mechanical Parameters

After analysing all wells intersecting the Upper Cretaceous in the area, we chose to use well GRL-01 as a type well for the area. Data from this well is used to estimate Young's Modulus for each of the layers, using empirical relationships selected according to lithology.

Although several other wells penetrate the Upper Cretaceous in the area of interest, there was insufficient data available to calculate mechanical properties accurately, preventing consideration of lateral variability in the Young's Modulus.

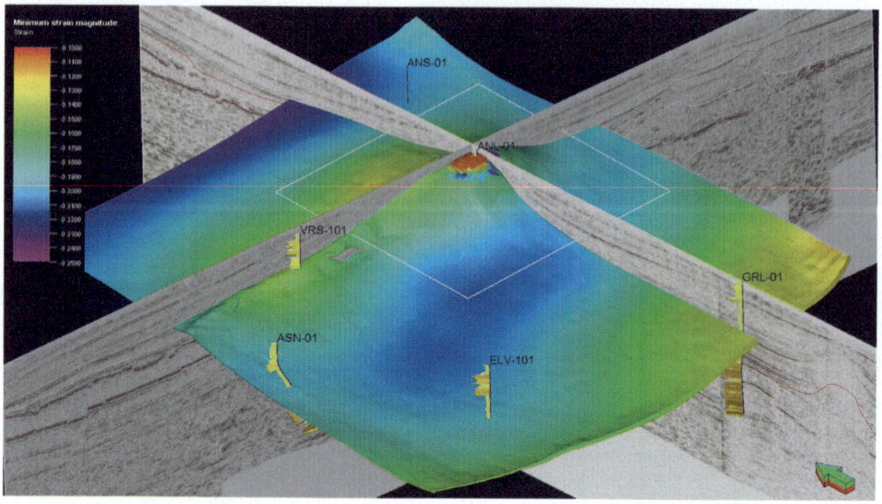

Fig. 10.11 Three-dimensional model of the magnitude of the minimum compressive principal strain from the backstripping algorithm

Stratigraphic tops were checked for quality against published wells and used for comparison between the stratigraphic tops and with the seismic data. Seismic horizons corresponding to the Top Lower North Sea Group, Top Upper Cretaceous and the Base of the Upper Germanic Group provide a good understanding of the general stratigraphic relationships throughout the area. The geometry of the seismic reflections combined with the relative proportions of stratigraphic units from across the area were subsequently used as constraints to control the vertical thickness and proportions of layers and lithologies within the geocellular grid. Each of the markers used in this study is named and described in Table 10.1.

By combining the stratigraphic observations of TNO and our well log observations, we classified the lithologies into three types, carbonate, shale and sandstone (Table 10.2). For each stratigraphic interval, we calculated the average sonic slowness which was then used to estimate the uniaxial compressive strength (Table 10.3), from which a second relation is used to calculate the Young's Modulus (Table 10.4). The well log for GRL-01 showing the measured gamma ray (GR) log, sonic slowness (DT) as well as the calculated Young's Modulus (Young's Mod.) averaged for each of the lithologies in Table 10.1.

From the log analysis, we identify three zones as more brittle and constrained by more ductile strata above and below (Fig. 10.12). We simulate the fracture development in the upper three of these, all deposited within the Cretaceous; the Texel Formation (CKTX), Lower Holland Marl (KNGLL) and the Bentheim Sandstone (KNNSP). The fact that these layers are more brittle and are also constrained by less brittle units above and below, offers the improved likelihood of the development of layer-bound fractures.

Table 10.1 Summary of the stratigraphy within the depth of interest present in the area with brief lithological descriptions (compiled from data released by TNO detailed in Hummelman et al. 2019)

Code	Formation	Lithology
CKGR	Ommelanden formation	Fine-grained limestone
CKTX	Texel formation	Limestones and marly chalks
KNGLU	Upper holland marl member	Marls
KNGLM	Middle holland claystone member	Shaley claystone
KNGLL	Lower holland marl member	Marl or calcareous claystone
KNNC	Vlieland claystone formation	Claystone
KNNSP	Bentheim sandstone member of Vlieland sandstone formation	Massive sandstones, calcareous with shell fragments
KNNCV	Bentheim claystone member of Vlieland claystone formation	Marly, silty to sandy claystone
SKCFM	Middle member of Coevorden formation	Silty/sandy claystones
SKCFL	Lower member of Coevorden formation	Calcareous claystones

Table 10.2 Summary of the average measured parameters gamma ray (GR av.) and sonic slowness (DT av.), the lithology (sandstone—sst, limestone—lst and shale—sh) and the modelled parameters uniaxial compressive strength (UCS) and Young's Modulus calculated per stratigraphic interval

Stratigraphic unit	GR av. (API units)	DT av. (μs/ft)	Litho.	UCS (MPa)	Young's Modulus (GPa)
CKGR	19	86	lst	24.4	3.1
CKTX	22	68	lst	37.4	7.1
KNGLU	52	90	sh	27.1	3.8
KNGLM	74	103	sh	18.4	2.5
KNGLL	54	83	sh	34.3	5.0
KNNC	76	94	sh	23.8	3.3
KNNSP	26	81	sst	66.1	67.8
KNNCV	79	105	sh	17.3	2.3
SKCFM	50	82	sh	35.7	5.2

Table 10.3 Equations to estimate the uniaxial compressive strength, C_0, of intervals based on the interpreted lithology

Lithotype	Uniaxial compressive strength, C_0	Citation
Carbonate	$C_0 = \frac{\left(\frac{7682}{DT}\right)^{1.82}}{145}$	Militzer (1973)
Shale	$C_0 = 0.77 * \left(\frac{304.8}{DT}\right)^{2.93}$	Horsrud (2001)
Sandstone	$C_0 = 1200 * e^{-0.036*DT}$	McNally (1987)

Source from compilations by Zoback (2007)

Table 10.4 Equations to estimate the Young's Modulus, E, of intervals based on the interpreted lithology and calculated uniaxial compressive strength, C_0

Lithotype	Young's Modulus, E	Citation
Carbonate	$E = \left(\frac{C_0}{13.8}\right)^{\frac{1}{0.51}}$	Militzer (1973)
Shale	$E = \left(\frac{C_0}{7.97}\right)^{\frac{1}{0.91}}$	Horsrud (2001)
Sandstone	$E = \left(\frac{C_0}{12}\right)^{\frac{1}{0.8}}$	Knipe et al. (2008)

Source from compilations by Mavko et al. (1998)

We model each of these layers as a single layer of uniform Young's Modulus and assume that the vertical proportion of different lithologies will remain constant across the area. Therefore, using the relative thickness of different layers from GRL-01, we distribute these evenly across the area, constrained by the Top Lower North Sea Group, Top Upper Cretaceous and Base Upper Germanic seismic horizon interpretations. A comparison of the output model horizons suggests that these layers fit with the geometry observed in the seismic volume.

10.2.3 Fracture Model

From the modelling result of the three stratigraphic units (Fig. 10.13), we can see immediately that the fracture density and orientations vary spatially. On closer inspection, one can observe that the average fracture length and the fracture network anisotropy also vary spatially around the structure (Fig. 10.13b, d, f). Maps of the implicit fracture data (Fig. 10.13a, c, e) show that the mean linear density of layer-bound macrofractures, $_{MF}P_{32}$, is greater in the thinner layers (Fig. 10.14), with a maximum of 0.25 fractures/m in the thickest formation, the Texel, and c.1.2 fractures/m in the thinnest formation, the Vlieland Sandstone.

This is consistent with the observation in Sect. 8.1.2 that the layer thickness forms one of the most significant controls on fracture spacing. Furthermore, we notice that in the Texel formation, the layer thickness is slightly thinner in the area where the fractures form earliest. This also correlates to the area above the salt diapir that has experienced the greatest total strain, illustrated in Fig. 10.15.

In the lower two formations, the northern area features a much higher fracture intensity than the southern area. The east-west line separating these two areas is the trend of the salt roller feeding the salt diapir at Anloo. The stress shadow width for a single fracture is proportional to its height, and, therefore, as the layer thickness increases so does the stress shadow width, resulting in more widely spaced fractures. We observe this in both the implicit and explicit results as a decrease in fracture intensity and fracture density towards the South. This variation is illustrated clearly when viewed in cross-section (Fig. 10.15).

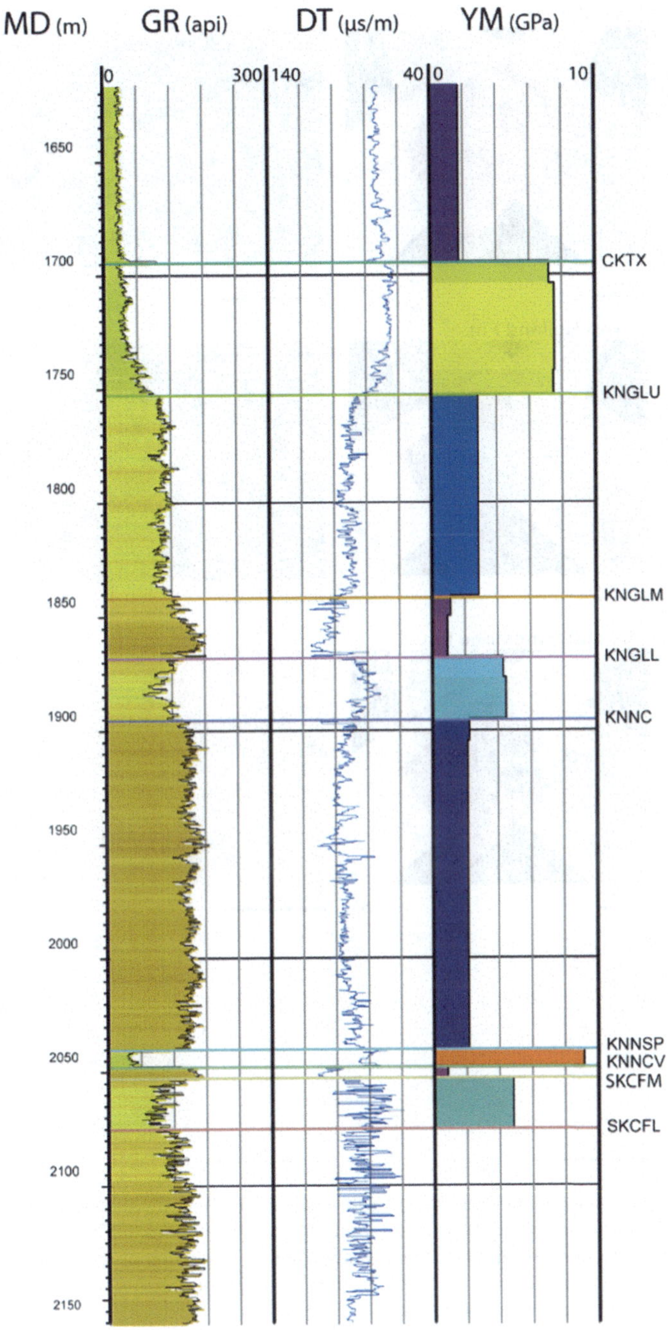

Fig. 10.12 Well log for GRL-01 illustrating measured depth along borehole (MD) with the measured gamma ray (GR) log, sonic slowness (DT) and calculated Young's Modulus (YM). Stratigraphic markers labelled to the right of the logs are named and described in Table 10.1

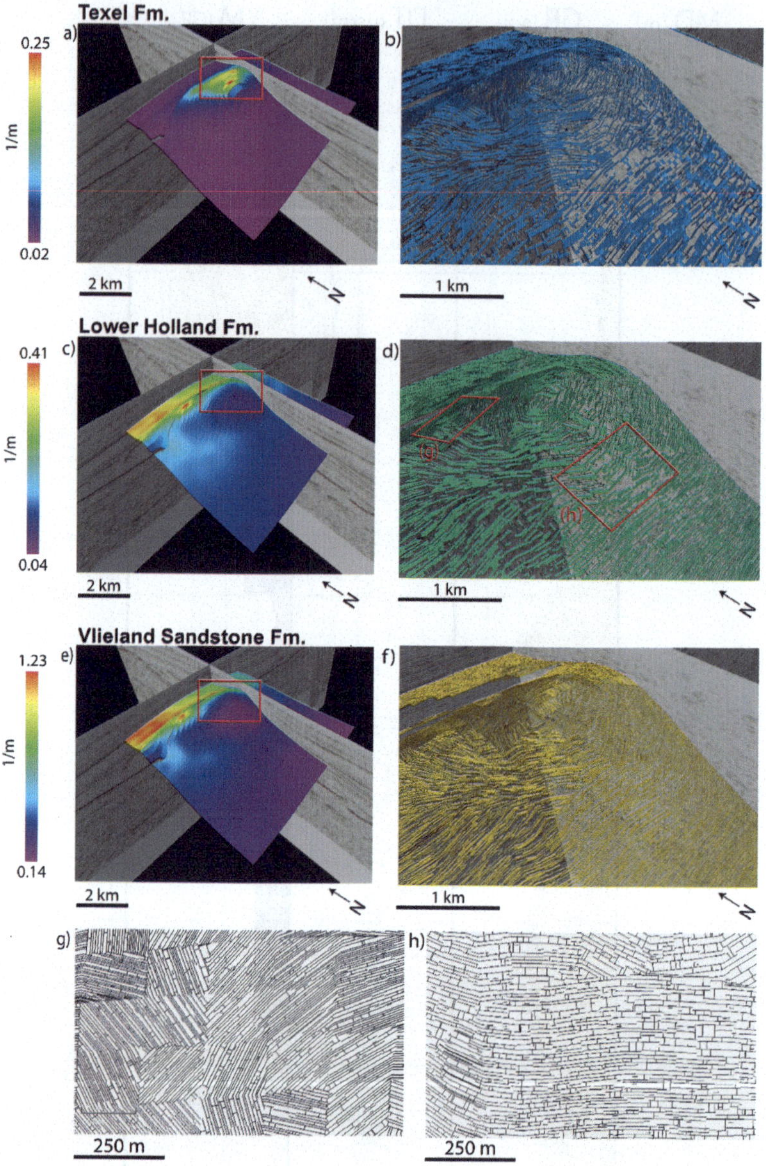

Fig. 10.13 Maps of the mean linear macrofracture density, $_{MF}P_{32}$, (fractures/m), representing the area of fractures per unit volume of rock (**a**, **c** and **e**), and the explicit DFN (**b**, **d** and **f**) for the three modelled layers at maximum fracture saturation. Close inspection of the explicit DFN (**g** and **h**) show contrasting areas of anisotropy caused by the relative proportion of fractures oriented in-line with the major and minor axes of strain rate

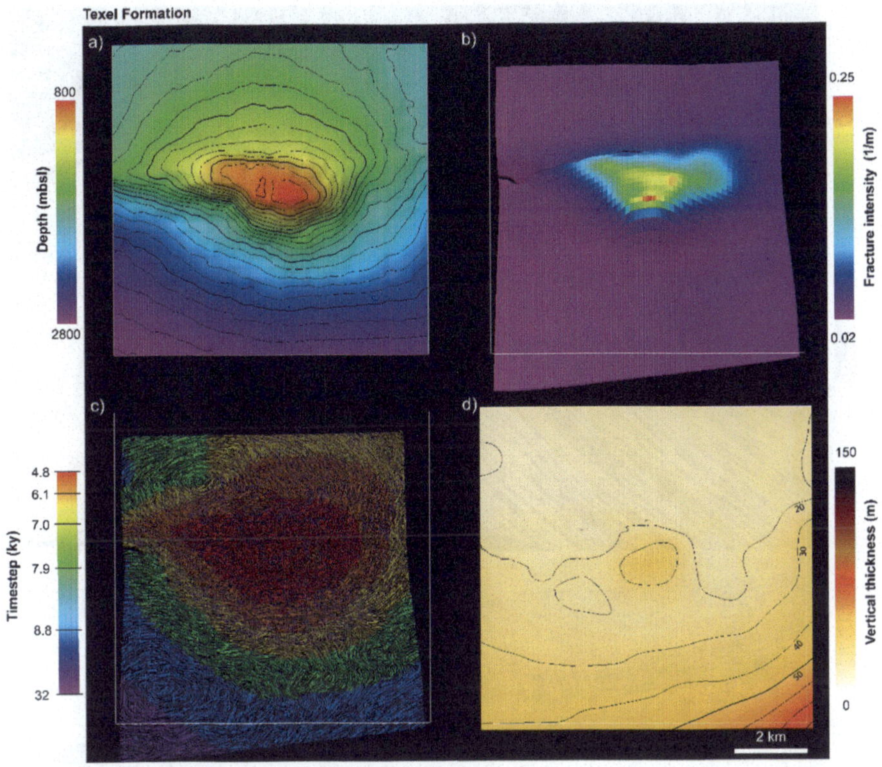

Fig. 10.14 Map view of **a** the depth to Top Zechstein Formation, **b** the strain distribution, **c** the explicit DFN and **d** an isochore of the Texel Formation. The DFN is coloured by the timing of fracture propagation. The area modelled in **b** shows the same area of modelling as the smaller area of interest highlighted in Fig. 10.9

Importantly, in the context of the discussion of controls on fracturing, we observe that fracture lengths of the layer-bound fracture population decrease with increasing fracture intensity. In the Lower Holland Formation fracture, intensity and anisotropy to the North of the salt diapir (Fig. 10.13g) are increased compared to those in the South (Fig. 10.13h). We also observe that fractures in the primary set tend to be longer, leading to increased anisotropy.

This observation could be critical in consideration of the implications for subsurface fluid flow. If an application requires the transfer of fluid from a matrix to fracture, increased volumetric fracture density may be preferable. If a system requires higher connectivity across greater distances, fewer but larger fractures may be preferred, and their orientation is likely to be critical.

Comparing different stages of the fracture network evolution for the Texel Formation (Figs. 10.15 and 10.16), we observe early fracturing on the crest of the Anloo diapir, where the strain is greatest, and a continued spread of the deformation front down the flanks of the high. As the deformation front spreads, the northern areas are

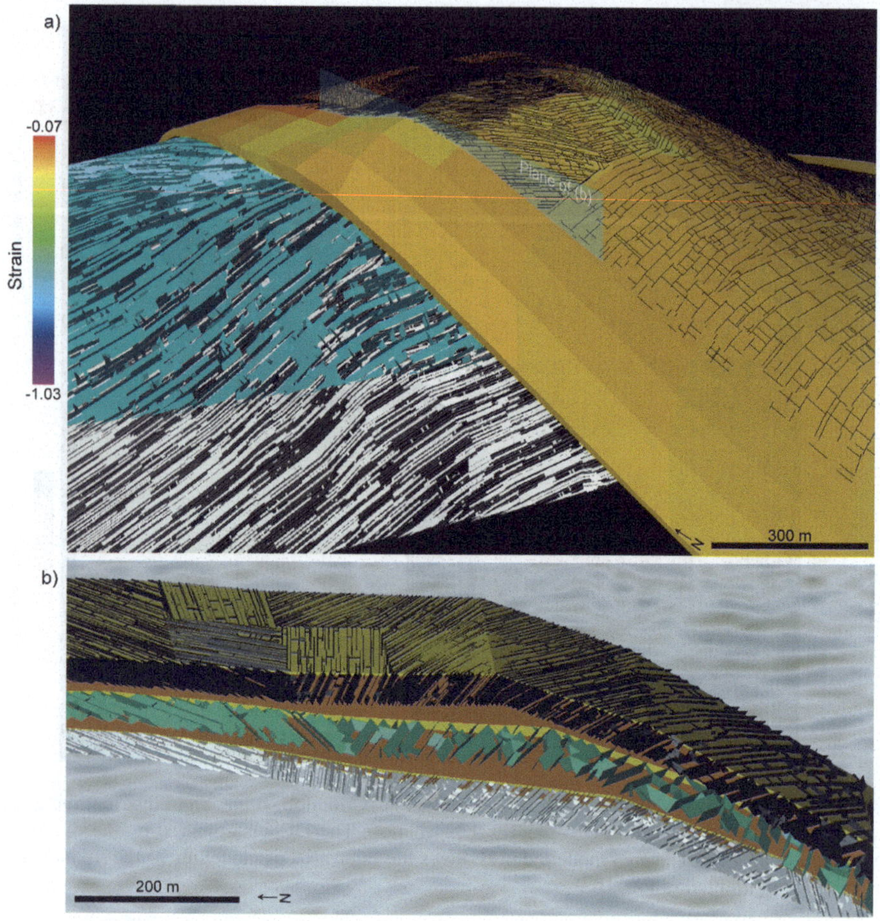

Fig. 10.15 Images focussing on the antiform above the Anloo salt diapir, in the red boxes on Fig. 10.13. Here we see the three stacked fracture networks in the Texel Formation (black), Lower Holland Marl Member (green) and Bentheim Sandstone Member (white). Inset image **a** shows the gridded strain property with the fracture networks. Areas of higher-strain magnitude are associated with increased fracture intensity. In **b**, the fracture height of the layer-bound fractures is equivalent to layer thickness. No definitive information on the fracture network can be discerned from the seismic dataset, although we may expect increased seismic attenuation and anisotropy may be a response to fracturing

fractured slightly earlier and to a greater intensity, consistent with the slight thinning of the formation in this direction (Fig. 10.14d).

Close inspection of the fracture network (Fig. 10.14c) shows that during continuing deformation, there is limited infilling of space between fractures initiated in earlier timesteps. This illustrates observations made in Sect. 8.2, whereby at any given location, the fracture network rapidly reaches saturation. During fracturing, the fracture network expands as a sharp front (Fig. 10.16).

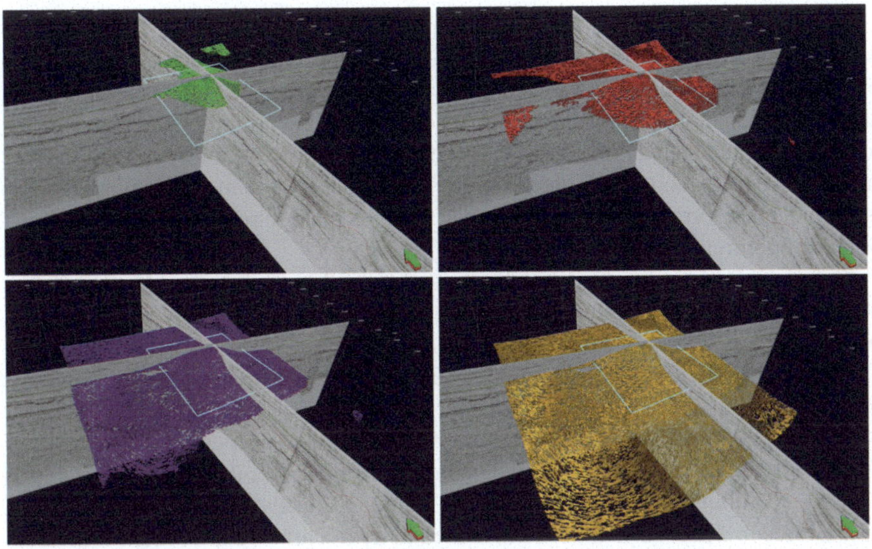

Fig. 10.16 Large-scale model run covering around 300 square km. Shows the evolution of the fracture network for the Texel Formation coloured coded to illustrate their timing of fracturing (green = 6.5 kyr, red = 8 kyr, purple = 9.0 kyr, orange = 37 kyr). Note the outwards expansion of the deformation front, consistent with observations of Sect. 8.2. The square shows the area modelled in Fig. 10.14

10.2.4 Industrial Application

When modelling fractures in an area with limited calibration data, such as this, it is important to adopt a workflow that encompasses the uncertainties and allows us to estimate the risks quantitatively. It is also important that the models can easily be refined and updated when new data becomes available. A common way to do this is by creating multiple models representing high, low and intermediate case scenarios. All these models should be geologically realistic and consistent with the data available but should encompass the range of uncertainties in the data.

We can do this very easily by using the intermediate stages in the development of the fracture network as low and intermediate scenarios and the final fracture network as the high scenario. This has the advantage that all intermediate stages are geologically realistic but cover a range of possible outcomes in a consistent manner. For example, we may be uncertain to what extent the fracture network has developed, but we can be certain that fractures will develop first in the high-strain areas and then propagate into the low-strain areas. Therefore, a model comprising fractures only in the high-strain areas would be a realistic low-case scenario, but a model comprising fractures only in the low-strain areas would not. This also allows us to narrow down the range of possible scenarios easily as more data becomes available, for example, from new wellbores.

Where there is uncertainty in the input parameters to the model, for example, the mechanical properties, this approach also allows the rapid generation of multiple models and easy reduction of the range of uncertainties as new data becomes available. Using sensitivity analysis, the impact of different uncertainties and their relative importance can be analysed. We can also examine lateral variability in the resulting fracture networks, to identify, for example, areas that will be more likely to develop isotropic or anisotropic fracture networks, or areas that are more likely to have long or short fractures.

Industrial applications are typically presented two key challenges when faced with natural fractures, are fractures present, and if so, what is the likely impact of that fracturing?

To evaluate whether fractures are present, within the framework of the approach presented in this book, one must consider the mechanical properties and thickness of layers in conjunction with an interpretation of the areas experiencing most accumulated strain where possible. Using examples from above, it would be possible to develop a combined common risk segment mapping framework for the consideration of fractures in a similar manner to that used in hydrocarbon exploration. Individual uncertainties may include layer thickness, Young's modulus, pore pressure and total strain.

Modelling the area with the presented framework could provide a map of which areas are likely to have fractured earliest in a phase of deformation. If fractures of a specific orientation or degree of anisotropy are desired, these may also be mapped directly from the modelling outputs and used to investigate targets.

To consider the potential implications of a known fracture network, our focus would be on the latter elements, seeking to understand the most likely trends in dominant orientation and the degree of anisotropy in the fracture network. In applications requiring transfer of a fluid from a matrix to fracture space, more intense fracturing may be welcome; however, in situations requiring heat transfer from rock to a transient fluid, a more connected fracture network offering high permeability may be preferable.

Critical to either of these scenarios, in this chapter, we demonstrate that multiple iterations of modelling may be completed over a large-scale area (c.60 km^2). This efficiency of modelling allows the use of multiple cases and potentially even full-cycle stochastic workflows to consider the likely variation of fracture networks dependent on the variation of geological controls with reasonable constraints.

References

Aabø, T.M., J.S. Dramsch, C.L. Würtzen, S. Seyum, and M. Welch. 2019. An integrated workflow for fracture characterization in chalk reservoirs, applied to the Kraka Field. *Marine and Petroleum Geology* 112: 104065.

Daniilidis, A., and R. Herber. 2017. Salt intrusions providing a new geothermal exploration target for higher energy recovery at shallower depths. *Energy*. https://doi.org/10.1016/j.energy.2016.10.094.

Elshehawi, S., E. Bregman, P. Schot, and A. Grootjans. 2019. Natural isotopes and ion compositions identify changes in groundwater flows affecting wetland vegetation in the Drentsche Aa Brook Valley, The Netherlands. *Journal of Ecological Engineering* 20 (3): 112–125.

Gunnink, J., D. Maljers, S. Van Gessel, A. Menkovic, and H. Hummelman. 2013. Digital Geological Model (DGM): A 3D raster model of the subsurface of the Netherlands. *Netherlands Journal of Geosciences—Geologie En Mijnbouw* 92 (1): 33–46. https://doi.org/10.1017/S0016774600000263.

Heijnen, L.J., G. Diephuis, P. Betts, A. Huck, A. Willemsen, and N.A. Buik. 2010. Feasibility for the application of geothermal energy for greenhouses in South-East Drenthe, The Netherlands, results of a multi-disciplinary study. In *Proceedings of the world geothermal congress.*

Horsrud, P. 2001. Estimating mechanical properties of shale from empirical correlations. *SPE Drilling & Completion* 16 (02): 68–73.

Hummelman, J., D. Maljers, A. Menkovic, R. Reindersma, J. Stafleu, and R. Vernes. 2019. Totstandkomingsrapport Digitaal Geologisch Model (DGM), Report: TNO 2019 11653.

Jorgensen, L., and P. Andersen. 1991. Integrated study of the Kraka field. In *SPE 23082, presented at Offshore Europe*, Aberdeen, United Kingdom.

Klinkby, L., L. Kristensen, E.B. Nielsen, K. Zinck-Jørgensen, and L. Stemmerik. 2005. Mapping and characterization of thin chalk reservoirs using data integration: the Kraka field, Danish North Sea. *Petroleum Geoscience* 11: 113–124.

Knipe, R.J., R.K. Davies, S. Freeman, S. Harris, A. Brown, C. Souque, C. Tueckmantel, H. Lickorish, J. Campbell, M.J. Welch, R. Burden, J. Miller, A. Li, and E. Stuart. 2008. Foundation fault project. RDR consortium research project. RDR project no. 9582.

Maerten, L. 1999. Mechanical interaction of intersecting normal faults: theory, field examples and applications. Ph.D. thesis, Stanford University.

Mavko, G., T. Mukerji, and J. Dvorkin. 1998. The rock physics handbook: tools for seismic analysis in porous media. Cambridge University Press.

McNally, G.H. 1987. Estimation of coal measures rock strength using sonic and neutron logs. *Geoexploration* 24 (4–5): 381–395.

Militzer, H. 1973. Einige Beiträge der Geophysik zur primärdatenerfassung im Bergbau.

Pederson, J. M., W.F. Stewart, and I.P. Yuwono. 2002. Coaxing information from petrophysics and integrated data analysis: the Menggala South Field Case History, Sumatra, Indonesia. SPE—Asia Pacific Oil and Gas Conference, 506–515. https://doi.org/10.2523/77885-ms.

Petersen, K.D., and O.R. Clausen. 2018. Forward and reverse dynamic and kinematic modeling of the evolution of the Kraka Field. Unpublished DHRTC report.

Rasmussen, E.S., O.V. Vejbæk, T. Bidstrup, S. Piasecki, and K. Dybkjær. 2005. Late Cenozoic depositional history of the Danish North Sea basin: Implications for the petroleum systems in the Kraka, Halfdan, Siri and Nini fields. *Geological Society London Petroleum Geology Conference Series* 6: 1347–1358.

Smit F., E. Bregman, and F. Magri. 2018. Coupling earth surface processes and geological structures to explain environmental features as observed onshore Northern Netherlands SEQ. https://doi.org/10.1190/segam2018-2995900.110.1190/segam2018-2995900.1

van Ansem I., C. Cronenberg, and R. Eenkhoorn. 2009. Carbon capture and storage—action plan for the Northern Netherlands. ISBN 978-90-79147-06-9.

Welch, M.J., M. Lüthje, and A.C. Glad. 2019. Influence of fracture nucleation and propagation rates on fracture geometry: insights from geomechanical modelling. *Petroleum Geoscience* 25: 470–489.

Zoback, M. 2007. Reservoir geomechanics. *Cambridge University Press*. https://doi.org/10.1017/CBO9780511586477.

Chapter 11
Conclusions and Further Work

Abstract In this chapter, we summarise the key insights we gain from the new method and discuss its potential applications in geomechanical and fluid flow modelling. We consider the value that it could add in various industrial contexts, including tunnel excavation and mining, groundwater management, hydrocarbon exploration and production, geothermal energy extraction and CO_2 sequestration. Finally, we discuss how the technique could be developed further, to model fractures that have been reactivated in multiple deformation episodes and to model fractures that propagate across multiple mechanical layers.

Keywords Fracture modelling · Discrete fracture networks · Naturally fractured reservoir · Subcritical crack propagation · Reservoir modelling · Geothermal energy · Carbon sequestration · Geomechanics

In this book, we have demonstrated that it is possible to develop an algorithm for simulating the growth of layer-bound fracture networks that is geomechanically robust enough to generate accurate and useful fracture models but also fast enough to model large networks containing hundreds of thousands of fractures across large geological structures. We do this by applying the geomechanical rules governing fracture propagation and interaction to cumulative distribution functions describing the entire fracture populations, rather than to individual fractures.

11.1 Insights from the Algorithm

This algorithm can give us insights into the controls on the development and resulting geometry of fracture networks. Some of these insights are fairly intuitive, or simply affirm or quantify relationships, that have been described previously from modelling or field studies. For example:

- There is a direct correlation between layer thickness and fracture spacing when the fractures have stress shadows.

© The Editor(s) (if applicable) and The Author(s), under exclusive license to Springer Nature Switzerland AG 2020

M. J. Welch et al., *Modelling the Evolution of Natural Fracture Networks*,
https://doi.org/10.1007/978-3-030-52414-2_11

- Fractures develop first in areas of high stiffness and/or high horizontal strain and then propagate outwards into areas of lower stiffness or lower strain.
- Fracture lengths will be limited by the presence of perpendicular fracture sets that can block fracture propagation: the longest fractures are likely to be found in uniaxial fracture networks, where there are no perpendicular fractures to intersect, while the shortest fractures are likely to be those in secondary fracture sets that propagate between pre-existing primary fractures.
- Linkage of parallel fracture segments that have terminated due to stress shadow interaction, by the development of relay segments, will significantly increase the mean fracture length.
- Fracture connectivity (measured by the ratio of connected R and Y nodes to unconnected I nodes) will increase as the fracture network develops.
- High fluid overpressure (or low lithostatic stress) promotes the growth of Mode 1 dilatant fractures rather than Mode 2 shear fractures.

However, some of these insights are unexpected or even counter-intuitive, for example:

- The duration of the deformation episode is often not a significant control of fracture density. This is because the fracture network grows very rapidly once the fractures start propagating and can reach local saturation in a very short time relative to the duration deformation episode (often tens of thousands rather than millions of years).
- This also means that the fracture network will tend to sweep across the geological structure as a sharp "deformation front". Lateral variation in the stiffness or the horizontal strain rate will, therefore, result in a sharp boundary between unfractured and fully saturated zones, rather than a gradual variation in fracture density across the structure.
- The subcritical fracture propagation index and the initial microfracture density are both key controls on the mean length of layer-bound fractures. This is because they both control the nucleation rate of layer-bound fractures: if the fracture nucleation rate is higher (due to a low subcritical fracture propagation index or a high initial microfracture density), then there is more likelihood of fractures terminating due to stress shadow interaction, and hence the mean fracture length will be shorter.
- The fracture network will often develop more quickly if the subcritical fracture propagation index is low, even though individual fractures will propagate more slowly. There are two possible reasons for this: first, as noted above, the fracture nucleation rate is higher when the subcritical fracture propagation index is lower; and second, a low subcritical fracture propagation index allows fractures to start propagating even when the stress concentration at the fracture tip is low, while a high subcritical fracture propagation index requires the stress concentration at the fracture tip to approach the critical value for the fracture to propagate. Even though subcritical fracture propagation at low stress is slow, over geological timescales it can be sufficient to cause the fracture network to reach saturation.
- In addition to a high fluid overpressure (or low lithostatic stress), a high subcritical fracture propagation index is also required to generate Mode 1 dilatant fractures;

if the subcritical fracture propagation index is low, Mode 2 shear fractures will tend to develop first.

This is by no means an exhaustive list of insights that we can obtain using the algorithm and many aspects are worth further investigation. One phenomenon that should certainly be studied further is the development of stress shadows around fractures. We have shown that stress shadows can explain the spacing of fractures observed in outcrops such as Nash Point, and many other examples; conversely, we have also shown that stress shadows are incompatible with the close spacing of fractures observed in fracture corridors, in outcrops such as Pegwell Bay. However, it is not clear what factors determine whether fractures will develop stress shadows or not. This is an important question, as it would allow us to predict whether we would expect to find regularly-spaced fracture or fracture corridors. As we discussed in Sect. 9.3.3, it may be related to the fluid overpressure or to the boundaries between the mechanical layers, but we do not yet have enough examples to come to any conclusions.

11.2 Applications of the Method

Both implicit and explicit fracture models built using this method can be useful in applications where it is necessary to understand the impact of natural fracture networks on geomechanics and rock failure or on fluid flow in the subsurface. Examples of the former application include tunnelling, mining, foundation stability and nuclear waste storage; examples of the latter include groundwater management, hydrocarbon exploration and production, geothermal energy extraction and CO_2 sequestration. In both cases, a more accurate and detailed characterisation of the natural fracture network will result in more accurate geomechanical or fluid flow modelling.

11.2.1 Applications in Geomechanical Modelling

In the case of geomechanical modelling, numerical models such as finite element models must be populated with bulk rock elastic moduli. The bulk rock elastic moduli for a naturally fractured rockmass can be calculated using the equations given in Chap. 6. This obviously requires implicit fracture data, especially the mean linear fracture density (P_{32}) of the different fracture sets, the fracture mode and orientation. Lateral variations in the fracture density can have a big impact on the bulk rock elastic moduli, and hence, the stress and strain distribution, around an engineering structure such as a tunnel or mineshaft. Furthermore, the constitutive relationships for fractured rock derived in Chap. 6 are clearly anisotropic (technically they are orthotropic), and hence, a numerical method that can handle anisotropy is required for optimal results.

The density, distribution, size and connectivity of naturally fractured networks will also be an important control on rock failure and is thus important for predicting, avoiding or mitigating tunnel or foundation instability, or geohazards such as cliff collapse and landslides.

11.2.2 Applications in Fluid Flow Modelling

In the case of fluid flow modelling, natural fracture networks often provide long-distance, high permeability flow conduits in the subsurface, and must, therefore, be included in any flow model. However, even if the fracture density is known, fracture permeability is very difficult to predict, as it is highly non-linear and very sensitive to factors such as fracture aperture, length and connectivity. It is also often highly anisotropic, reflecting anisotropy in the fracture network, and thus needs to be represented by a permeability tensor rather than a scalar value. In smaller fracture networks, it may be possible to simulate flow through an explicit fracture network directly. A detailed discussion of the different methods of modelling fluid flow through fracture networks is beyond the scope of this book. However, we will observe that the more accurately we want to model fluid flow through the fractures, the more detailed fracture characterisation data we will require to do so; and conversely, if we have more detailed fracture characterisation data available, we will require a more advanced method of modelling fluid flow through the fractures to fully utilise it. For this reason, we believe that further research into modelling fluid flow through fractures would complement this work.

In tight rocks, such as shales, or heavily-cemented sandstones and limestones, fractures will also provide a significant addition to porosity, and in igneous or metamorphic rocks they will often provide the only connected porosity. Modelling fluid flow through fracture networks is thus especially important in unconventional hydrocarbon production and geothermal energy extraction from basement rocks. In higher porosity rocks, most fluid storage will be in the matrix pores, and hence, the transfer of fluid between the matrix pores and the fractures is key. This is especially the case in multi-phase flow systems such as hydrocarbon reservoirs or CO_2 sequestration sites; the fractures may have very different relative permeabilities and capillary entry pressures to the host rock, and hence, the saturation of the different phases may vary considerably between the fractures and the matrix. The rate of fluid transfer between the matrix pores and the fractures will be dependent on a number of properties of the fracture network such as fracture spacing, volumetric fracture density, fracture length and fracture connectivity, which will control the mean "block size" of fracture-bound blocks of the host rock. Finally, in some cases cemented or cataclastic fractures may act as fluid barriers in permeable rocks. In these cases, the fracture density and connectivity will control the reduction in permeability caused by the fractures.

11.2.3 Applications in Risk Analysis

As we saw in Sect. 10.2, this method is particularly useful for quickly generating a preliminary fracture model, especially where we have very limited data. The only requirements for generating a fracture model using this method are the geometry of the layer of interest and some idea of the strain history, which can be approximated based on the regional tectonic history or estimated from the curvature of the layer if no other data is available. Direct mechanical property data is useful but if it is not available, the model can be populated with standard mechanical properties for the lithologies in question, or with mechanical properties calculated from well log data using empirical algorithms. The preliminary model can then be refined as more data becomes available.

The method also makes it easy to produce multiple, geologically consistent models for use in risk analysis. One way to do this is to generate multiple models representing different stages in the evolution of the fracture network. The early stages of fracture network evolution can be used for the low case scenarios, the middle stages for the mid-case scenarios and the final fracture network for the high-case scenario. This will generate a series of geologically consistent models that accurately reflect the probability of the fracture network being present at different locations around the structure, as well as demonstrating the evolution of fracture connectivity and anisotropy through time at those locations. Furthermore, if there is uncertainty in input parameters such as mechanical properties or horizontal strain rate, it is easy to run multiple models varying the parameter in question to determine the potential impact of this uncertainty on the resulting fracture network. In some cases, the parameter in question may have little impact on the final fracture network; in other cases, it may have a significant impact, suggesting that further work may be required to reduce the uncertainty.

11.2.4 Comparison with Conventional Methods

To summarise, this method of generating fracture models has advantages over the conventional stochastic methods in four key aspects: accuracy, insights gained, ease of use and risk analysis. These are summarised in Table 11.1.

11.3 Future Developments

In this book, we have focussed on modelling the growth of layer-bound fractures formed during a single deformation episode. This is applicable to many outcrop and subsurface examples of natural fracture systems, including those examined in Chaps. 9 and 10. However, we also see examples of natural fracture systems resulting

Table 11.1 Advantages of using this method of generating fracture models compared with conventional stochastic techniques

	Stochastic methods	This method
Accuracy	Since they are calibrated against limited and incomplete data, conventional methods often generate geologically unrealistic fracture models, and at best, there will be a high level of uncertainty in the results	Since it simulates the physical processes of fracture nucleation and propagation, the output will automatically honour the geology, geomechanical properties and structural evolution of the reservoir. The detail of the output will reflect the detail of the input data, but even using basic input data the outputs will be geologically consistent
Insights gained	Key parameters such as fracture density, size and connectivity must be specified in the input data. Since the fracture models merely reflect this input data, they are only able to provide limited help to quantify parameters of interest (e.g. connected fracture volume, permeability anisotropy, etc.)	Key parameters such as fracture density, size and connectivity emerge from the simulation, and cannot be calculated directly from the input data. The method, therefore, adds genuine additional information and insight to the model
Ease of use	Since there is very little constraint on the input parameters, a "trial and error" approach with a lot of tweaking is required to get reasonable results. It can, therefore, be very time-consuming to generate fracture models	Since this is a deterministic method, it is easy to set up and run. It can typically be run with data already available in most static geomodels (e.g. curvature of the top horizon and default mechanical properties for the known lithology). However, it is easy to build more detailed models, for example, taking into account lateral variability in mechanical property due to facies variations
Risk analysis	Since the input parameters are very poorly constrained and often do not represent geological attributes at all, it is difficult to systematically generate multiple realistic models or to rank them in likelihood. It is of course easy to generate multiple models by changing input parameters but it is difficult to analyse these systematically to quantify the range of possible outcomes	Since this method simulates the growth of the fracture network, it is easy to output models at different stages of fracture development, which will accurately reflect the fracture networks we would expect for different intensities of deformation. Where there is uncertainty in, for example, mechanical properties, we can vary these to generate multiple models very quickly

from multiple deformation episodes or spanning multiple mechanical layers. In this section, we will consider the enhancements to the model that would be necessary to model these.

11.3.1 Multiple Deformation Episodes

When a natural fracture system develops in an isotropic rockmass during a single deformation episode, we expect it to comprise either one or two fracture sets that strike perpendicular to the principle horizontal stress orientations. With an extensional applied strain, the fractures will either have Mode 1 dilatant displacement or Mode 2 normal shear displacement and, hence, the horizontal component of displacement will always be parallel to the respective principal stress orientation. We can, therefore, use the orthotropic form of Hooke's Law derived in Chap. 6, and the stress and strain tensors will be coaxial.

However, if a layer is subjected to multiple deformation episodes, the orientation of the applied horizontal strain in the later episodes may vary, and, therefore, the principal stresses during these episodes may not be orthogonal to the fractures developed in earlier deformation episodes. In the simplest case, where the earlier fractures are not reactivated during later deformation episodes (e.g. due to healing or cementation), we will develop new fracture sets striking perpendicular to the new principal stresses. These can be simulated by generating multiple fracture models corresponding to the different deformation episodes and overlaying the results.

In many cases, however, the fractures developed during earlier deformation episodes may be reactivated during the later episodes. In this case, there will be an element of strike-slip displacement on the fractures and their horizontal displacement will not be parallel to the horizontal principal stress orientations. We will need to use an anisotropic form of Hooke's Law to relate the stress and strain tensors, which will not be coaxial.

There may also be scenarios in which strike-slip fractures develop during single deformation episodes, for example, in response to a strike-slip deformation episode in which the minimum horizontal strain is extensional but the maximum horizontal strain is compressional. Equally, a biaxial horizontal strain in which the minimum and maximum horizontal strains are approximately equal can sometimes generate polygonal fracture systems, which comprise multiple fracture sets striking in different orientations (Cartwright et al. 2003).

An algorithm that allowed for reactivation of pre-existing oblique fractures and non-coaxial stress and strain tensors would be useful for investigating the circumstances in which we would expect growth of existing fractures compared with the formation of new fractures. It would also be useful for predicting the likely orientation of fracture sets developed in strike-slip deformation episodes and for predicting the development of polygonal fracture systems. We hope to develop and publish more in this respect in future.

Fig. 11.1 Examples of Mode 1 fractures crossing multiple mechanical layers from Kilve, southeast England. Note that the majority of the fractures seen here are confined to the brittle limestone layer, but two of the fractures propagate across the underlying ductile shale and connect to the next limestone layer. Note also that the aperture of these fractures is much greater than that of the layer-bound fractures

11.3.2 Multiple Mechanical Layers

Most fractures observed in natural fracture systems are confined to a single mechanical layer and terminate at the layer boundaries. However, we do also see fractures that span multiple mechanical layers, often crossing ductile layers to connect to other brittle layers (e.g. Fig. 11.1). Although these fractures are usually rare, they are particularly important for two reasons:

- They tend to be larger than the layer-bound fractures, with greater aperture or displacement. This is because the equilibrium fracture displacement is proportional to fracture height, and thus will increase as the fractures grow. Indeed, it has been suggested that larger fault systems often develop from fractures that propagate across mechanical boundaries to link adjacent brittle layers and gradually accumulate and localise displacement (e.g. Davies et al. 2012).
- They can provide important fluid flow pathways between reservoir units that might otherwise be isolated by the intervening ductile layers. This can provide pressure connectivity between different aquifer or reservoir units but also cause breaching of sealing units for hydrocarbon or CO_2 accumulations.

The algorithms required to model the development of fracture networks spanning multiple mechanical layers are more complex than those presented in this book, for several reasons:

- The analytical expressions for stress concentration around fractures used in Sects. 3.14.1 and 4.1 only apply in a homogeneous elastic medium. We would, therefore, need to develop more complex expressions to calculate the stress concentration around the tips of fractures spanning multiple mechanical layers.

- The propagation rate of layer-bound fractures is independent of the fracture length since it is controlled by the fracture height which is constant. This simplifies the expressions for growth of the layer-bound fracture population derived in Sects. 4.2 and 4.3. This would not be the case for fractures spanning multiple mechanical layers, which will grow in height as they grow in length.
- Since the stress shadow width is also proportional to fracture height, this will also grow through time for fractures spanning multiple mechanical layers. At any given time, we would, therefore, expect fractures of different sizes to have different stress shadow widths, which will make a calculation of the exclusion zone volume much more complex.

It may be necessary to use numerical approximations to address some of these issues. However, previous studies have shown that by making some approximations, it is possible to derive simple but useful mechanical models of fractures spanning multiple mechanical layers (e.g. Rijken and Cooke 2001; Welch et al. 2009). It is, therefore, hoped that a similar approach will yield fruit in this case.

11.4 Concluding Remarks

With this book, we have shown in detail the physics behind our unique method to calculate both implicit and explicit fracture networks. We have shown how it is used and calibrated on three outcrops and applied with success to two subsurface examples. With this method, we envisage that it will be possible to create more accurate models and lower the uncertainty in the prediction of natural fracture networks, making it possible to plan and execute projects in tunnel excavation and mining, groundwater management, hydrocarbon exploration and production, geothermal energy extraction and CO_2 sequestration in a safer, faster and more economically viable way.

References

Cartwright, J., D. James, and A. Bolton. 2003. The genesis of polygonal fault systems: A review. In *Subsurface sediment mobilization*, ed. P. Van Rensbergen, R.R. Hillis, A.J. Maltman, and C.K. Morley, 216, 223–243. Geological Society of London Special Publication.

Davies, R.K., R.J. Knipe, and M.J. Welch. 2012. The role of vertical mechanical heterogeneity in predicting fault zone architecture. In *Proceedings of the 3rd EAGE International Conference on Fault and Top Seals*, Montpelier, France.

Rijken, P., and M.L. Cooke. 2001. Role of shale thickness on vertical connectivity of fractures: Application of crack-bridging theory to the Austin Chalk, Texas. *Tectonophysics* 337: 117–133.

Welch, M.J., R.K. Davies, R.J. Knipe, and C. Tueckmantel. 2009. A dynamic model for fault nucleation and propagation in a mechanically layered section. *Tectonophysics* 474: 473–492.

Appendix A
Calculating Fracture Populations When Subcritical Index $b \leq 2$

Abstract In the main text, we derive expressions for fracture propagation where the subcritical fracture propagation index is greater than 2. This will be the case in most realistic geological scenarios. However in this appendix, we show how to modify these expressions for scenarios where the subcritical fracture propagation index is less than or equal to 2

Keywords Fracture modelling · Discrete fracture networks · Naturally fractured reservoir · Subcritical crack propagation

The equations for the cumulative fracture density distribution functions given in the main text are all based on the solution of the differential equation (3.7) in Sect. 3.2, which is valid when the subcritical fracture propagation index, b, is greater than 2. This will be the case in most realistic scenarios, as b typically has a value of 10 or more in most rock types (Atkinson 1984; Holder et al. 2001). However for completeness, we include here the solutions to (3.7) and resulting cumulative fracture density distribution functions for $b < 2$ and $b = 2$.

Solutions for $b < 2$

Equation (3.7) relates the rate of microfracture growth to the microfracture radius:

$$\frac{dr}{dt} = A\left(\frac{2}{\sqrt{\pi}K_c}\right)^b \sigma_d^b r^{\frac{b}{2}} \tag{A.1}$$

When $b < 2$, the solution to (A.1) is given by

$$r(t) = \gamma(|t_i + t|)^\beta \tag{A.2}$$

Equation (3.9) is still valid, but note that β is now positive. (3.10) now becomes

$$r_0 = \left(r_1^{\frac{1}{\beta}} - \gamma^{\frac{1}{\beta}} t_1\right)^\beta \tag{A.3}$$

© The Editor(s) (if applicable) and The Author(s), under exclusive license to Springer Nature Switzerland AG 2020

M. J. Welch et al., *Modelling the Evolution of Natural Fracture Networks*,
https://doi.org/10.1007/978-3-030-52414-2

The form of (A.3) for variable driving stress can still be written as

$$r_0 = \left(r_{\mathbf{N}}^{\frac{1}{\beta}} + \Gamma_{\mathbf{N}} \right)^{\beta} \tag{A.4}$$

but the cumulative microfracture growth factor, $\Gamma_{\mathbf{N}}$, must be redefined so that

$$\Gamma_{\mathbf{N}} = -\sum_{\mathbf{J}=1}^{\mathbf{N}} \left[\gamma_{\mathbf{J}}^{\frac{1}{\beta}} (\mathrm{t}_{\mathbf{J}} - \mathrm{t}_{\mathbf{J}-1}) \right] \tag{A.5}$$

If we substitute (A.5) for (3.12), the subsequent cumulative fracture density distribution equations for ${}_{\mu F}P_{30}$, ${}_{\mu F}P_{32}$, ${}_{MF}P_{30}$, ${}_{MF}P_{32}$, ${}^{a}_{\mu F}P_{30}$, ${}^{a}_{\mu F}P_{32}$, ${}^{a}_{MF}P_{30}$, ${}^{s}_{MF}P_{30}$, ${}^{a}_{MF}P_{32}$ and ${}^{s}_{MF}P_{32}$ (3.14, 3.18, 4.10, 4.16, 5.29, 5.33, 5.35, 5.41–5.47, 5.50, 5.55, and 5.67–5.72, respectively), which are all based on Γ, will still be valid. The only exceptions are the equations for ${}^{s}_{\mu F}P_{30}$ and ${}^{s}_{\mu F}P_{32}$ (5.32 and 5.34), which must be modified to:

$$\begin{aligned} &{}^{s}_{\mu F}P_{30}(\mathrm{r}_N, \mathrm{t}_N) \\ &= B \sum_{K=1}^{N} \left[\begin{array}{c} \frac{\overline{q_K}\theta_{K-1}}{(1-\beta c)\gamma_K^{\frac{1}{\beta}}} \left(\left(\left(\frac{h}{2}\right)^{\frac{1}{\beta}} + \Gamma_K \right)^{1-\beta c} - \left(\left(\frac{h}{2}\right)^{\frac{1}{\beta}} + \Gamma_{K-1} \right)^{1-\beta c} - \left(\mathrm{r}_N^{\frac{1}{\beta}} + \Gamma_K \right)^{1-\beta c} + \left(\mathrm{r}_N^{\frac{1}{\beta}} + \Gamma_{K-1} \right)^{1-\beta c} \right) \\ + (\theta_{K-1} - \theta'_{K-1}) \left(\left(\left(\frac{h}{2}\right)^{\frac{1}{\beta}} + \Gamma_K \right)^{-\beta c} - \left(\left(\frac{h}{2}\right)^{\frac{1}{\beta}} + \Gamma_{K-1} \right)^{-\beta c} \right) \end{array} \right] \end{aligned} \tag{A.6}$$

and

$$\begin{aligned} &{}^{s}_{\mu F}P_{32}(0, \mathrm{t}_N) \\ &\approx \frac{\pi B}{(1-\beta c)} \left(\frac{h}{2}\right)^2 \sum_{i=1}^{I-1} \left[\left(\frac{i}{2I}\right)^2 \sum_{K=1}^{N} \left[\frac{\overline{q_K}\theta_{K-1}}{\gamma_K^{\frac{1}{\beta}}} \left(\begin{array}{c} \left(\left(\frac{ih}{2I}\right)^{\frac{1}{\beta}} + \Gamma_{K-1} \right)^{1-\beta c} - \left(\left(\frac{ih}{2I}\right)^{\frac{1}{\beta}} + \Gamma_K \right)^{1-\beta c} \\ - \left(\left(\frac{(i+1)h}{2I}\right)^{\frac{1}{\beta}} + \Gamma_{K-1} \right)^{1-\beta c} + \left(\left(\frac{(i+1)h}{2I}\right)^{\frac{1}{\beta}} + \Gamma_K \right)^{1-\beta c} \end{array} \right) \right] \right] \end{aligned} \tag{A.7}$$

Solutions for $b = 2$

When $b = 2$, the solution to (A.1) takes on an exponential form:

$$r(t) = \mathrm{r}_0 \mathrm{e}^{\gamma t} \tag{A.8}$$

where

$$\gamma = A \left(\frac{2\sigma_d}{\sqrt{\pi} K_c} \right)^b = A \frac{4\sigma_d^2}{\pi K_c^2} \tag{A.9}$$

Note that β is undefined when $b = 2$. Rearranging (A.8) gives:

$$r_0 = r_1 e^{-\gamma t_1} \tag{A.10}$$

Table A.1 Equivalent forms for various factors in the equations for the cumulative fracture density distribution when $b = 2$ and when $b \neq 2$. Note that Γ_* can signify any function

Component when $b \neq 2$		Component when $b = 2$
$\left(\frac{h}{2}\right)^{\frac{b}{2}}$	$\Leftrightarrow$	$\frac{h}{2}$
$\left(\frac{2}{h}\right)^{\frac{b}{2}}\frac{1}{2\beta}$	$\Leftrightarrow$	$\frac{1}{h}$
$\left(\left(\frac{h}{2}\right)^{\frac{1}{\beta}}+\Gamma_*\right)^{-\beta c}$	$\Leftrightarrow$	$e^{-c\left(\ln\left(\frac{h}{2}\right)+\Gamma_*\right)}=\left(\frac{h}{2}\right)^{-c}e^{-c\Gamma_*}$
$\frac{-\beta}{(1-\beta c)}\left(\left(\frac{h}{2}\right)^{\frac{1}{\beta}}+\Gamma_*\right)^{1-\beta c}$	$\Leftrightarrow$	$\frac{1}{c}e^{-c\left(\ln\left(\frac{h}{2}\right)+\Gamma_*\right)}=\frac{1}{c}\left(\frac{h}{2}\right)^{-c}e^{-c\Gamma_*}$
$\frac{(2-\beta c)}{-\beta}\left(\left(\frac{h}{2}\right)^{\frac{1}{\beta}}+\Gamma_*\right)^{1-\beta c}$	$\Leftrightarrow$	$ce^{-c\left(\ln\left(\frac{h}{2}\right)+\Gamma_*\right)}=c\left(\frac{h}{2}\right)^{-c}e^{-c\Gamma_*}$
$\frac{\beta^2}{(1-\beta c)(1-2\beta c)}\left(\left(\frac{h}{2}\right)^{\frac{1}{\beta}}+\Gamma_*\right)^{2-\beta c}$	$\Leftrightarrow$	$\frac{1}{c^2}e^{-c\left(\ln\left(\frac{h}{2}\right)+\Gamma_*\right)}=\frac{1}{c^2}\left(\frac{h}{2}\right)^{-c}e^{-c\Gamma_*}$
$\left(\frac{h}{2}\right)^{\frac{1-\beta c}{\beta}}$	$\Leftrightarrow$	$\left(\frac{h}{2}\right)^{-c}$

and the form of (A.9) for variable driving stress is

$$r_0 = r_{\mathbf{N}} e^{\Gamma_{\mathbf{N}}} \tag{A.11}$$

where

$$\Gamma_{\mathbf{N}} = -\sum_{\mathbf{J}=1}^{\mathbf{N}}\left[\gamma_{\mathbf{J}}^{\frac{1}{\beta}}\left(t_{\mathbf{J}} - t_{\mathbf{J}-1}\right)\right] \tag{A.12}$$

The subsequent cumulative fracture density distribution equations for ${}_{\mu F}P_{30}$, ${}_{\mu F}P_{32}$, ${}_{MF}P_{30}$, ${}_{MF}P_{32}$, ${}^{a}_{\mu F}P_{30}$, ${}^{s}_{\mu F}P_{30}$, ${}^{a}_{\mu F}P_{32}$, ${}^{s}_{\mu F}P_{32}$, ${}^{a}_{MF}P_{30}$, ${}^{s}_{MF}P_{30}$, ${}^{a}_{MF}P_{32}$ and ${}^{s}_{MF}P_{32}$ will all take on exponential forms, equivalent to the power-law forms described by (3.14, 3.18, 4.10, 4.16, 5.29, 5.33, 5.35, 5.41–5.47, 5.50, 5.55, and 5.67–5.72), respectively. We can convert from the power-law forms to the exponential forms by making the substitutions listed in Table A.1. For example, (3.14) becomes

$${}_{\mu F}P_{30}(r_N, t_N) = Br_0^{-c} = Br_{\mathbf{N}}^{-c}e^{-c\Gamma_{\mathbf{N}}} \tag{A.13}$$

Equation (4.10) becomes

$${}_{MF}P_{30}(L_N, t_N) = B\left(\frac{h}{2}\right)^{-c} e^{-c\left(\Gamma_N + \frac{L_N}{h}\right)} \tag{A.14}$$

and (4.16) becomes

$$ {}_{MF}P_{32}(L_N, t_N) = 2Bh\left(\frac{h}{2}\right)^{-c}\left(\frac{h}{2c}\left(e^{-c\left(\Gamma_N+\frac{L_N}{h}\right)}-1\right)+\frac{L_N}{2}e^{-c\left(\Gamma_N+\frac{L_N}{h}\right)}\right) \tag{A.15} $$

Note that when $b = 2$, we can derive an analytical solution to the integral for the mean linear microfracture density, ${}_{\mu F}P_{32}$.

$$ \begin{aligned} {}_{\mu F}P_{32}(r_N, t_N) &= \int_{r_N}^{\frac{h}{2}} cBr^{-(c+1)}e^{-c\Gamma_N}\pi r^2 dr \\ &= \pi cBe^{-c\Gamma_N}\left[\frac{r^{2-c}}{2-c}\right]_{r_N}^{\frac{h}{2}} = \pi Be^{-c\Gamma_N}\frac{c}{2-c}\left(\left(\frac{h}{2}\right)^{2-c} - r_N^{2-c}\right) \end{aligned} \tag{A.16} $$

when $c \neq 2$, or

$$ {}_{\mu F}P_{32}(r_N, t_N) = \pi cBe^{-c\Gamma_N}[\ln(r)]_{r_N}^{\frac{h}{2}} = \pi Be^{-c\Gamma_N}c\left(\ln\left(\frac{h}{2}\right) - \ln(r_N)\right) \tag{A.17} $$

when $c = 2$.

Appendix B
Calculating the Total Exclusion Zone Volume

Abstract In this appendix we show how to derive an expression for the exclusion zone volume, representing the volume in which new macrofractures cannot nucleate due to the stress shadows of adjacent macrofractures. Since the exclusion zones around individual fractures can partially overlap, the total exclusion zone volume depends on the spatial distribution of the fractures, and will evolve through time as the fracture population grows. However, it is an important parameter as it controls not only the macrofracture nucleation rate but also the probability of fracture intersection, the fracture connectivity and the length distribution of the secondary fracture set. We show how to derive numerical expressions that also describe the evolution of the fracture spacing distribution through time.

Keywords Fracture modelling · Discrete fracture networks · Naturally fractured reservoir · Stress shadow

In this appendix we show how to derive an expression for the exclusion zone volume, representing the volume in which new macrofractures cannot nucleate due to the stress shadows of adjacent macrofractures. Since the exclusion zones around individual fractures can partially overlap, the total exclusion zone volume depends on the spatial distribution of the fractures, and will evolve through time as the fracture population grows. However, it is an important parameter as it controls not only the macrofracture nucleation rate but also the probability of fracture intersection, the fracture connectivity and the length distribution of the secondary fracture set. We show how to derive numerical expressions that also describe the evolution of the fracture spacing distribution through time.

As we saw in Sect. 5.1, when calculating the probability of fracture deactivation, it is necessary to calculate the combined volume of the "exclusion zones" surrounding a set of macrofractures. The exclusion zone around an individual macrofracture represents the zone in which no other parallel macrofracture may nucleate or propagate, because its stress shadow would overlap that of the first macrofracture. Therefore,

© The Editor(s) (if applicable) and The Author(s), under exclusive license to Springer Nature Switzerland AG 2020

M. J. Welch et al., *Modelling the Evolution of Natural Fracture Networks*,
https://doi.org/10.1007/978-3-030-52414-2

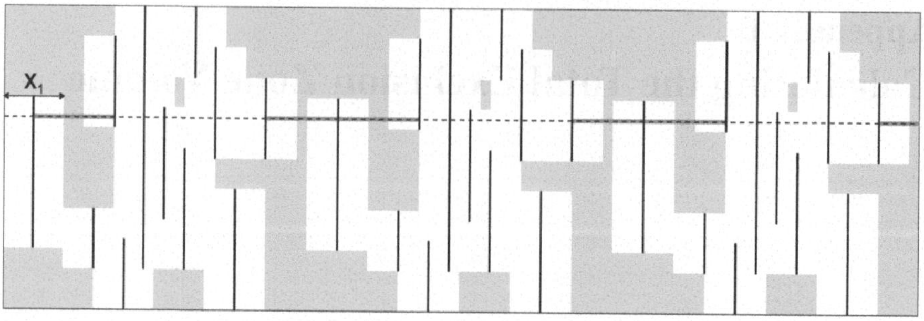

Fig. B.1 Schematic illustration of the stress shadows (shaded boxes) and exclusion zones (striped boxes) around a set of N-S striking macrofractures (solid black lines). The stress shadows around adjacent fractures cannot overlap, but the exclusion zones, which represent the minimum separation between the fractures, can partially overlap. The dashed line represents a 1D transect perpendicular to the fracture strike, which we can use to calculate the fracture spacing distribution

total exclusion zone volume $\chi(W)$ represents the proportion of the total volume in which new macrofractures with stress shadow width W cannot nucleate, and into which existing macrofractures with stress shadow width W cannot propagate. The inverse of the total exclusion zone volume is the clear zone volume, $1 - \chi(W)$, which represents the proportion of the total volume in which new macrofractures can nucleate.

If all macrofractures in the set have the same stress shadow width, the exclusion zone around each macrofracture is, therefore, twice as wide as the stress shadows (see Fig. B.1). However, calculating the total volume of the exclusion zones around all the macrofractures is more complicated than calculating the total volume of all the stress shadows because, unlike the stress shadows, the exclusion zones can partially overlap. The degree of overlap will depend on the spacing between adjacent macrofractures, but because the macrofractures nucleate in random locations, this spacing will vary. To derive a rigorous formula for the total exclusion zone volume, we must, therefore, consider the spacings between each pair of adjacent macrofractures as comprising a population of macrofracture spacings and examine how this population will change when we add additional macrofractures.

We can simplify the calculation by reducing it to a one-dimensional transect along a line perpendicular to the fracture strike (as shown on Fig. B.1). In this case, the density of fractures intersected by the transect will be given by the total mean linear macrofracture density, ${}_{MF}P_{32}$. We can represent the spacings between adjacent macrofractures intersected by the transect by a cumulative density distribution function $S(x)$, which gives the number of adjacent macrofractures separated by a distance x or greater (measured along the transect), divided by the total transect length. If the transect is sufficiently long that we can ignore end effects, or if we connect both ends of the transect to form an infinitely repeating loop, as shown in Fig. B.2, then the total density of all macrofracture spacings, $S(0)$, will equal the total mean linear macrofracture density.

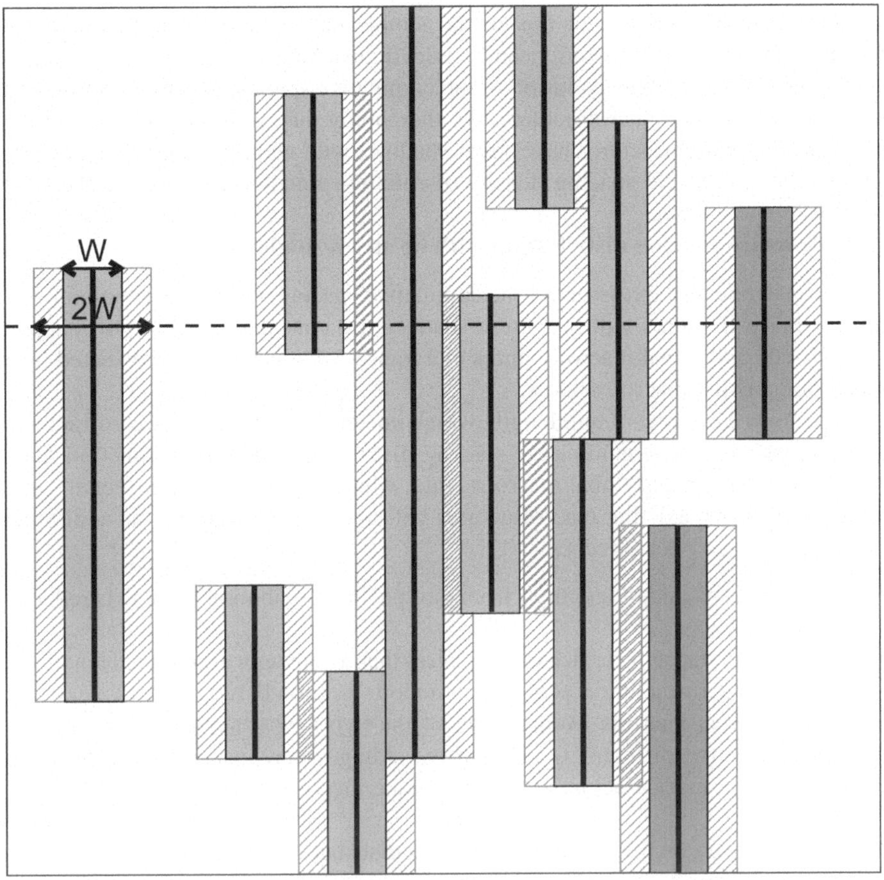

Fig. B.2 Schematic illustration showing a finite transect across a periodically repeating pattern of macrofractures to avoid end effects. The portion of the transect shaded grey represents the total length occupied by macrofracture spacings of length x_1 or greater

We can also calculate the proportion of the transect occupied by macrofracture spacings of a given length (e.g. x_1) or greater, which we denote $\mathbb{L}(\mathrm{x}_1)$. This is calculated by differentiating the cumulative spacing distribution function, $S(x)$, with respect to x, multiplying the result by the spacing length x, and then integrating from x_1 to infinity:

$$\mathbb{L}(\mathrm{x}_1) = \int_{\mathrm{x}_1}^{\infty} -\frac{dS}{dx} x dx \tag{B.1}$$

Note that if we can ignore end effects, the proportion of the transect occupied by all macrofracture spacings, $L(0)$, must be 1.

We start by demonstrating this in the simple case where there are no stress shadows, so new macrofractures can be added anywhere on the transect. We will show how to calculate the evolution of the cumulative spacing distribution function as we add additional macrofractures. We then apply the same methodology to the case where all macrofractures have stress shadows with width W, and show how we can use the cumulative spacing density distribution function to calculate the total exclusion zone volume.

Macrofracture spacing distribution with no stress shadows

We can derive an expression for the cumulative spacing density distribution function for macrofractures with no stress shadows by considering what will happen when we add additional macrofractures to a transect that is already intersected by a number of existing macrofractures.

We assume a transect of unit length, which intersects a number of macrofractures given by ${}_{MF}P_{32}$, with a cumulative spacing distribution given by a function $S(x)$. We will now add an additional macrofracture, which intersects the transect at some point between two existing macrofractures with spacing x_n. Adding this additional macrofracture has three effects:

- It removes a spacing of length x_n from the spacing distribution, so $S(x)$ is reduced by 1 for all $x \leq x_n$.
- It adds two new spacings, each of length less than x_n but with a combined length x_n. If the new macrofracture is placed at a random location between the two existing macrofractures, then the average size of the two new spacings will be evenly distributed between 0 and x_n. Therefore, adding a new macrofracture between these existing macrofractures will on average increase $S(x)$ by a factor $2–2x/x_n$ for all $x \leq x_n$.
- It has no impact on the cumulative spacing distribution function for $x > x_n$.

Combining these effects, the increment in $S(x)$ from adding the new macrofracture can be described by:

$$\Delta S(x) = \begin{cases} 1 - \frac{2x}{x_n} & \text{for} x \leq x_n \\ 0 & \text{for} x > x_n \end{cases} \tag{B.2}$$

We can generalise (B.2) to obtain an expression for the rate of change of $S(x)$ for any value of x, as we add new macrofractures (and thereby increase ${}_{MF}P_{32}$):

$$\frac{dS(x)}{d_{MF}P_{32}} = \int_x^{\infty} F(x_n)\left(1 - \frac{2x}{x_n}\right)dx_n \tag{B.3}$$

where $F(x_n)$ represents the probability that any new macrofracture is placed between two existing macrofractures with spacing x_n. This is proportional to the total length of all macrofracture spacings of length x_n:

$$F(x_n) = -\frac{dS(x_n)}{dx_n}x_n \tag{B.4}$$

Combining (B.3) and (B.4) gives

$$\begin{aligned}\frac{dS(x)}{d_{\mathrm{MF}}\mathrm{P}_{32}} &= \int_x^\infty -\frac{dS(x_n)}{dx_n}x_n\left(1-\frac{2x}{x_n}\right)dx_n \\ &= -\int_x^\infty \frac{dS(x_n)}{dx_n}x_n dx_n + \int_x^\infty 2x\frac{dS(x_n)}{dx_n}dx_n \\ &= -\int_x^\infty \frac{dS(x_n)}{dx_n}x_n dx_n - 2xS(x)\end{aligned} \tag{B.5}$$

since $S(\infty) = 0$.

A solution to (B.5) can be obtained using an exponential function for $S(x)$, of the form

$$S(x) = \mathbb{A}e^{-\mathbb{B}x} \tag{B.6}$$

where $\mathbb{A}$ and $\mathbb{B}$ are functions of ${}_{\mathrm{MF}}\mathrm{P}_{32}$ but are independent of x. In this case, we can show that

$$\begin{aligned}\frac{dS(x)}{dx} &= -\mathbb{A}\mathbb{B}e^{-\mathbb{B}x} \\ \Rightarrow \frac{dS(x)}{d_{\mathrm{MF}}\mathrm{P}_{32}} &= -\int_x^\infty -\mathbb{A}\mathbb{B}e^{-\mathbb{B}x_n}x_n dx_n - 2\mathbb{A}xe^{-\mathbb{B}x} \\ &= -\left[\mathbb{A}x_n e^{-\mathbb{B}x_n} + \frac{\mathbb{A}}{\mathbb{B}}e^{-\mathbb{B}x_n}\right]_x^\infty - 2\mathbb{A}xe^{-\mathbb{B}x} \\ &= \frac{\mathbb{A}}{\mathbb{B}}e^{-\mathbb{B}x} - \mathbb{A}xe^{-\mathbb{B}x}\end{aligned} \tag{B.7}$$

However, differentiating (B.6) directly with respect to ${}_{\mathrm{MF}}\mathrm{P}_{32}$ (remembering that $\mathbb{A}$ and $\mathbb{B}$ are both functions of ${}_{\mathrm{MF}}\mathrm{P}_{32}$) gives

$$\frac{dS(x)}{d_{\mathrm{MF}}\mathrm{P}_{32}} = \frac{d\mathbb{A}}{d_{\mathrm{MF}}\mathrm{P}_{32}}e^{-\mathbb{B}x} - \mathbb{A}x\frac{d\mathbb{B}}{d_{\mathrm{MF}}\mathrm{P}_{32}}e^{-\mathbb{B}x} \tag{B.8}$$

Comparing (B.7) and (B.8), we can see that a solution is obtained when $d\mathbb{A}/d_{\mathrm{MF}}\mathrm{P}_{32} = d\mathbb{B}/d_{\mathrm{MF}}\mathrm{P}_{32} = 1$ and $\mathbb{A} = \mathbb{B} = {}_{\mathrm{MF}}\mathrm{P}_{32}$. Therefore,

$$S(x) = {}_{\mathrm{MF}}\mathrm{P}_{32}e^{-{}_{\mathrm{MF}}\mathrm{P}_{32}x} \tag{B.9}$$

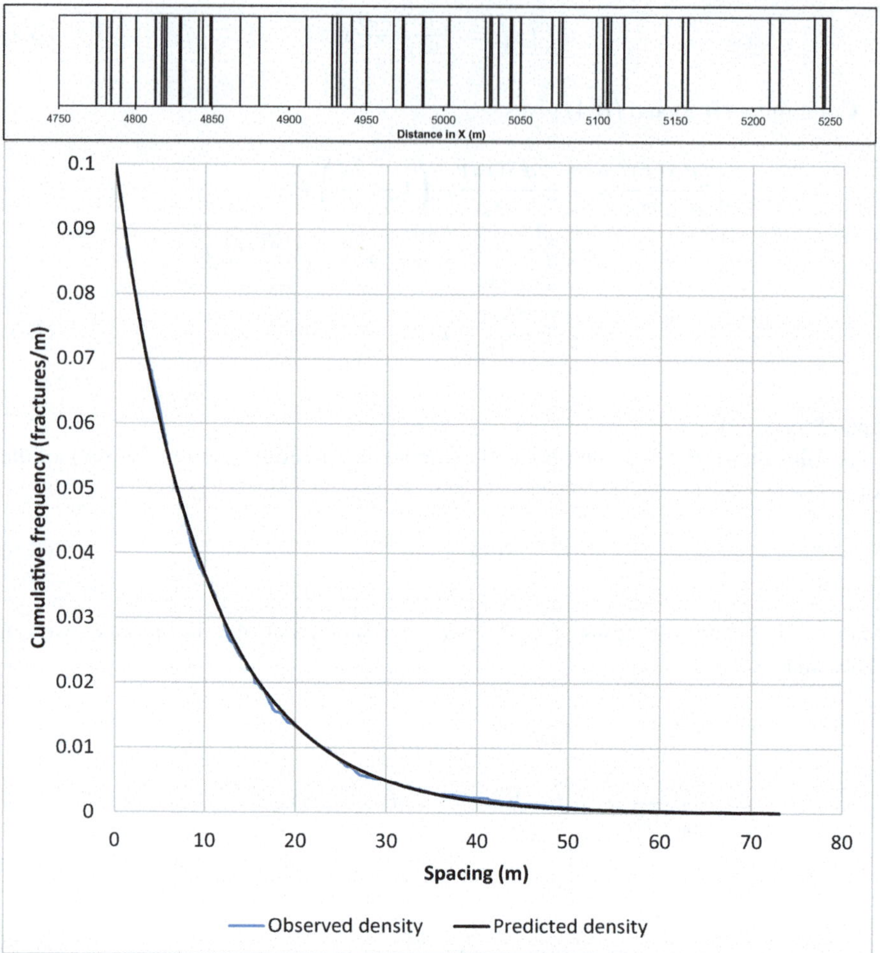

Fig. B.3 Cumulative macrofracture spacing distributions obtained by adding 1000 macrofractures with no stress shadows at random locations along a transect of length 10000 m. The blue line represents the empirical cumulative spacing distribution of the randomly placed macrofractures, while the black line represents the theoretical cumulative spacing distribution derived from (B.9). Above is shown a small portion of the transect; the vertical lines represent the macrofractures crossing the transect

i.e. the macrofracture spacings will follow an exponential distribution if macrofractures are placed at random locations with no constraints. We can demonstrate this by plotting the cumulative macrofracture spacing distribution function $S(x)$ alongside the actual spacings obtained empirically by adding macrofractures at random locations on a long transect. This is illustrated in Fig. B.3.

Macrofracture spacing distribution with stress shadows

We can use a similar technique to determine the spacing distribution for macrofractures with stress shadows. We can then use this to calculate the total exclusion zone volume. However, the calculation is more complicated in this scenario, as there are restrictions on where we can add new macrofractures. One consequence of this is that macrofracture spacing will no longer follow a purely exponential distribution.

In this calculation, we define the macrofracture spacing as the distance between the edges of the stress shadows surrounding adjacent macrofractures, rather than the distance between the macrofractures themselves; as we will see later, this makes it easier to calculate the total macrofracture exclusion zone volume. Therefore. the proportion of the transect occupied by all macrofracture spacings, $\mathbb{L}(0)$, will be $1 - \psi$, where ψ is the proportion of the transect occupied by the macrofracture stress shadows.

As before, we consider the effect on the cumulative spacing distribution function $S(x)$ of adding additional macrofractures to a transect that is already intersected by a number of existing macrofractures. Again we assume a transect of unit length which intersects a number of macrofractures given by ${}_{MF}P_{32}$. However in this case, we can only add new macrofractures between the stress shadows of the existing macrofractures, and the new macrofractures must be placed a distance of at least $W/2$ from the edges of the existing stress shadows (where W is the stress shadow width) to avoid stress shadow overlap. This means that we cannot place new macrofractures into spacings with length less than W; such spacings form a "dead zone" that can never be filled by macrofracture stress shadows. It is, therefore, impossible in this scenario to reach 100% saturation of the transect with macrofracture stress shadows.

We will now add an additional macrofracture at some point between two existing macrofractures, which have a spacing between their respective stress shadows of x_n. Adding this new macrofracture has three effects:

- It removes a spacing of length x_n from the spacing distribution, so $S(x)$ will be reduced by 1 for all $x \leq x_n$.
- It adds two new spacings, with a combined length $x_n - W$. If the new macrofracture is placed at a random location between the two existing macrofractures, then the average size of the two new spacings will be evenly distributed between 0 and $x_n - W$. Therefore, adding a new macrofracture between these existing macrofractures will on average increase $S(x)$ by a factor $2–2x/(x_n - W)$ for all $x \leq x_n - W$.
- It has no impact on the cumulative spacing distribution function for $x > x_n$.

Combining these effects, the increment in $S(x)$ from adding the new macrofracture can be described by:

$$\Delta S(x) = \begin{cases} 1 - \frac{2x}{x_n - W} & \text{for } x \leq x_n - W \\ -1 & \text{for } x_n - W < x \leq x_n \\ 0 & \text{for } x > x_n \end{cases} \tag{B.10}$$

Generalising (B.10) to obtain an expression for the rate of change of the cumulative spacing distribution, $S(x)$, for any value of x gives:

$$\frac{dS(x)}{d_{MF}P_{32}} = \int_{x}^{x+W} -F(x_n)dx_n + \int_{x+W}^{\infty} F(x_n)\left(1 - \frac{2x}{x_n - W}\right)dx_n \tag{B.11}$$

As before, $F(x_n)$ represents the probability that any new macrofracture is placed between two existing macrofractures with spacing x_n. However, this must also be modified, since (as noted above) the new macrofracture can only be placed in spacings with length greater than W, and even in these spacings it cannot be placed within a distance $W/2$ of each endpoint. Therefore $F(x_n)$ is now given by

$$F(x_n) = \begin{cases} \dfrac{-\frac{dS(x_n)}{dx_n}(x_n - W)}{\int_W^{\infty} -\frac{dS(x_n)}{dx_n}(x_n - W)dx_n} & \text{for } x_n \geq W \\ 0 & \text{for } x_n < W \end{cases} \tag{B.12}$$

and hence

$$\frac{dS(x)}{d_{MF}P_{32}} = \begin{cases} \dfrac{\int_x^{x+W} \frac{dS(x_n)}{dx_n}(x_n - W)dx_n - \int_{x+W}^{\infty} \frac{dS(x_n)}{dx_n}(x_n - W)\left(1 - \frac{2x}{x_n - W}\right)dx_n}{\int_W^{\infty} -\frac{dS(x_n)}{dx_n}(x_n - W)dx_n} & \text{for } x \geq W \\ \dfrac{\int_W^{x+W} \frac{dS(x_n)}{dx_n}(x_n - W)dx_n - \int_{x+W}^{\infty} \frac{dS(x_n)}{dx_n}(x_n - W)\left(1 - \frac{2x}{x_n - W}\right)dx_n}{\int_W^{\infty} -\frac{dS(x_n)}{dx_n}(x_n - W)dx_n} & \text{for } x < W \end{cases} \tag{B.13}$$

We start by looking at the form of (B.13) for $x \geq W$. Using the substitution $x_w = x_n - W$, so that $dx_w = dx_n$, we can write this as

$$\begin{aligned} \frac{dS(x)}{d_{MF}P_{32}} &= \frac{\int_{x-W}^{x} \frac{dS(x_w + W)}{dx_w} x_w dx_w - \int_x^{\infty} \frac{dS(x_w + W)}{dx_w} x_w \left(1 - \frac{2x}{x_w}\right) dx_w}{\int_0^{\infty} -\frac{dS(x_w + W)}{dx_w} x_w dx_w} \\ &= \frac{\int_{x-W}^{x} \frac{dS(x_w + W)}{dx_w} x_w dx_w - \int_x^{\infty} \frac{dS(x_w + W)}{dx_w} x_w dx_w - 2xS(x + W)}{\int_0^{\infty} -\frac{dS(x_w + W)}{dx_w} x_w dx_w} \end{aligned} \tag{B.14}$$

We assume that, as before, $S(x)$ follows an exponential function, at least for $x \geq W$, of the form

$$S(x) = \mathbb{A}e^{-\mathbb{B}(x-W)} \tag{B.15}$$

where $\mathbb{A}$ and $\mathbb{B}$ are functions of $_{MF}P_{32}$, but are independent of x. In this case, we can show that

$$\begin{aligned} \frac{dS(x)}{dx} &= -\mathbb{A}\mathbb{B}e^{-\mathbb{B}(x-W)} \\ \Rightarrow \frac{dS(x+W)}{dx} &= -\mathbb{A}\mathbb{B}e^{-\mathbb{B}x} \end{aligned}$$

$$
\begin{aligned}
\Rightarrow \frac{dS(x)}{d_{\mathrm{MF}}\mathrm{P}_{32}} &= \frac{\int_{x-W}^{x} -\mathbb{A}\mathbb{B}e^{-\mathbb{B}x_w} x_w dx_w - \int_x^{\infty} -\mathbb{A}\mathbb{B}e^{-\mathbb{B}x_w} x_w dx_w - 2x\mathbb{A}e^{-\mathbb{B}x}}{\int_0^{\infty} \mathbb{A}\mathbb{B}e^{-\mathbb{B}x_w} x_w dx_w} \\
&= \frac{\left[\mathbb{A}x_w e^{-\mathbb{B}x_w} + \frac{\mathbb{A}}{\mathbb{B}}e^{-\mathbb{B}x_w}\right]_{x-W}^{x} - \left[\mathbb{A}x_w e^{-\mathbb{B}x_w} + \frac{\mathbb{A}}{\mathbb{B}}e^{-\mathbb{B}x_w}\right]_x^{\infty} - 2x\mathbb{A}e^{-\mathbb{B}x}}{-\left[\mathbb{A}x_w e^{-\mathbb{B}x_w} + \frac{\mathbb{A}}{\mathbb{B}}e^{-\mathbb{B}x_w}\right]_0^{\infty}} \\
&= \frac{2\frac{\mathbb{A}}{\mathbb{B}}e^{-\mathbb{B}x} - \mathbb{A}(x-W)e^{-\mathbb{B}(x-W)} - \frac{\mathbb{A}}{\mathbb{B}}e^{-\mathbb{B}(x-W)}}{\frac{\mathbb{A}}{\mathbb{B}}} \\
&= \left(2e^{-\mathbb{B}W} - 1\right)e^{-\mathbb{B}(x-W)} - \mathbb{B}(x-W)e^{-\mathbb{B}(x-W)}
\end{aligned}
\tag{B.16}
$$

Differentiating (B.15) directly with respect to ${}_{\mathrm{MF}}\mathrm{P}_{32}$ (remembering that $\mathbb{A}$ and $\mathbb{B}$ are both functions of ${}_{\mathrm{MF}}\mathrm{P}_{32}$) gives

$$
\frac{dS(x)}{d_{\mathrm{MF}}\mathrm{P}_{32}} = \frac{d\mathbb{A}}{d_{\mathrm{MF}}\mathrm{P}_{32}}e^{-\mathbb{B}(x-W)} - \mathbb{A}(x-W)\frac{d\mathbb{B}}{d_{\mathrm{MF}}\mathrm{P}_{32}}e^{-\mathbb{B}(x-W)} \tag{B.17}
$$

Since (B.16) and (B.17) are valid for any value of x, we can equate the $e^{-\mathbb{B}(x-W)}$ terms to get

$$
\frac{d\mathbb{A}}{d_{\mathrm{MF}}\mathrm{P}_{32}} = \left(2e^{-\mathbb{B}W} - 1\right) \tag{B.18}
$$

and we can equate the $(x-W)e^{-\mathbb{B}(x-W)}$ terms to get

$$
\frac{d\mathbb{B}}{d_{\mathrm{MF}}\mathrm{P}_{32}} = \frac{\mathbb{B}}{\mathbb{A}} \tag{B.19}
$$

Although there is no obvious analytical solution to (B.18) and (B.19), it is fairly easy to derive values for $\mathbb{A}$ and $\mathbb{B}$ numerically, as shown in Fig. B.4. Initially, both $\mathbb{A}$ and $\mathbb{B}$ increase in proportion to ${}_{\mathrm{MF}}\mathrm{P}_{32}$ (i.e. $\mathbb{A} \approx \mathbb{B} \approx {}_{\mathrm{MF}}\mathrm{P}_{32}$) so the macrofracture spacing distribution is similar to that obtained with no stress shadows; however, as more fractures are added, the gradient of $\mathbb{A}$ levels off while the gradient of $\mathbb{B}$ increases. Eventually $\mathbb{A}$ starts to drop back towards zero, which it reaches when the total stress shadow volume is about 0.75. This represents the point at which there are no macrofracture spacings with length greater than W, and hence it is not possible to add any more macrofractures, even though approximately 25% of the transect is still outside the macrofracture stress shadows.

We can now look at the form of (B.13) for spacings where $x < W$. Again using the substitution $x_w = x_n - W$, so that $dx_w = dx_n$, we can write this as

$$
\frac{dS(x)}{d_{\mathrm{MF}}\mathrm{P}_{32}} = \frac{\int_0^{x} \frac{dS(x_w+W)}{dx_w} x_w dx_w - \int_x^{\infty} \frac{dS(x_w+W)}{dx_w} x_w dx_w - 2xS(x+W)}{\int_0^{\infty} -\frac{dS(x_w+W)}{dx_w} x_w dx_w} \tag{B.20}
$$

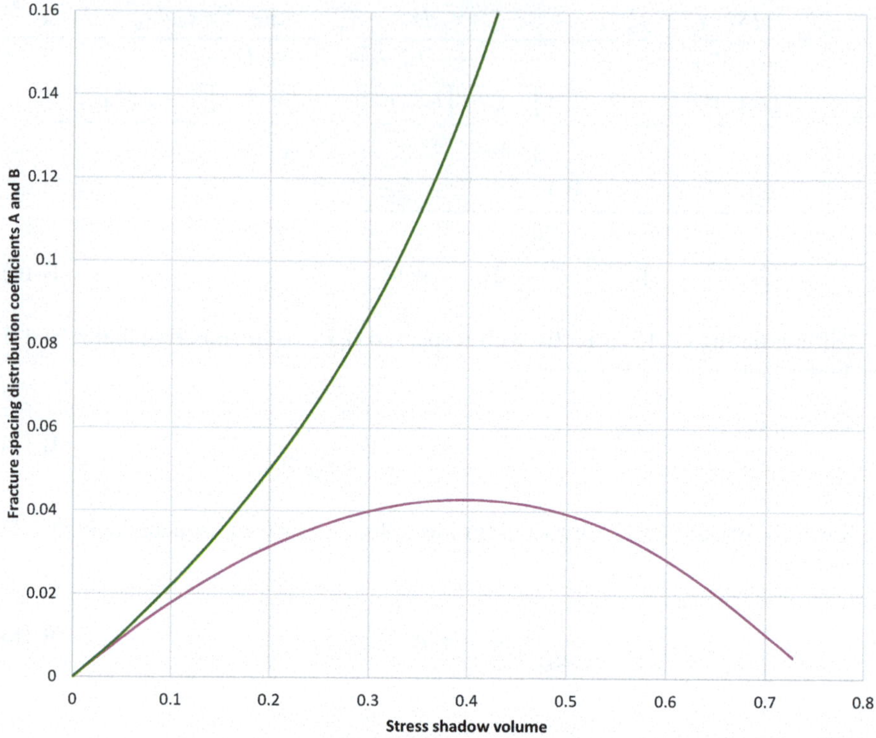

Fig. B.4 Chart showing the cumulative spacing distribution parameters $\mathbb{A}$ (in pink) and $\mathbb{B}$ (in green) as a function of total stress shadow volume, derived numerically from (B.18) and (B.19). NB The total stress shadow volume $\psi = W_{\mathrm{MF}}\mathrm{P}_{32}$

Note that, although (B.20) is valid for $x < W$, it only contains expressions for $S(x + W)$ and $dS(x + W)/dx$, and since $x + W \geq W$, we can substitute (B.15) for these terms to obtain

$$\begin{aligned}
\frac{dS(x)}{d_{\mathrm{MF}}\mathrm{P}_{32}} &= \frac{\int_0^x -\mathbb{A}\mathbb{B}e^{-\mathbb{B}x_w}x_w dx_w - \int_x^\infty -\mathbb{A}\mathbb{B}e^{-\mathbb{B}x_w}x_w dx_w - 2x\mathbb{A}e^{-\mathbb{B}x}}{\int_0^\infty \mathbb{A}\mathbb{B}e^{-\mathbb{B}x_w}x_w dx_w} \\
&= \frac{\left[\mathbb{A}x_w e^{-\mathbb{B}x_w} + \frac{\mathbb{A}}{\mathbb{B}}e^{-\mathbb{B}x_w}\right]_0^x - \left[\mathbb{A}x_w e^{-\mathbb{B}x_w} + \frac{\mathbb{A}}{\mathbb{B}}e^{-\mathbb{B}x_w}\right]_x^\infty - 2x\mathbb{A}e^{-\mathbb{B}x}}{-\left[\mathbb{A}x_w e^{-\mathbb{B}x_w} + \frac{\mathbb{A}}{\mathbb{B}}e^{-\mathbb{B}x_w}\right]_0^\infty} \\
&= \frac{2\frac{\mathbb{A}}{\mathbb{B}}e^{-\mathbb{B}x} - \frac{\mathbb{A}}{\mathbb{B}}}{\frac{\mathbb{A}}{\mathbb{B}}} \\
&= \left(2e^{-\mathbb{B}x} - 1\right)
\end{aligned} \tag{B.21}$$

Again there is no obvious analytical solution to (B.21), but we can integrate $dS(x)/d_{\mathrm{MF}}\mathrm{P}_{32}$ numerically to obtain the cumulative macrofracture spacing $S(x)$ for any $x < W$. We can thus plot the full cumulative macrofracture spacing distribution

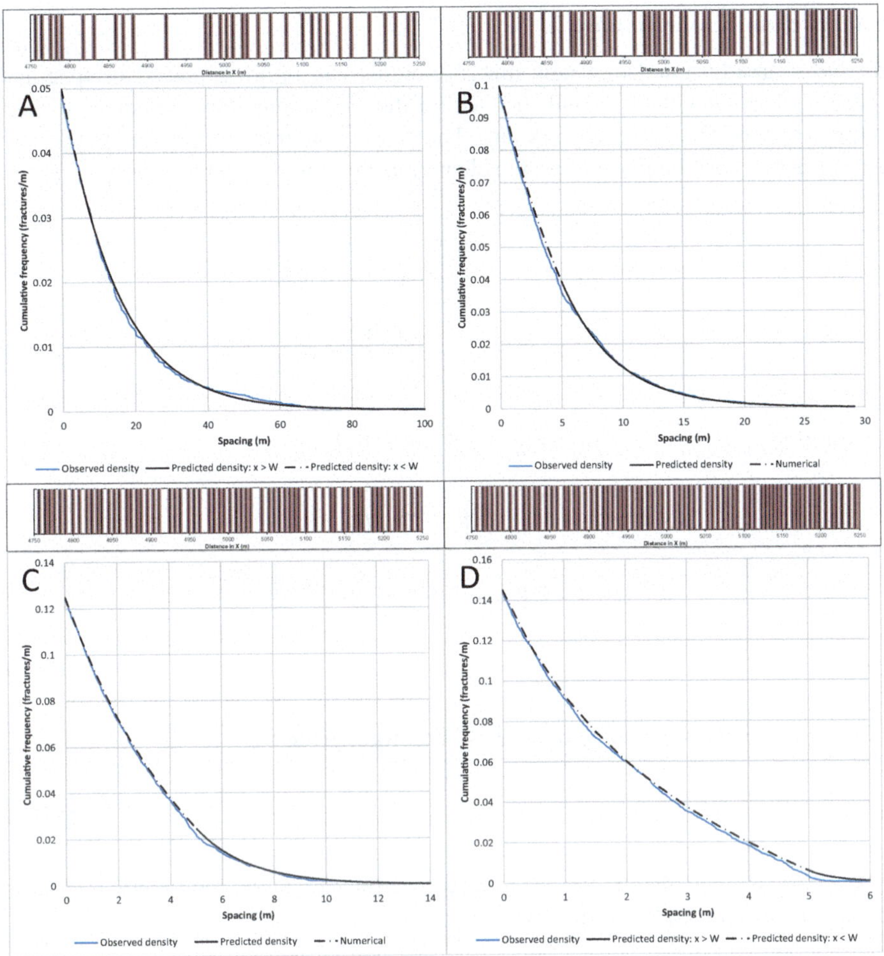

Fig. B.5 Cumulative macrofracture spacing distributions obtained by adding macrofractures with stress shadows of width $W = 5$ m at random locations along a transect of length 10000 m. New macrofractures were added until there was no space to add any more without overlapping the stress shadows of existing macrofractures. **a** Spacing distributions are shown after adding 500 fractures, $\psi = 0.25$ (**a**); after adding 1000 fractures, $\psi = 0.5$ (**b**); after adding 1250 fractures, $\psi = 0.625$ (**c**); after adding 1451 fractures, $\psi = 0.726$ (**d**). The blue lines represent the empirical cumulative spacing distribution of the randomly placed macrofractures, while the black lines represent the theoretical cumulative spacing distribution derived from (B.15) (solid line) and (B.21) (dashed line). Above is shown a small portion of each transect; the vertical black lines represent the macrofractures crossing the transect and the red shading represents the stress shadows around each of these

function $S(x)$ and compare this with the actual spacings obtained when we add macrofractures at random locations on a long transect. This is illustrated in Fig. B.5.

Total exclusion zone volume

We have already noted that new macrofractures can only be placed in spacings with length greater than W and that within these spacings they cannot be placed within a distance $W/2$ of each endpoint. The proportion of the transect in which new macrofractures can be placed (i.e. the clear zone volume, $1 - \chi$) is, therefore, given by

$$\begin{aligned} 1 - \chi &= \int_W^\infty -\frac{dS(x_n)}{dx_n}(x_n - W)dx_n \\ &= \int_0^\infty -\frac{dS(x_w + W)}{dx_w}x_w dx_w \end{aligned} \tag{B.22}$$

substituting $x_w = x_n - W$. Inserting (B.15) into (B.22) gives

$$\begin{aligned} 1 - \chi &= \int_0^\infty -\mathbb{A}\mathbb{B}e^{-\mathbb{B}x_w}x_w dx_w \\ &= -\left[\mathbb{A}x_w e^{-\mathbb{B}x_w} + \frac{\mathbb{A}}{\mathbb{B}}e^{-\mathbb{B}x_w}\right]_0^\infty \\ &= \frac{\mathbb{A}}{\mathbb{B}} \end{aligned} \tag{B.23}$$

The total exclusion zone volume χ is, therefore, given by

$$\chi = 1 - \frac{\mathbb{A}}{\mathbb{B}} \tag{B.24}$$

We can also calculate the rate of change of the exclusion zone volume as new macrofractures are added to the transect:

$$\begin{aligned} \frac{d\chi}{d_{\mathrm{MF}}\mathrm{P}_{32}} &= \frac{\mathbb{A}\frac{d\mathbb{B}}{d_{\mathrm{MF}}\mathrm{P}_{32}} - \frac{d\mathbb{A}}{d_{\mathrm{MF}}\mathrm{P}_{32}}\mathbb{B}}{\mathbb{B}^2} \\ &= \frac{\mathbb{A}\frac{\mathbb{B}}{\mathbb{A}} - \left(2e^{-\mathbb{B}W} - 1\right)\mathbb{B}}{\mathbb{B}^2} \\ &= \frac{2}{\mathbb{B}}\left(1 - e^{-\mathbb{B}W}\right) \end{aligned} \tag{B.25}$$

Equation (B.25) can also be expressed in terms of the total stress shadow volume, ψ:

$$\frac{d\chi}{d\psi} = \frac{2}{\mathbb{B}W}\left(1 - e^{-\mathbb{B}W}\right) \tag{B.26}$$

We can validate this method by comparing the values for χ calculated using (B.24) or (B.26) with the actual exclusion zone volume measured when macrofractures are added at random locations to a long transect. This is illustrated in Figure 70, which shows a good fit between the two datasets, indicating that (B.24) and (B.26) give reliable and accurate results. However, they may be difficult to use in practice, since the values for $\mathbb{A}$ and $\mathbb{B}$ must be calculated numerically. This is most easily done using the timestep discretisation; new values for $\mathbb{A}$ and $\mathbb{B}$ can be recalculated at the end of each timestep using (B.18) and (B.19), based on the previous values for $\mathbb{A}$ and $\mathbb{B}$ and the calculated increase in total mean linear macrofracture density during the timestep.

Ideally, however, we would like to derive an expression for χ that can be calculated analytically, based on parameters that we have exact values for, e.g. total mean linear macrofracture density, ${}_{\mathrm{MF}}\mathrm{P}_{32}$, or total stress shadow volume, ψ. Although it is not possible to derive such an expression that gives exact values for χ, it is possible to derive an approximation that is valid in the early stages of fracture development.

As we noted previously, in the early stages of fracture development, when the total macrofracture density is low, the macrofracture spacings can be approximated by a simple exponential distribution of the form

$$S(x) \approx \tilde{\mathbb{A}} e^{-\tilde{\mathbb{B}} x} \tag{B.27}$$

We can use (B.27) to derive an approximation for the clear zone volume, $1-\chi$, expressed purely in terms of the total stress shadow volume, ψ. Firstly, we can see that

$$\tilde{\mathbb{A}} = S(0) = {}_{\mathrm{MF}}\mathrm{P}_{32} \tag{B.28}$$

We also know that the total volume of all spacings, of any length, is equal to the inverse of the total stress shadow volume:

$$\begin{aligned} &\int_0^\infty -\frac{dS(x_n)}{dx_n} x_n dx_n = 1-\psi \\ &\Rightarrow \int_0^\infty \tilde{\mathbb{A}}\tilde{\mathbb{B}} e^{-\tilde{\mathbb{B}}x} x_n dx_n = 1-\psi \\ &\Rightarrow -\left[\tilde{\mathbb{A}} x_n e^{-\tilde{\mathbb{B}}x} + \frac{\tilde{\mathbb{A}}}{\tilde{\mathbb{B}}} e^{-\tilde{\mathbb{B}}x}\right]_0^\infty = \frac{\tilde{\mathbb{A}}}{\tilde{\mathbb{B}}} = 1-\psi \end{aligned} \tag{B.29}$$

and since $\tilde{\mathrm{A}} = {}_{\mathrm{MF}}\mathrm{P}_{32}$,

$$\tilde{\mathbb{B}} = \frac{{}_{\mathrm{MF}}\mathrm{P}_{32}}{1-\psi} \tag{B.30}$$

We can now derive an expression for $1-\chi$ using a variant of the method used previously:

$$
\begin{aligned}
1-\chi &\approx \int_W^\infty -\frac{dS(x_n)}{dx_n}(x_n-W)dx_n = \int_W^\infty \frac{dS(x_n)}{dx_n}Wdx_n - \int_W^\infty \frac{dS(x_n)}{dx_n}x_ndx_n \\
&= -{}_{\mathrm{MF}}\mathrm{P}_{32}e^{-\frac{{}_{\mathrm{MF}}\mathrm{P}_{32}W}{1-\psi}}W - \int_W^\infty -\frac{{}_{\mathrm{MF}}\mathrm{P}_{32}^2}{1-\psi}e^{-\frac{{}_{\mathrm{MF}}\mathrm{P}_{32}x_n}{1-\psi}}x_ndx_n \\
&= -{}_{\mathrm{MF}}\mathrm{P}_{32}We^{-\frac{{}_{\mathrm{MF}}\mathrm{P}_{32}W}{1-\psi}} - \left[{}_{\mathrm{MF}}\mathrm{P}_{32}x_ne^{-\frac{{}_{\mathrm{MF}}\mathrm{P}_{32}x_n}{1-\psi}} + (1-\psi)e^{-\frac{{}_{\mathrm{MF}}\mathrm{P}_{32}x_n}{1-\psi}}\right]_W^\infty \\
&= -{}_{\mathrm{MF}}\mathrm{P}_{32}We^{-\frac{{}_{\mathrm{MF}}\mathrm{P}_{32}W}{1-\psi}} + {}_{\mathrm{MF}}\mathrm{P}_{32}We^{-\frac{{}_{\mathrm{MF}}\mathrm{P}_{32}W}{1-\psi}} + (1-\psi)e^{-\frac{{}_{\mathrm{MF}}\mathrm{P}_{32}W}{1-\psi}}
\end{aligned}
\tag{B.31}
$$

Since $\psi = {}_{\mathrm{MF}}\mathrm{P}_{32}W$, (B.31) can be reduced to

$$1-\chi \approx (1-\psi)e^{-\frac{\psi}{1-\psi}} \tag{B.32}$$

and we avoid the need for numerical approximations. We can also derive an analytical expression for the differential of the exclusion zone volume χ with respect to ψ:

$$
\begin{aligned}
\chi &\approx 1-\left((1-\psi)e^{-\frac{\psi}{1-\psi}}\right) \\
\Rightarrow \frac{d\chi}{d\psi} &\approx e^{-\frac{\psi}{1-\psi}} + (1-\psi)\frac{(1-\psi)+\psi}{(1-\psi)^2}e^{-\frac{\psi}{1-\psi}} \\
&= \frac{2-\psi}{1-\psi}e^{-\frac{\psi}{1-\psi}}
\end{aligned}
\tag{B.33}
$$

As Fig. B.6 shows, (B.32) gives a good approximation for the clear zone volume when the total stress shadow volume is less than c.0.45. However as the total stress shadow volume exceeds c.0.5, (B.32) significantly overestimates the clear zone volume. Therefore, this approximation is useful when simulating the early stages of fracture growth but should not be used as the fracture network approaches saturation.

Proximity zone volume

There may also be circumstances in which we want to calculate the proportion of the transect that lies within an arbitrary distance X_d of the nearest macrofracture, where X_d is unrelated to the macrofracture stress shadow width W. We can nevertheless use a variant of the exclusion zone formula to calculate the total volume of these "proximity zones". In this case, the width of each proximity zone (on both sides of the macrofracture) will be given by $X_p = 2X_d$. The method for calculating the total proximity zone volume $\chi'(X_p)$ will depend on the proximity zone width:

- If the proximity zone width is less than the stress shadow width, then the proximity zones around individual macrofractures cannot overlap, since the stress shadows cannot overlap. The total proximity zone volume is, therefore, given by ${}_{\mathrm{MF}}\mathrm{P}_{32}X_p$.

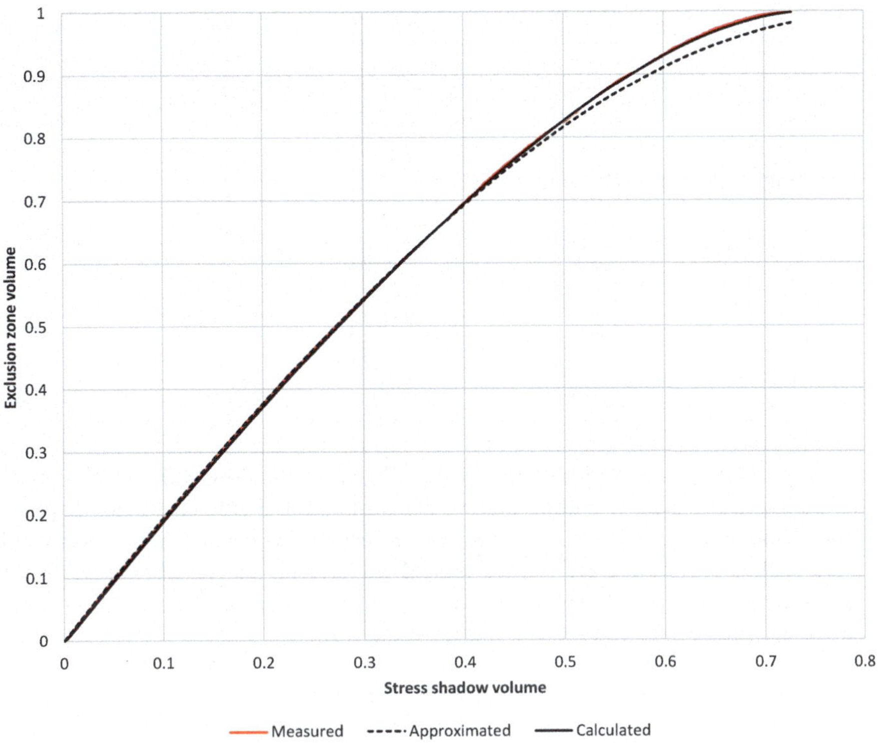

Fig. B.6 Total exclusion zone volume as a function of total stress volume, calculated from (B.24) (solid black line), compared with empirical values for exclusion zone volume measured as 1451 macrofractures with stress shadow width 5 m were added one by one to a transect of length 10000 m (see Fig. B.5) (solid red line). Also shown is an estimate for the exclusion zone volume calculated using the approximation (B.32) (dashed black line). This gives a good approximation at low to moderate macrofracture densities ($\psi <$ c.0.45) but a poor approximation as we approach saturation, where the clear zone volume predicted by the approximation may be many times greater than the calculated the clear zone volume

- If the proximity zone width is greater than twice the stress shadow width, then the entire stress shadow volume will lie within the proximity zones, as will all the spacings between the macrofracture stress shadow with length $x < (X_p - W)$, and a portion of the larger spacings lying within a distance of $(X_p - W)/2$ of either end. The easiest way to calculate this volume is to calculate the total volume not lying within the proximity zone, $1 - \chi'(X_p)$. This can be calculated in the same way as the clear zone volume $1 - \chi$, but substituting $(X_p - W)$ for W:

$$1 - \chi'(X_p) = \int_{X_p - W}^{\infty} -\frac{dS(x_n)}{dx_n}\left(x_n - (X_p - W)\right)dx_n$$
$$= \int_0^{\infty} -\frac{dS\left(x_w + (X_p - W)\right)}{dx_w} x_w dx_w \quad \text{(B.34)}$$

where $x_w = x_n - (X_p - W)$. Since $(X_p - W) > W$, we can solve (B.34) by noting that

$$S(x) = \mathbb{A}e^{-\mathbb{B}(x-W)} = \mathbb{A}e^{-\mathbb{B}(X_p-2W)}e^{-\mathbb{B}(x-(X_p-W))} \tag{B.35}$$

Inserting (B.35) into (B.34) gives us

$$\begin{aligned} 1-\chi'(X_p) &= \int_0^\infty \mathbb{A}\mathbb{B}e^{-\mathbb{B}(X_p-2W)}e^{-\mathbb{B}x_w}x_w dx_w \\ &= -\left[\mathbb{A}x_w e^{-\mathbb{B}(X_p-2W)}e^{-\mathbb{B}x_w} + \frac{\mathbb{A}}{\mathbb{B}}e^{-\mathbb{B}(X_p-2W)}e^{-\mathbb{B}x_w}\right]_0^\infty \\ &= \frac{\mathbb{A}}{\mathbb{B}}e^{-\mathbb{B}(X_p-2W)} \end{aligned} \tag{B.36}$$

- If the proximity zone width is greater than the stress shadow width but less than twice the stress shadow width, then we can still use (B.34) to calculate the proximity zone volume, but we cannot simply insert (B.35) because this is not valid for $x < W$ (i.e. for $x_w < 2W - X_p$). Indeed we have no analytical expression for S in the range $0 < x < W$. However, we can derive an approximate solution using an approximation for $S(x)$ where $x < W$ of the form

$$S(x) \approx \mathbb{A}_2 e^{-\mathbb{B}_2 x} - \mathbb{C}_2 = \mathbb{A}_2 e^{-\mathbb{B}_2(X_p-W)}e^{-\mathbb{B}_2(x-(X_p-W))} - \mathbb{C}_2 \tag{B.37}$$

Since $S(0) = {}_{\mathrm{MF}}\mathrm{P}_{32}$ we can see that

$$\mathbb{A}_2 - \mathbb{C}_2 = {}_{\mathrm{MF}}\mathrm{P}_{32} \tag{B.38}$$

and by comparison with (B.35), we can see that

$$\begin{aligned} S(W) &= \mathbb{A}_2 e^{-\mathbb{B}_2 W} - \mathbb{C}_2 = \mathbb{A} \\ \Rightarrow e^{-\mathbb{B}_2 W} &= \frac{\mathbb{A}+\mathbb{C}_2}{\mathbb{A}_2} \\ \Rightarrow \mathbb{B}_2 &= \frac{\ln(\mathbb{A}_2) - \ln(\mathbb{A}+\mathbb{C}_2)}{W} \end{aligned} \tag{B.39}$$

The total volume not lying within the proximity zone, $1 - \chi'(X_p)$, will, therefore, be given by

$$\begin{aligned} 1-\chi'(X_p) &\approx \int_0^{2W-X_p} -\frac{dS(x_w+(X_p-W))}{dx_w}x_w dx_w + \int_{2W-X_p}^\infty -\frac{dS(x_w+(X_p-W))}{dx_w}x_w dx_w \\ &= \int_0^{2W-X_p} \mathbb{A}_2\mathbb{B}_2 e^{-\mathbb{B}_2(X_p-W)}e^{-\mathbb{B}_2 x_w}x_w dx_w + \int_{2W-X_p}^\infty \mathbb{A}\mathbb{B}e^{-\mathbb{B}(X_p-2W)}e^{-\mathbb{B}x_w}x_w dx_w \\ &= -\left[\mathbb{A}_2 x_w e^{-\mathbb{B}_2(X_p-W)}e^{-\mathbb{B}_2 x_w} + \frac{\mathbb{A}_2}{\mathbb{B}_2}e^{-\mathbb{B}_2(X_p-W)}e^{-\mathbb{B}_2 x_w}\right]_0^{2W-X_p} \end{aligned}$$

$$
\begin{aligned}
&-\left[\mathbb{A}x_w e^{-\mathbb{B}(X_p-2W)}e^{-\mathbb{B}x_w} + \frac{\mathbb{A}}{\mathbb{B}}e^{-\mathbb{B}(X_p-2W)}e^{-\mathbb{B}x_w}\right]_{2W-X_p}^{\infty} \\
&= \frac{\mathbb{A}_2}{\mathbb{B}_2}e^{-\mathbb{B}_2(X_p-W)} - \mathbb{A}_2(2W-X_p)e^{-\mathbb{B}_2 W} - \frac{\mathbb{A}_2}{\mathbb{B}_2}e^{-\mathbb{B}_2 W} + \frac{\mathbb{A}}{\mathbb{B}} + \mathbb{A}(2W-X_p) \\
&= \frac{\mathbb{A}_2}{\mathbb{B}_2}e^{-\mathbb{B}_2(X_p-W)} + \left(\frac{\mathbb{A}}{\mathbb{B}} - \frac{\mathbb{A}+\mathbb{C}_2}{\mathbb{B}_2}\right) - \mathbb{C}_2(2W-X_p)
\end{aligned} \tag{B.40}
$$

Since we know that $X_p(W) = 1 - \psi$, then by setting $X_p = W$ and substituting (B.38) we can show that

$$
1-\chi'(W) = \frac{\mathbb{A}_2}{\mathbb{B}_2} - \frac{\mathbb{A}+\mathbb{C}_2}{\mathbb{B}_2} + \frac{\mathbb{A}}{\mathbb{B}} - \mathbb{C}_2 W = \frac{{}_{\mathrm{MF}}\mathrm{P}_{32}}{\mathbb{B}_2} - \frac{\mathbb{A}}{\mathbb{B}_2} + \frac{\mathbb{A}}{\mathbb{B}} - \mathbb{C}_2 W = 1-\psi
$$

$$
\Rightarrow \mathbb{B}_2 = \frac{{}_{\mathrm{MF}}\mathrm{P}_{32} - \mathbb{A}}{(1-\psi) - \frac{\mathbb{A}}{\mathbb{B}} + \mathbb{C}_2 W} \tag{B.41}
$$

From (B.39), we can also show that

$$
\begin{aligned}
&\left({}_{\mathrm{MF}}\mathrm{P}_{32} + \mathbb{C}_2\right)e^{-\mathbb{B}_2 W} - \mathbb{C}_2 = \mathbb{A} \\
&\Rightarrow \mathbb{C}_2 = \frac{{}_{\mathrm{MF}}\mathrm{P}_{32}e^{-\mathbb{B}_2 W} - \mathbb{A}}{1-e^{-\mathbb{B}_2 W}}
\end{aligned} \tag{B.42}
$$

Values for $\mathbb{B}_2$ and $\mathbb{C}_2$ can then be determined iteratively using (B.41) and (B.42).

A general expression for the total proximity zone volume, $\chi'(X_p)$, is, therefore, given by

$$
\chi'(X_p) = \begin{cases} {}_{\mathrm{MF}}\mathrm{P}_{32}X_p & \text{if} X_p \le W \\ \sim 1 - \frac{\mathbb{A}_2}{\mathbb{B}_2}e^{-\mathbb{B}_2(X_p-W)} - \left(\frac{\mathbb{A}}{\mathbb{B}} - \frac{\mathbb{A}}{\mathbb{B}_2}\right) & \text{if} W < X_p \le 2W \\ 1 - \frac{\mathbb{A}}{\mathbb{B}}e^{-\mathbb{B}(X_p-2W)} & \text{if} X_p > 2W \end{cases} \tag{B.43}
$$

If we want to calculate the proportion of the transect that lies within an arbitrary distance X_d on one side (e.g. the southernmost side) of the nearest macrofracture (the "one-sided proximity zone volume"), we can use the same formula (B.43) but with the proximity zone width equal to the specified maximum distance from the nearest macrofracture (i.e. $X_p = X_d$). This is still valid because the minimum distance between two adjacent macrofractures is still given by the stress shadow width W, and so where $X_p > W$, the proximity zone comprises the entire stress shadow width between the two macrofractures, plus a portion of the spacing between the proximity zones of width $X_p - W$.

Appendix C
Calculating the Optimal Timestep Duration

Abstract In this appendix we show how to optimise the execution of the workflow by calculating a timestep duration consistent with a specified maximum increment in the growth of the fracture network. This allows us to determine the duration of each timestep dynamically, as the calculation is progressing, in a way that balances the trade-off between runtime and accuracy. We can thus avoid both overly long timesteps, that lead to inaccuracy, and overly short timesteps, that increase runtime.

Keywords Fracture modelling · Discrete fracture networks · Naturally fractured reservoir · Fracture growth · Subcritical crack propagation · Optimisation

Setting the timestep duration involves a trade-off between accuracy and calculation time. The principal source of inaccuracy comes from the approximation that the fracture populations remain constant during the timestep for the purposes of calculating stress-strain relationships and probability of fracture deactivation. These are mostly a function of the mean linear macrofracture density, ${}_{MF}P_{32}$, and the stress shadow volume, ψ_{MF}.

Therefore, in order to calculate a timestep duration to optimise the trade-off between accuracy and calculation time, we must derive an expression relating the increase in stress shadow volume $\Delta\psi_{MF}$ to the timestep duration Δt. We can then use this to calculate an optimal duration for each timestep, to ensure that the increase in stress shadow volume within the timestep remains within an acceptable limit. Note that this limit can also be applied in the evenly distributed strain scenario, even though fractures in the same set do not interact. In both scenarios, however, an increase in ψ_{MF} per timestep in the range 0.002–0.01 will usually give a reasonable trade-off between runtime and accuracy. Where there are multiple active fracture sets, we can calculate the optimal timestep duration for each set as described below and adopt the shortest of these estimates as the timestep duration.

© The Editor(s) (if applicable) and The Author(s), under exclusive license to Springer Nature Switzerland AG 2020

M. J. Welch et al., *Modelling the Evolution of Natural Fracture Networks*,
https://doi.org/10.1007/978-3-030-52414-2

Calculating $\Delta\psi_{MF}$

We can derive an expression relating $\Delta\psi_{MF}$ to Δt by integrating the total volumetric active half-macrofracture density, ${}^{a}_{\underline{MF}}P_{30}(0, t_n)$, multiplied by the half-macrofracture propagation rate, $d\ell/dt$, and the stress shadow width, W:

$$\Delta\psi_{MF} = \int_{t_{N-1}}^{t_N} {}^{a}_{\underline{MF}}P_{30}(0, t)\frac{d\ell}{dt} W dt \tag{C.1}$$

where the timestep duration $\Delta t = t_N - t_{N-1}$.

As we saw in Sect. 4.2, the total volumetric macrofracture density at any time t_n during timestep **N** is equal to the total volumetric density of microfractures with radius greater than half the layer thickness, ${}_{\mu F}P_{30}(h/2, t_n)$. If we ignore macrofracture deactivation within the timestep, we can obtain an approximate estimate of the volumetric density of active half-macrofractures at time t_n by multiplying the total volumetric macrofracture density at time t_n by the ratio of active half-macrofractures to all full macrofractures at the end of the previous timestep:

$$\underline{MF}{}^{a}P_{30}(0, t_n) \approx \frac{{}^{a}_{\underline{MF}}P_{30}(0, t_{N-1})}{{}_{\mu F}P_{30}\left(\frac{h}{2}, t_{N-1}\right)} {}_{\mu F}P_{30}\left(\frac{h}{2}, t_n\right) \tag{C.2}$$

Substituting (C.2) into (C.1) will give us an overestimate of the actual rate of increase in volumetric macrofracture ratio during the timestep, since it ignores fracture deactivation during the timestep. Therefore, we will underestimate the timestep duration required to generate the specified increase in the stress shadow volume ψ_{MF}; however, this discrepancy is likely to be small and should not significantly increase the model runtime.

The exact form of the solution to (C.2) will depend on whether the fracture driving stress is constant or increasing.

Timesteps with constant driving stress

If the fracture driving stress, σ_d, is constant throughout the timestep, the half-macrofracture propagation rate will also be constant, as given by (5.1). The stress shadow width is always constant, given by (4.27) or (4.28) for Mode 1 or Mode 2 macrofractures, respectively.

The total volumetric density of microfractures with radius greater than half the layer thickness at time t_n during timestep **N** is given by modifying (3.14) to get:

$${}_{\mu F}P_{30}\left(\frac{h}{2}, t_n\right) = B\left(\left(\frac{h}{2}\right)^{\frac{1}{\beta}} + \Gamma_{N-1} + \gamma_N^{\frac{1}{\beta}}(t_n - t_{N-1})\right)^{-\beta c} \tag{C.3}$$

Substituting (C.2), (C.3), (5.1) and W into (C.1) gives

$$\Delta\psi_{MF} = \int_{t_{N-1}}^{t_N} \frac{{}^{a}_{\underline{MF}}P_{30}(0, t_{N-1})}{{}_{\mu F}P_{30}\left(\frac{h}{2}, t_{N-1}\right)} B\left(\left(\frac{h}{2}\right)^{\frac{1}{\beta}} + \Gamma_{N-1} + \gamma_N^{\frac{1}{\beta}}(t - t_{N-1})\right)^{-\beta c} \alpha_{MF}\sigma_d^b W dt$$
$$= \frac{{}^{a}_{\underline{MF}}P_{30}(0, t_{N-1})}{{}_{\mu F}P_{30}\left(\frac{h}{2}, t_{N-1}\right)} B\alpha_{MF}\sigma_d^b W \int_{t_{N-1}}^{t_N} \left(\left(\frac{h}{2}\right)^{\frac{1}{\beta}} + \Gamma_{N-1} + \gamma_N^{\frac{1}{\beta}}(t - t_{N-1})\right)^{-\beta c} dt \quad \text{(C.4)}$$

This integral can be solved analytically to give

$$\Delta\psi_{MF} = \frac{{}^{a}_{\underline{MF}}P_{30}(0, t_{N-1})}{{}_{\mu F}P_{30}\left(\frac{h}{2}, t_{N-1}\right)} B\alpha_{MF}\sigma_d^b W \left[\frac{\left(\left(\frac{h}{2}\right)^{\frac{1}{\beta}} + \Gamma_{N-1} + \gamma_N^{\frac{1}{\beta}}(t - t_{N-1})\right)^{1-\beta c}}{(1-\beta c)\gamma_N^{\frac{1}{\beta}}}\right]_{t_{N-1}}^{t_N}$$
$$= \frac{{}^{a}_{\underline{MF}}P_{30}(0, t_{N-1})B\alpha_{MF}\sigma_d^b W}{(1-\beta c)\gamma_N^{\frac{1}{\beta}}\,{}_{\mu F}P_{30}\left(\frac{h}{2}, t_{N-1}\right)} \left(\left(\left(\frac{h}{2}\right)^{\frac{1}{\beta}} + \Gamma_{N-1} + \gamma_N^{\frac{1}{\beta}}(t_N - t_{N-1})\right)^{1-\beta c} - \left(\left(\frac{h}{2}\right)^{\frac{1}{\beta}} + \Gamma_{N-1}\right)^{1-\beta c}\right)$$
$$= \frac{-\beta^{a}_{\underline{MF}}P_{30}(0, t_{N-1})BW}{(1-\beta c)_{\mu F}P_{30}\left(\frac{h}{2}, t_{N-1}\right)} \left(\frac{h}{2}\right)^{\frac{b}{2}} \left(\left(\left(\frac{h}{2}\right)^{\frac{1}{\beta}} + \Gamma_{N-1} + \gamma_N^{\frac{1}{\beta}}(t_N - t_{N-1})\right)^{1-\beta c} - \left(\left(\frac{h}{2}\right)^{\frac{1}{\beta}} + \Gamma_{N-1}\right)^{1-\beta c}\right) \quad \text{(C.5)}$$

Rearranging (C.5) gives us

$$\frac{(1-\beta c)_{\mu F}P_{30}\left(\frac{h}{2}, t_{N-1}\right)}{-\beta^{a}_{\underline{MF}}P_{30}(0, t_{N-1})BW} \left(\frac{2}{h}\right)^{\frac{b}{2}} \Delta\psi_{MF} = \left(\left(\left(\frac{h}{2}\right)^{\frac{1}{\beta}} + \Gamma_{N-1} + \gamma_N^{\frac{1}{\beta}}(t_N - t_{N-1})\right)^{1-\beta c} - \left(\left(\frac{h}{2}\right)^{\frac{1}{\beta}} + \Gamma_{N-1}\right)^{1-\beta c}\right)$$
$$\Rightarrow \frac{(1-\beta c)_{\mu F}P_{30}\left(\frac{h}{2}, t_{N-1}\right)}{-\beta^{a}_{\underline{MF}}P_{30}(0, t_{N-1})BW} \left(\frac{2}{h}\right)^{\frac{b}{2}} \Delta\psi_{MF} + \left(\left(\frac{h}{2}\right)^{\frac{1}{\beta}} + \Gamma_{N-1}\right)^{1-\beta c} = \left(\left(\frac{h}{2}\right)^{\frac{1}{\beta}} + \Gamma_{N-1} + \gamma_N^{\frac{1}{\beta}}(t_N - t_{N-1})\right)^{1-\beta c}$$
$$\Rightarrow \left(\frac{(1-\beta c)_{\mu F}P_{30}\left(\frac{h}{2}, t_{N-1}\right)}{-\beta^{a}_{\underline{MF}}P_{30}(0, t_{N-1})BW} \left(\frac{2}{h}\right)^{\frac{b}{2}} \Delta\psi_{MF} + \left(\left(\frac{h}{2}\right)^{\frac{1}{\beta}} + \Gamma_{N-1}\right)^{1-\beta c}\right)^{\frac{1}{1-\beta c}} = \left(\frac{h}{2}\right)^{\frac{1}{\beta}} + \Gamma_{N-1} + \gamma_N^{\frac{1}{\beta}}(t_N - t_{N-1})$$
$$\Rightarrow -\gamma_N^{\frac{1}{\beta}}(t_N - t_{N-1}) = \left(\left(\frac{h}{2}\right)^{\frac{1}{\beta}} + \Gamma_{N-1}\right) - \left(\frac{(1-\beta c)_{\mu F}P_{30}\left(\frac{h}{2}, t_{N-1}\right)}{-\beta^{a}_{\underline{MF}}P_{30}(0, t_{N-1})BW} \left(\frac{2}{h}\right)^{\frac{b}{2}} \Delta\psi_{MF} + \left(\left(\frac{h}{2}\right)^{\frac{1}{\beta}} + \Gamma_{N-1}\right)^{1-\beta c}\right)^{\frac{1}{1-\beta c}} \quad \text{(C.6)}$$

and finally

$$\Delta t = (t_N - t_{N-1}) = \frac{1}{\gamma_N^{\frac{1}{\beta}}} \left(\begin{array}{c} \left(\frac{(1-\beta c)_{\mu F}P_{30}\left(\frac{h}{2}, t_{N-1}\right)}{-\beta^{a}_{\underline{MF}}P_{30}(0, t_{N-1})BW} \left(\frac{2}{h}\right)^{\frac{b}{2}} \Delta\psi_{MF} + \left(\left(\frac{h}{2}\right)^{\frac{1}{\beta}} + \Gamma_{N-1}\right)^{1-\beta c}\right)^{\frac{1}{1-\beta c}} \\ -\left(\left(\frac{h}{2}\right)^{\frac{1}{\beta}} + \Gamma_{N-1}\right) \end{array}\right) \quad \text{(C.7)}$$

Timesteps with increasing driving stress

If the driving stress increases during the timestep, the half-macrofracture propagation rate will also increase. In Chap. 6, we introduced the concept of a weighted mean driving stress for the timestep that takes account of this increase, and we can adopt a similar approach here. We can thus rewrite (C.4) as

$$\Delta\psi_{MF} = \frac{{}_{MF}^{a}P_{30}(0,t_{N-1})}{{}_{\mu F}P_{30}\left(\frac{h}{2},t_{N-1}\right)} B\alpha_{MF} W \int_{t_{N-1}}^{t_N} \left(\left(\frac{h}{2}\right)^{\frac{1}{\beta}} + \Gamma_{N-1} - \frac{A}{\beta}\left(\frac{2}{\sqrt{\pi}K_c}\right)^b \sigma_d^b (t - t_{N-1})\right)^{-\beta c} \sigma_d^b dt \quad \text{(C.8)}$$

Equation (C.8) can be solved by substitution. If we define $\Lambda_n = \sigma_d^b(t_n - t_{N-1})$, then we can make the substitutions

$$\begin{aligned}
&\sigma_d^b t \rightarrow \Lambda_n \\
&\sigma_d^b dt \rightarrow d\Lambda_n \\
&t_N \rightarrow \overline{\sigma_{dN}^b}(t_N - t_{N-1}) = \frac{\left(\sigma_{dN-1} + \frac{d\sigma_d}{dt}(t_N - t_{N-1})\right)^{b+1} - \sigma_{dN-1}^{b+1}}{(b+1)\frac{d\sigma_d}{dt}} \\
&t_{N-1} \rightarrow 0
\end{aligned} \quad \text{(C.9)}$$

Equation (C.8) can thus be written as

$$\Delta\psi_{MF} = \frac{{}_{MF}^{a}P_{30}(0,t_{N-1})}{{}_{\mu F}P_{30}\left(\frac{h}{2},t_{N-1}\right)} B\alpha_{MF} W \int_{0}^{\overline{\sigma_{dN}^b}(t_N - t_{N-1})} \left(\left(\frac{h}{2}\right)^{\frac{1}{\beta}} + \Gamma_{N-1} - \frac{A}{\beta}\left(\frac{2}{\sqrt{\pi}K_c}\right)^b \Lambda_n\right)^{-\beta c} d\Lambda_n \quad \text{(C.10)}$$

This can be solved in a similar manner to (C.4), to give

$$\begin{aligned}
\Delta\psi_{MF} &= \frac{{}_{MF}^{a}P_{30}(0,t_{N-1})}{{}_{\mu F}P_{30}\left(\frac{h}{2},t_{N-1}\right)} B\alpha_{MF} W \left[\frac{\left(\left(\frac{h}{2}\right)^{\frac{1}{\beta}} + \Gamma_{N-1} - \frac{A}{\beta}\left(\frac{2}{\sqrt{\pi}K_c}\right)^b \Lambda_n\right)^{1-\beta c}}{-(1-\beta c)\frac{A}{\beta}\left(\frac{2}{\sqrt{\pi}K_c}\right)^b}\right]_0^{\overline{\sigma_{dN}^b}(t_N - t_{N-1})} \\
&= \frac{-\beta\, {}_{MF}^{a}P_{30}(0,t_{N-1}) B\alpha_{MF} W}{(1-\beta c)A\left(\frac{2}{\sqrt{\pi}K_c}\right)^b {}_{\mu F}P_{30}\left(\frac{h}{2},t_{N-1}\right)} \left(\begin{array}{c}\left(\left(\frac{h}{2}\right)^{\frac{1}{\beta}} + \Gamma_{N-1} - \frac{A}{\beta}\left(\frac{2}{\sqrt{\pi}K_c}\right)^b \frac{\left(\sigma_{dN-1} + \frac{d\sigma_d}{dt}(t_N - t_{N-1})\right)^{b+1} - \sigma_{dN-1}^{b+1}}{(b+1)\frac{d\sigma_d}{dt}}\right)^{1-\beta c} \\ -\left(\left(\frac{h}{2}\right)^{\frac{1}{\beta}} + \Gamma_{N-1}\right)^{1-\beta c}\end{array}\right) \\
&= \frac{-\beta\, {}_{MF}^{a}P_{30}(0,t_{N-1}) BW}{(1-\beta c)\,{}_{\mu F}P_{30}\left(\frac{h}{2},t_{N-1}\right)} \left(\frac{h}{2}\right)^{\frac{b}{2}} \left(\begin{array}{c}\left(\left(\frac{h}{2}\right)^{\frac{1}{\beta}} + \Gamma_{N-1} - \frac{A}{\beta}\left(\frac{2}{\sqrt{\pi}K_c}\right)^b \frac{\left(\sigma_{dN-1} + \frac{d\sigma_d}{dt}(t_N - t_{N-1})\right)^{b+1} - \sigma_{dN-1}^{b+1}}{(b+1)\frac{d\sigma_d}{dt}}\right)^{1-\beta c} \\ -\left(\left(\frac{h}{2}\right)^{\frac{1}{\beta}} + \Gamma_{N-1}\right)^{1-\beta c}\end{array}\right)
\end{aligned} \quad \text{(C.11)}$$

Rearranging (C.11) gives us

$$\begin{aligned}
&\frac{(1-\beta c)\,{}_{\mu F}P_{30}\left(\frac{h}{2},t_{N-1}\right)}{-\beta_{MF^a} P_{30}(0,t_{N-1})BW} \left(\frac{2}{h}\right)^{\frac{b}{2}} \Delta\psi_{MF} \\
&= \left(\begin{array}{c}\left(\left(\frac{h}{2}\right)^{\frac{1}{\beta}} + \Gamma_{N-1} - \frac{A}{\beta}\left(\frac{2}{\sqrt{\pi}K_c}\right)^b \frac{\left(\sigma_{dN-1} + \frac{d\sigma_d}{dt}(t_N - t_{N-1})\right)^{b+1} - \sigma_{dN-1}^{b+1}}{(b+1)\frac{d\sigma_d}{dt}}\right)^{1-\beta c} \\ -\left(\left(\frac{h}{2}\right)^{\frac{1}{\beta}} + \Gamma_{N-1}\right)^{1-\beta c}\end{array}\right)
\end{aligned}$$

$$\Rightarrow \frac{(1-\beta c)_{\mu F} P_{30}\left(\frac{h}{2}, t_{N-1}\right)}{-\beta_{\underline{MF}^a} P_{30}(0, t_{N-1}) BW}\left(\frac{2}{h}\right)^{\frac{b}{2}} \Delta\psi_{MF} + \left(\left(\frac{h}{2}\right)^{\frac{1}{\beta}} + \Gamma_{N-1}\right)^{1-\beta c}$$

$$= \left(\begin{array}{c} \left(\frac{h}{2}\right)^{\frac{1}{\beta}} + \Gamma_{N-1} \\ -\frac{A}{\beta}\left(\frac{2}{\sqrt{\pi} K_c}\right)^b \frac{\left(\sigma_{dN-1} + \frac{d\sigma_d}{dt}(t_N - t_{N-1})\right)^{b+1} - \sigma_{dN-1}^{b+1}}{(b+1)\frac{d\sigma_d}{dt}} \end{array}\right)^{1-\beta c}$$

$$\Rightarrow \left(\frac{(1-\beta c)_{\mu F} P_{30}\left(\frac{h}{2}, t_{N-1}\right)}{-\beta_{\underline{MF}^a} P_{30}(0, t_{N-1}) BW}\left(\frac{2}{h}\right)^{\frac{b}{2}} \Delta\psi_{MF} + \left(\left(\frac{h}{2}\right)^{\frac{1}{\beta}} + \Gamma_{N-1}\right)^{1-\beta c}\right)^{\frac{1}{1-\beta c}}$$

$$= \left(\frac{h}{2}\right)^{\frac{1}{\beta}} + \Gamma_{N-1} - \frac{A}{\beta}\left(\frac{2}{\sqrt{\pi} K_c}\right)^b \frac{\left(\sigma_{dN-1} + \frac{d\sigma_d}{dt}(t_N - t_{N-1})\right)^{b+1} - \sigma_{dN-1}^{b+1}}{(b+1)\frac{d\sigma_d}{dt}}$$

$$\Rightarrow \frac{A}{\beta}\left(\frac{2}{\sqrt{\pi} K_c}\right)^b \frac{\left(\sigma_{dN-1} + \frac{d\sigma_d}{dt}(t_N - t_{N-1})\right)^{b+1} - \sigma_{dN-1}^{b+1}}{(b+1)\frac{d\sigma_d}{dt}}$$

$$= \left(\left(\frac{h}{2}\right)^{\frac{1}{\beta}} + \Gamma_{N-1}\right) - \left(\frac{(1-\beta c)_{\mu F} P_{30}\left(\frac{h}{2}, t_{N-1}\right)}{-\beta_{\underline{MF}^a} P_{30}(0, t_{N-1}) BW}\left(\frac{2}{h}\right)^{\frac{b}{2}} \Delta\psi_{MF} + \left(\left(\frac{h}{2}\right)^{\frac{1}{\beta}} + \Gamma_{N-1}\right)^{1-\beta c}\right)^{\frac{1}{1-\beta c}}$$

$$\Rightarrow \left(\sigma_{dN-1} + \frac{d\sigma_d}{dt}(t_N - t_{N-1})\right)^{b+1}$$

$$= \frac{\beta}{A}\left(\frac{2}{\sqrt{\pi} K_c}\right)^{-b}(b+1)\frac{d\sigma_d}{dt}\left(\begin{array}{c} \left(\left(\frac{h}{2}\right)^{\frac{1}{\beta}} + \Gamma_{N-1}\right) \\ -\left(\frac{(1-\beta c)_{\mu F} P_{30}\left(\frac{h}{2}, t_{N-1}\right)}{-\beta_{\underline{MF}^a} P_{30}(0, t_{N-1}) BW}\left(\frac{2}{h}\right)^{\frac{b}{2}} \Delta\psi_{MF} + \left(\left(\frac{h}{2}\right)^{\frac{1}{\beta}} + \Gamma_{N-1}\right)^{1-\beta c}\right)^{\frac{1}{1-\beta c}} \end{array}\right) + \sigma_{dN-1}^{b+1} \quad \text{(C.12)}$$

and finally

$$\Delta t = \frac{dt}{d\sigma_d}\left(\begin{array}{c} \frac{\beta}{A}\left(\frac{2}{\sqrt{\pi} K_C}\right)^{-b}(b+1)\frac{d\sigma_d}{dt}\left(\begin{array}{c} \left(\left(\frac{h}{2}\right)^{\frac{1}{\beta}} + \Gamma_{N-1}\right) \\ -\left(\frac{(1-\beta c)_{\mu F} P_{30}\left(\frac{h}{2}, t_{N-1}\right)}{-\beta_{\underline{MF}}^a P_{30}(0, t_{N-1}) BW}\left(\frac{2}{h}\right)^{\frac{b}{2}} \Delta\psi_{MF} + \left(\left(\frac{h}{2}\right)^{\frac{1}{\beta}} + \Gamma_{N-1}\right)^{1-\beta c}\right)^{\frac{1}{1-\beta c}} \end{array}\right) \\ +\sigma_{dN-1}^{b+1} \end{array}\right)^{\frac{1}{b+1}} - \sigma_{dN-1}\frac{dt}{d\sigma_d} \quad \text{(C.13)}$$

Note that where $\sigma_{d\mathbf{N-1}} >> d\sigma_d/dt$, (C.13) may be difficult to calculate because $(\sigma_{d\mathbf{N-1}} + d\sigma_d/dt\ \Delta t)^{b+1} \approx \sigma_{d\mathbf{N-1}}^{b+1}$. In this case, it may be helpful to rewrite (C.13) as

$$\Delta t = \frac{dt}{d\sigma_d} A^{-\frac{1}{b+1}}\left(\frac{2}{\sqrt{\pi} K_C}\right)^{-\frac{b}{b+1}}\left(\begin{array}{c} \beta(b+1)\frac{d\sigma_d}{dt}\left(\begin{array}{c} \left(\left(\frac{h}{2}\right)^{\frac{1}{\beta}} + \Gamma_{N-1}\right) \\ -\left(\frac{(1-\beta c)_{\mu F} P_{30}\left(\frac{h}{2}, t_{N-1}\right)}{-\beta_{\underline{MF}}^a P_{30}(0, t_{N-1}) BW}\left(\frac{2}{h}\right)^{\frac{b}{2}} \Delta\psi_{MF} + \left(\left(\frac{h}{2}\right)^{\frac{1}{\beta}} + \Gamma_{N-1}\right)^{1-\beta c}\right)^{\frac{1}{1-\beta c}} \end{array}\right) \\ +A\left(\frac{2}{\sqrt{\pi} K_c}\right)^b \sigma_{dN-1}^{b+1} \end{array}\right)^{\frac{1}{b+1}} - \sigma_{dN-1}\frac{dt}{d\sigma_d} \quad \text{(C.14)}$$

We can then eliminate some of the extreme values by substituting:

$$A\left(\frac{2}{\sqrt{\pi}K_c}\right)^b \sigma_{dN-1}^{b+1} = \sigma_{dN-1} A\left(\frac{2\sigma_{dN-1}}{\sqrt{\pi}K_C}\right)^b \tag{C.15}$$

If we still cannot calculate (C.13) accurately, we can assume that $d\sigma_d/dt = 0$ and use the expression for constant driving stress given by (C.7).

The manufacturer's authorised representative in the EU is Springer Nature Customer Service Centre GmbH, Europaplatz 3, 69115 Heidelberg, Germany. If you have any concerns regarding our products, please contact ProductSafety@springernature.com

Printed and bound by CPI Group (UK) Ltd, Croydon, CR0 4YY
23/09/2026
02221546-0004